To The Student

Michael Hacker

David Burghardt

Today, we are surrounded by engineered systems. We have become dependent upon the products and services of modern engineering and technology. There is a lot to learn about how people and industries will continue to be affected by technological change.

School is the ideal place to begin learning about engineering and technology and the role they play in our culture. This book will introduce you to the study of engineering and technology and help you understand how they affect your life. Engineering and technology enable us to change the world to meet our needs and to improve our lives.

As learners, you will come to understand engineering and technology as interdisciplinary subjects. As you have fun participating in the design activities, you will integrate other subject matter such as Math, Science, Social Studies, and Language Arts.

The content of this book will provide you with engineering and technological literacy that will enable you to make informed choices about technology issues and developments as you become a responsible citizen. Most important, be safe and have fun.

Sincerely,

Michael Hacker

Dave Burghardt

Engineering AND Technology Education

Learning by Design

Second Edition

Michael Hacker
Hofstra University
Co-Director, Center for Technological Literacy

David Burghardt
Hofstra University
Professor of Engineering
Co-Director, Center for Technological Literacy

Prentice Hall

Boston   Columbus   Indianapolis   New York   San Francisco   Upper Saddle River
Amsterdam   Cape Town   Dubai   London   Madrid   Milan   Munich   Paris   Montreal   Toronto
Delhi   Mexico City   São Paulo   Sydney   Hong Kong   Seoul   Singapore   Taipei   Tokyo

Editorial Director: Vernon R. Anthony
Senior Acquisitions Editor: Wyatt Morris
Development Editor: Dan Trudden
Editorial Assistant: Yvette Schlarman
Director of Marketing: David Gesell
Marketing Assistant: Les Roberts
School Marketing Manager: Laura Cutone
Senior Managing Editor: Joellen Gohr
Associate Managing Editor: Alexandrina B. Wolf
Senior Operations Supervisor: Pat Tonneman

Text and Cover Designer: PreMediaGlobal
Cover Art: sgame/Shutterstock.com
Media Director: Ally Graesser
Media Project Manager: Karen Bretz
Full-Service Project Management:
  PreMediaGlobal
Composition: PreMediaGlobal
Printer/Binder: Courier/Kendallville
Cover Printer: Lehigh-Phoenix Color/Hagerstown
Text Font: 12.5/15 NewCenturySchlbkLTStd

Credits and acknowledgments borrowed from other sources and reproduced, with permission, in this textbook appear on pages 601–602.

Microsoft® and Windows® are registered trademarks of the Microsoft Corporation in the U.S.A. and other countries. Screen shots and icons reprinted with permission from the Microsoft Corporation. This book is not sponsored or endorsed by or affiliated with the Microsoft Corporation.

5 6 7 8 9 10 V011 14

Prentice Hall
is an imprint of

www.PearsonSchool.com/CTE

ISBN 10:   0-13-237874-4
ISBN 13: 978-0-13-237874-1

About the Authors

In their work together, Michael Hacker and David Burghardt have consistently focused on helping students improve their engineering and technological literacy. The development of the informed design procedure that is at the heart of this textbook is an outgrowth of their research. Other efforts have involved gaining approval for and helping to implement a New York State mandate for a middle-school technology education program; developing learning standards for technology education; conceiving, developing, and implementing a high-school pre-engineering program; and conducting large-scale projects funded by the National Science Foundation that are focused on teaching and learning in science, technology, engineering, and mathematics (STEM).

 **Michael Hacker** serves as the Co-Director of the Center for Technological Literacy at Hofstra University. For over 45 years, engineering and technology education have been at the core of his professional life. He has been a classroom teacher, department supervisor, and university professor. He rose to local, state, and national prominence as an advocate for contemporary engineering and technology education programs in the nation's schools.

As the former New York State Education Department Supervisor for Technology Education, Hacker has managed the development and implementation of innovative middle- and high-school curricula that have served as national models. He was named State Supervisor of the Year in 1993 by the ITEEA, received the ITEEA Award of Distinction in 1995, and was inducted into the prestigious ITEEA Academy of Fellows in 2004. He was honored with the Epsilon Pi Tau Distinguished Service Award in 2001.

Hacker is the author of five secondary-school textbooks and dozens of journal articles. He served as a co-editor of three sets of NATO proceedings, as well as a major international publication focused on the role of engineering and technology education in fostering human development. He has served nationally as a consultant to the American Association for the Advancement of Science, as a National Science Foundation expert reviewer, and as a writer for the National Standards for Technological Literacy.

 **David Burghardt** is a Distinguished Professor of Engineering, Co-Director of the Center for Technological Literacy, and the former chair of Computer Science at Hofstra University. He is the author of ten texts in engineering and technology as well as many articles in engineering and technology education. He is a professional engineer in New York State, a chartered engineer in the United Kingdom, a member of the IEEE committee on linking colleges of engineering and education, an NSF reviewer, a principal investigator on NSF and NASA projects, and a writer for the New York State math, science, and technology learning standards.

At the Center for Technological Literacy, Burghardt has helped create a graduate program in integrated math, science, and technology for elementary and middle-school teachers. In addition, he designed a course that places informed design at the center of learning.

Acknowledgments

The authors dedicate this book to the technology teachers who have forged a vision for change into a viable and exciting Technology Education program, and we offer grateful thanks to the following individuals:

Cory Booth, Jonesboro Middle School, Sharpsburg, Georgia

Patricia Hunsucker, Evans Middle School, Evans, Georgia

Jim Kiggens, CEO, Course Games, Santa Barbara, California

Virgina King, Helfrich Park STEM Academy, Evansville, IN

Alan Horowitz, an exceptionally talented technology teacher, who contributed the quick activities that so significantly enhanced our work.

Shai Hacker for his help in developing the content on social media.

Robert Barden, former Head of Global Technology, Deutschebank, for his unwavering friendship, support, and conceptual thinking that helped frame many of the approaches taken in this book.

Mrs. Rose Ambrosino, Mathematics Consultant

Dr. Mary Dery, who helped with the chemical technology chapter

George Granlund, for the teacher annotations

Patricia Murphy, Open University, Milton Keyes, England

Dr. Colleen Hill, California State University at Long Beach

Dr. Bruce Tulloch, Director of Science, Bethlehem, NY CSD

We also wish to express our appreciation to the Pearson team that helped us create the Second Edition. Special thanks to Wyatt Morris, Senior Acquisitions Editor, Dan Trudden, Developmental Editor, and Alex Wolf, Associate Managing Editor.

Thanks to Victoria Hrdina for her excellent work on the new "Green Living" feature.

To all our friends and colleagues in Technology Education, we offer this work in the hope that it will contribute to the technological capability of our students and to the growth of our discipline.

Brief Contents

Contents

Special Features

How Technology Works...

Investigate technology principles through 3-D illustrations and interactive Web animations.

Inclined Plane

It is easier to slide a heavy box up a ramp than it is to lift it straight up.

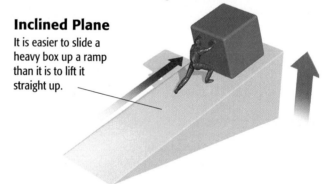

Connecting to STEM

Make direct links between Science, Technology, Engineering, and Math

People in Technology

Read about people working on the leading edge of technology.

Technology in the Real World

Read about how everyday technology works.

Design Activities

Plan, design, and build working projects that bring technology principles to life.

Using Your Text

You can use this book as a tool to master Technology Education. Spend a few minutes to become familiar with the way the book is set up, and see how you can unlock the secrets of Technology Education.

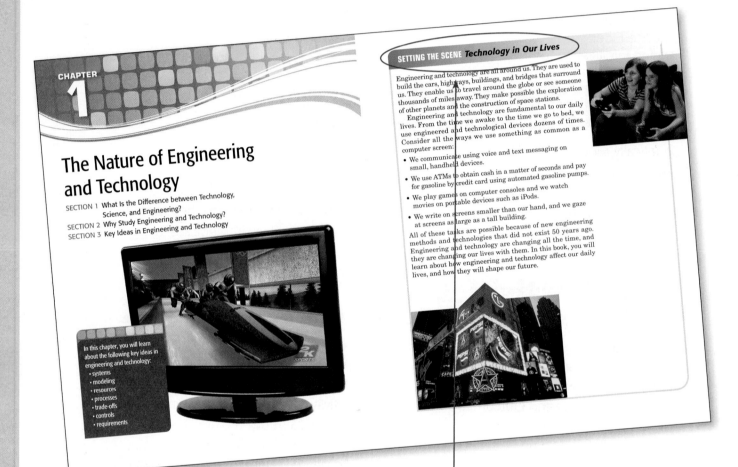

CHAPTER OPENER

Each chapter opens with **"Setting the Scene,"** which offers a real-world connection.

SECTION OPENER

Each section opener begins with a set of **"Benchmarks for Learning."** These keys prepare you to be aware of important concepts as you read.

READING STRATEGY

Each section opener provides a **reading strategy** to help guide you through the chapter material.

CRITICAL THINKING SKILLS

Each chapter contains many features designed to encourage you to apply **critical thinking skills.** Critical thinking allows you to identify and understand material as you read and relate concepts to your own life.

SECTION 1
Technology and Early History

Benchmarks for Learning
- Technology has influenced history, and it continues to have a major impact on society.
- An invention or innovation was often developed without scientific knowledge.
- The design and construction of modern structures evolved from prior technologies and techniques.

Reading Strategy
Outlining What were the important periods of early history? Which developments changed people's lives? Consider these questions and use an outline to organize the information you learn.

I. Early Historical Periods
  A. Stone Age
    1.
    2.
  B. Bronze Age
    1.
    2.

Vocabulary

Stone Age	Iron Age	Agricultural Era
Bronze Age	smelting	aqueduct
alloy		

Figure 2.1 Early humans used stone, bone, and wood to make many simple tools.

Describing How did early humans use technology?

Early Historical Periods

When did technology actually begin? People often think of our times as technological times and earlier eras as "before technology." In fact, people have always used technology. Recall that technology is simply using the knowledge we have to create things that people need. Because our knowledge has grown, technology has grown with it—from the simple spears and grinding stones of our earliest ancestors to the complex machinery and systems of today.

Much of our history has been defined by the materials that people used in a particular era. They used these materials to make tools that helped them carry out everyday tasks. The names of these materials have given us a way to classify historical periods. Three important periods of early history are the Stone Age, the Bronze Age, and the Iron Age.

Stone Age

The **Stone Age,** the period during which people used stone to make tools, lasted for almost 1 million years—from about 1 million B.C. to about 3000 B.C. (Other common materials used during this time were bones and wood.)

During this period, people lived very different lives from ours (Figure 2.1). Early humans used stone, bone, and wood to make axes, spears, scrapers, and even tools for starting a fire. They used the bow for hunting animals. They made needles from bone and used them to sew clothing from animal skins.

30 UNIT 1 The Nature of Technology

SECTION 1 Assessment

Recall and Comprehension

1. What are some of the earliest historical examples of technology?
2. Identify the first material that people used to create tools. For what purposes were these tools used?
3. Describe the advantages of iron over bronze.
4. Name the invention that made agriculture much more efficient.

Critical Thinking

1. **Analyzing** Explain how technologies such as writing and printing were responsible for other great advances in technology.
2. **Generalizing** Explain why the rate of technological change has increased.

QUICK ACTIVITY
Technology can be used to satisfy our desire for safety and security. Using a spring-type clothespin, string, a tongue depressor, insulated wire, aluminum foil, tape, a battery, and a buzzer, assemble a basic security system. The buzzer should sound when the tripwire string is pulled.
For more related Design Activities, see pages 52–55.

...ociety **35**

Technology in the Real World

Genetically Modified Foods

There is no question that genetic engineering lets us do amazing things. We can add vitamins in rice, producing better nutrition for millions of people. We can control the process of ripening in fruits—from tomatoes to coffee beans. We can even produce plants that "glow in the dark" when they need to be watered. All of these changes are produced by copying part of the genetic material from one organism to another.

Genetic engineering is used widely to control the process of ripening in fruits, such as these tomatoes.

Controlling a Sticky Pest
The bollworm, shown below, is a caterpillar that causes serious damage to crops such as cotton, corn, and potatoes. For decades, organic farmers have been treating their crops with a naturally occurring pesticide produced by the bacteria (small, single-celled organisms) *Bacillus thuringiensis,* or *Bt* for short. Bt is deadly to certain plant-eating insects, including the bollworm. Even better, it is harmless to most other insects.

Using genetic engineering, scientists made copies of the Bt gene that is harmful to the bollworm. They inserted this gene into common crop plants. The genetically modified crops remained pest free.

Some Unwanted Results
End of story? Not quite. The pests that were initially controlled by the Bt gene are starting to show signs of resistance to it. A greater percentage of each new generation of bollworms can tolerate the Bt gene. In addition, the Bt gene appears to be toxic to some beneficial insects, such as the monarch butterfly.

Scientists are quickly learning that genetic engineering can produce unanticipated and often unwanted results. Does this mean that we should stop using genetic engineering? Should we limit its use to certain types of organisms, such as plants and bacteria? Or should scientists be able to explore all the possibilities of genetic engineering, controlling and creating new life-forms at will? What do you think?

A bollworm munches its way through a cotton plant.

Critical Thinking

1. **Summarizing** What benefits did the new Bt crops produce for farmers?
2. **Extending** Given the problems that scientists encountered with Bt crops, what solutions might they explore?

CHAPTER 2 Technology and Society **49**

Applying Your Knowledge

1. Describe a problem in society or the environment that needs to be corrected.
2. Give an example of how a person's values might affect a decision about the kind of car to buy.
3. Suppose that you are an architect hired to design a house for this neighborhood. Planes fly overhead approximately every 7 to 10 minutes. What types of solutions might you explore?

4. You are in charge of market research for a new backpack. Design a questionnaire you will use to gather information from your classmates.
5. Draw simple sketches of three different designs for an ergonomic bottle. How is each one ergonomic?
6. Participate in and assume a leadership role in a design or science competition (e.g., A Technology Student Association, Lego League, First Robotics, Science Olympiad, or Science fair competition).

Critical Thinking

1. **Applying** Think of a problem that involves a personal issue. Propose a workable and economical solution to the problem. Include a sketch with explanations.
2. **Analyzing** Designers create a prototype or a smaller model after they have completed the preliminary design. Why is a prototype especially important when many final products will be made?
3. **Inferring** Your city has run out of land for a landfill (refuse disposal). The city government has chosen to build an expensive incinerator to handle the garbage problem. What might be some of the trade-offs that were made in reaching this decision?
4. **Analyzing Cause and Effect** Why are products sometimes recalled? How might designers help to prevent product recalls?
5. **Extending** Advertisers spend a lot of time and money trying to appeal to people's values. What types of values might be important when advertising high-quality, expensive sports gear?

Connecting to STEM
science • technology • engineering • math

Determining Risk
Risk is a part of life. Designers must keep this in mind as they create new products, such as a car. We know that the risk of a fatal accident is very low—about 24 deaths per billion miles, or 0.000000024. If you were to take a trip of 100 miles, what is the probability of your being involved in a fatal accident? How might designers reduce this number as they design new cars?

CHAPTER 3 Design and Problem Solving **79**

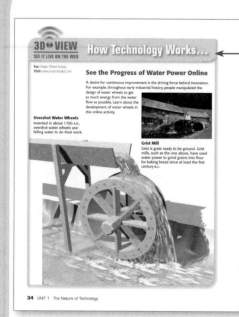

HOW TECHNOLOGY WORKS

How Technology Works features throughout the text incorporate 3-D Web activities with learning concepts.

Learn about a technical concept in your text and then watch it come to life on the Web in a 3-D, hands-on multimedia format.

PEOPLE IN TECHNOLOGY

Throughout the text, you'll learn about fascinating people who have made, or are currently making, an impact on technology.

LIVING GREEN

Each chapter contains a "Green Living" feature that introduces an environmental issue, offers suggestions for living a "greener" life, and an activity to complete on the topic.

CONNECTING TO STEM

You'll engage in fun activities that show how technology relates to other curriculum areas, specifically mathematics and science.

TECHNOLOGY IN THE REAL WORLD

Discover new and exciting technical innovations happening in the world today, and how today's technology affects your life.

CHAPTER REVIEW AND ASSESSMENT

Each chapter provides a point-by-point **summary** of key concepts, with explanations that allow you to reinforce material as you read each section.

A **Crossword Puzzle** found at the end of each chapter will help you reinforce key concepts throughout the text by means of a game that is fun and challenging to solve.

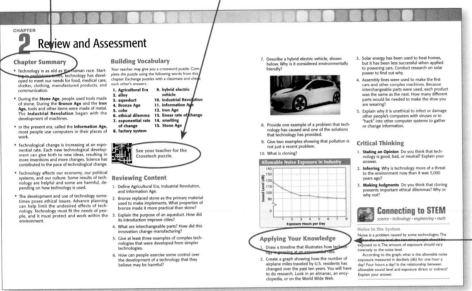

Apply your Knowledge questions end every chapter. These questions offer hands-on practice for using what you've just learned. They encourage you to think critically as you identify and evaluate technical situations.

UNDERSTANDING HOW TECHNOLOGY WORKS, AND APPLYING IT...BY DESIGN

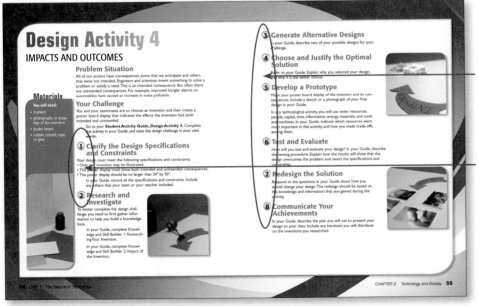

Design Activity features in every chapter offer hands-on practice of the technical concepts learned. Each exercise presents interesting technical challenges you can relate to and requires that you create a well-developed design solution.

In doing these activities, you will apply what you've learned to real-life situations and be able to understand the process by which the most effective design solutions are conceived.

Engineering AND Technology Education
Learning by Design

UNIT 1

The Nature of Technology

Unit Outline

Chapter 1
The Nature of Engineering and Technology

Chapter 2
Technology and Society

"Technology feeds on itself. Technology makes more technology possible."

—*Alvin Toffler,* Future Shock, 1970

The Nature of Engineering and Technology

In this chapter, you will learn about the following key ideas in engineering and technology:
- systems
- modeling
- resources
- processes
- trade-offs
- controls
- requirements

Engineering and technology are all around us. They are used to build the cars, highways, buildings, and bridges that surround us. They enable us to travel around the globe or see someone thousands of miles away. They make possible the exploration of other planets and the construction of space stations.

Engineering and technology are fundamental to our daily lives. From the time we awake to the time we go to bed, we use engineered and technological devices dozens of times. Consider all the ways we use something as common as a computer screen:

- We communicate using voice and text messaging on small, handheld devices.

- We use ATMs to obtain cash in a matter of seconds and pay for gasoline by credit card using automated gasoline pumps.

- We play games on computer consoles and we watch movies on portable devices such as iPods.

- We write on screens smaller than our hand, and we gaze at screens as large as a tall building.

All of these tasks are possible because of new engineering methods and technologies that did not exist 50 years ago. Engineering and technology are changing all the time, and they are changing our lives with them. In this book, you will learn about how engineering and technology affect our daily lives, and how they will shape our future.

What Is the Difference between Technology, Science, and Engineering?

Benchmarks for Learning

- Technology uses knowledge to turn resources into goods and services.
- Technology extends our natural abilities.
- Inventions and innovations have changed our society.
- Scientists study and explain the natural world, whereas technologists design new products and systems.
- Tools have been improved over time to do more difficult tasks and to do simple tasks more efficiently, accurately, or safely. Tools further the reach of hands, voices, memory, and the five human senses.
- Engineering is the process of creating or modifying technologies.

Reading Strategy

Listing As you read about the different aims of scientists and engineers, write down the important features of both occupations.

Scientists
- Study the natural world.

Engineers
- Design the human-made world.

Vocabulary

engineering	innovation	designing
technology	natural world	investigating
invention	human-made world	scientific inquiry

Technology in Our Daily Lives

Your alarm clock buzzes. You switch it off and check the large blinking numbers that show the time. You turn on the radio so you can listen to music. Once you get up, you switch on the lights in the bathroom and turn on the hot water for a shower. You prepare breakfast from foods that are kept fresh in a refrigerator. Maybe you get a ride to school in a bus or car. All of these activities and products are made possible by technology.

For most of us, nearly every part of our daily routine—from the time we get up to the time we go to sleep—depends on, or is affected by, technology (Figure 1.1). Artificial lighting makes it possible for us to be up and busy before dawn. Television and the Internet bring us live news from around the world.

We wear clothes that were designed on a computer and made from both natural and synthetic fabrics. Thermostats wired to modern furnaces and air conditioners keep our homes at a comfortable temperature. Technology shapes our lives in all these ways.

Some cultures, such as certain native South Pacific islanders or African tribes, are less dependent on advanced technology. People awake when the sun rises. They wash up when they can in nearby lakes or streams. They travel relatively short distances and live off the land. Our lives are very different from theirs because of our more advanced technology.

This textbook will help you learn how technology has become a part of our lives. As you read, you will learn how our routines—the ways we move about, work, learn, and communicate—are all affected by technology.

Defining Technology and Engineering

Technology makes use of human knowledge to convert resources into the goods and services that people need and want. Raw materials such as stone and wood are used to construct buildings. Solar energy is converted into electricity. Natural gas powers the furnaces that heat some of our homes.

Technology extends people's natural abilities. We can reach higher with ladders. We can swim underwater with scuba tanks. Like the skydivers in Figure 1.2, we can float through the air. We can communicate over great distances using telephones. We can even fly to the moon, circle it for months, and then fly back to Earth again.

Sometimes, technology allows the development of an entirely new device or product. This is called an **invention**. An invention can be very simple, such as a toothbrush, or it can be complex, such as a computer or cell phone.

Over time, an invention may be improved. This improvement is called an **innovation**. Because of innovations, technological devices used today are smaller, cheaper, and more powerful than the first models developed by inventors (Figure 1.3). We spend less time than our ancestors did on tasks such as cleaning and cooking. The innovations of technology have changed our daily lives dramatically.

Engineering is the process of creating or modifying technologies. Engineering is constrained by physical laws (laws

Figure 1.1 Most of our daily routines depend on, or are affected by, technology.

Contrasting *Which technology products did your parents not have when they were growing up?*

Technological Advances
Figure 1.2 Technology extends our capabilities.

of nature, such as energy cannot be created nor destroyed, or $E=mc^2$. Engineering is generally conducted by teams, and each member of a team contributes his or her special knowledge to the solution of a problem. Engineers generally design devices or structures or systems, and their designs must meet given performance specifications, or "specs." (Think about what the specifications might be for a bridge or for a passenger elevator.) **Designing** is the process of creating and planning a product or system. Engineers design under a set of constraints (limitations) related to the amount of time they have to complete the design, how much the design can cost, and limits on the kinds of materials that can be used. Engineers use math and science knowledge in creating design solutions to problems.

Comparing Technology and Science

Although people often talk about science and technology together, each field has a different focus. Science studies the **natural world**—Earth, space, and all living things not created by humans. Technology deals with the **human-made world**—buildings, bridges, communication systems, manufactured goods, and all the things that are the products of human creativity.

Figure 1.3 Technology has become much more portable.

Predicting *What innovations in communication technologies do you think might happen in the next ten years?*

Technologists

Technologists do their work by constructing and maintaining products, structures, and systems (Figure 1.4). Unlike scientists, technologists are not looking for answers to questions

Figure 1.4 Technologists, such as these architects, design the structures, products, and systems that make up the human-made world.

Classifying *Eyeglasses are one example of technology and science being used together to create a product. Name other examples.*

about the natural world. Rather, they are looking for ways to make the natural world easier to live in.

Technologists do not expect to find "correct" answers. Instead, they look for the best solutions to problems when there may be several good solutions. Technologists use their knowledge and tools to turn resources into goods and services that society needs or wants. They deal with the human-made world and its products and systems. Architects, product designers, and technicians are all technologists.

Scientists

Scientists try to understand things that occur in nature. They do their work by **investigating** the world around them (Figure 1.5). They explain the natural world based on the evidence they find in their investigations.

This process of examination is called **scientific inquiry**. In scientific inquiry, scientists observe and record the events around them. They perform careful experiments to test the effect of one thing (such as sunlight) on another (such as plant growth). Technology helps scientists in their experiments. Devices such as the microscope and the telescope help scientists learn more about the natural world. Biologists, chemists, and physicists are all scientists.

Engineers

Engineers create the human-made world and design artifacts and processes that never existed before. This is in

Figure 1.5 Scientists seek answers to questions about the natural world by collecting data and making observations.

Predicting *What forms of technology does this scientist use in his job?*

contrast to scientists, who study the natural world. Most often engineers do not literally construct the artifacts; they provide plans and directions for how the artifacts are to be constructed. Artifacts may be as small as a hand calculator or as large as a bridge. Engineers also design processes, such as those used in chemical and pharmaceutical industries to create chemicals and drugs, or those that direct how components are put together on an assembly line or those that control how checks are processed in banking.

Working Together

Engineers, scientists, and technologists work together. What scientists discover becomes new knowledge. Engineers add that knowledge to what they already know to design new technologies. Technologists construct and maintain the products, structures, and systems that engineers design. Scientific discoveries have brought about the invention of new technologies. For example, when X-rays were discovered, scientists noticed that this form of energy had the power to pass through human flesh, producing images of the insides of people. This knowledge led to the development of the X-ray machine, which technologists use, for example, to treat medical patients and to photograph passengers' suitcases at airports. Conversely, technology helps the development of science. X-rays have been harnessed for research in geology, chemistry, biology, and physics, and devices such as the microscope and the telescope help scientists learn more about the natural world.

SECTION 1 Assessment

Recall and Comprehension

1. What is technology?
2. How are the aims of science and technology different?
3. What is one example of a recent invention?
4. What is scientific inquiry?
5. What are innovations?

Critical Thinking

1. **Hypothesizing** Without modern technology, how would some of your daily routines be different?
2. **Evaluating** Think of some technological device that you use—anything from a toaster to a computer. How might an innovation improve it?

QUICK ACTIVITY

Keeping accurate time has been important since ancient times. Using a narrow 4-inch-long stick, tape, glue, and an 8-inch circular piece of cardboard, construct a sundial. Calibrate the sundial by placing it outside on a sunny day. Mark off the location of the stick's shadow every hour. To which compass direction does the shadow point at 12 noon? Why?
For more related Design Activities, see pages 24–27.

Why Study Engineering and Technology?

Reading Strategy

Mapping Use a concept map to help you keep track of all the different effects of technology.

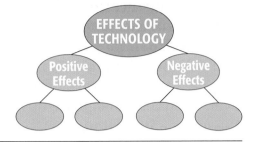

Vocabulary

technologically literate

Benchmarks for Learning

- People solve problems by using technology to develop new products and systems.
- Using technology to solve one problem sometimes creates new problems.
- Technology by itself is neither good nor bad, but its use may affect others. Therefore, decisions about products, processes, and systems must consider possible consequences.
- How technology is used determines whether it produces good or bad results.
- Technological literacy helps us make sense of our world.

Technology Has Improved Our Lives

Technology affects our lives in a variety of ways. Telephones, televisions, high-speed trains, space exploration, new medicines, synthetic clothing, and buildings under construction—all of these are the result of technology. Technology can help us satisfy our needs, making our lives more comfortable, healthy, and productive.

Many years ago, travel across the United States took weeks or even months. Handwritten letters were the only way of communicating with people in different towns. The Pony Express was the quickest way to deliver mail. Supplies were delivered slowly and usually came from local sources.

Today, jets can take us across the country in hours. We can write to, talk with, and see each other instantaneously using such devices as portable computers and digital video cameras (Figure 1.6). Using the latest technology, supplies can be ordered over the Internet and delivered overnight.

Figure 1.6 Technology has changed how and where we do our work.

Classifying *How has technology made your life easier?*

These changes are just a few examples of how dramatically technology has improved our lives. Years of invention and human creativity have produced changes in every part of our lives. New products and systems help us solve problems and accomplish tasks more easily.

Water-purification systems contribute to healthier living. Genetic engineering produces foods that stay fresh longer. Modern machinery allows us to manufacture goods cheaply and reliably. Computers are able to access, organize, and interpret huge amounts of data and information faster than ever. With technology, we have created a different world from the one we inherited.

Technology Creates New Challenges

The effects of technology are complex. A technological solution to a problem sometimes creates new problems (Figure 1.7). Convenient disposable products, such as hospital supplies, can create more waste. Although they allow foods to be stored longer, some food preservatives can harm our health. Jet planes and road traffic create noise and pollution (Figure 1.8). New technologies can have unforeseen or unwanted consequences.

Technology also presents us with many new choices. Will certain products harm the environment? Will our privacy be protected if we use the Internet? Are new technologies causing people to lose their jobs? Technology can create new challenges and may complicate our lives.

Given these challenges and complications, should we get rid of technology? If we went back to a simpler way of life, would

Challenges of Technology

Figure 1.7 All inventions may have unintended consequences.

Figure 1.8 This sound barrier prevents some of the traffic noise pollution from reaching nearby houses.

Making Judgments *What steps can society take to make sure technology is used wisely?*

Pluses and Minuses of Technology		
Invention	**Technological Plus**	**Technological Minus**
Cell phones	Ability to speak to someone from any location	Using cell phones while driving has increased the number of car accidents.
Highways	Ability to travel greater distances in less time	Cars create noise and pollution, particularly for homes near highways.
Air conditioning	Comfort at home and at work	Temperatures are 10°F higher in subways than outside due to exhaust from subway air conditioning units.
Internet	Global communication	Privacy is a major concern of computer users.

Figure 1.7

Figure 1.8

LIVING GREEN
Turn Your Junk into Gems

Most teens today are far more familiar with DVDs than VHS tapes, and many people download digital movies from the Internet or other media providers rather than rent from a video store. Similarly, very few people go out and buy CDs when MP3 downloads are readily available and very affordable. So what do we do with all our old CDs, DVDs, and VHS tapes? Remember that these products are manufactured from plastics and metal resins. When we toss them in the garbage they end up in landfills, taking thousands of years to degrade.

MAKE A CHANGE
There are tons of uses for your out-of-date, unwanted, or unusable compact discs, DVDS, even old software and computer games. Start by not throwing them in the trash. Then, check out the life cycle of a CD or DVD at the Environmental Protection Agency's website http://www.epa.gov/wastes/education/pdfs/finalposter.pdf. A great way to get the word out would be to hold a school-wide collection drive!

TRY THIS
Find a way to use your old computer or music discs in your next technology design project. If you've collected a bunch through a school drive you'll have a lot on hand. All these discs come with pre-drilled holes for axles and make great wheels for model vehicles. What other uses can you think of?

problems such as pollution disappear? Probably not. The fact is, we depend on technology. Our lives and our daily routines are built around it. We would not want to go back to a life without cars, flush toilets, telephones, or modern medical care.

Technology itself is not harmful, but the way we use it can sometimes cause problems. For example, computers have made many everyday tasks easier to complete. Yet many computer users complain of problems such as eyestrain as a result of too much time spent looking at the monitor. Yet technology can also often solve those problems. Therefore, we must learn how to use technology wisely. Knowledge gained from past experience can help us be more technologically literate, so we can better deal with unexpected problems.

Technological Literacy

To truly make sense of our world, we must understand what technology can and cannot do for us—we must become technologically literate.

People who are literate in English understand what is written or spoken in the English language. People who are **technologically literate** understand technology. They understand how processes work and how products are made.

New technologies in all areas are being distributed to the public at an ever-increasing pace. With some basic knowledge of technology, people can benefit both at work and at home by being able to choose the best products for their wants and needs. This basic knowledge then helps them to use and maintain products properly.

Technologically literate people know that technology is created to meet human needs. They also understand that the laws of nature impose limits on what technology can do. For example, we cannot reach speeds greater than the speed of light.

Technologically literate people can make informed decisions about how technology is used. As citizens in a democracy, we can affect how technology is used. We can write letters, vote, and speak out about what is happening. We can control technology.

SECTION 2 Assessment

Recall and Comprehension

1. What are three benefits of technology? What are three potentially negative impacts of technology?
2. What does it mean to be technologically literate?
3. What are some examples of new products or systems?

Critical Thinking

1. **Defending** Is technology good or bad? Defend your point of view.
2. **Making a Judgment** In what ways are you technologically literate? What area of technology would you like to know more about?
3. **Evaluating** Think of some technological device that you use. Does it have any drawbacks? How might it be improved?

QUICK ACTIVITY

Almost all inventions produce both desirable and undesirable results. On a large chart, make a list of ten inventions or discoveries. Next to each, list one benefit and one negative impact it has had on people or the environment. Compare your list with those of other students in the class.
For more related Design Activities, see pages 24–27.

People in Technology

Dean Kamen
Inventor of Battery-Powered Walking

"[The Segway will] be to the car what the car was to the horse and buggy. . . . A car simply doesn't belong in the city."

—Dean Kamen

Dean Kamen has given a lot of serious thought to people's everyday routines. As a young boy, he devised a way to change the sheets on his bed by using a system of ropes and pulleys. This contraption saved him the hassle of having to run from one side of his bed to the other.

The Segway—No Accelerator, No Brake Pedal

Now that he's an entrepreneur and engineer, Kamen's inventions have become a bit more

The Segway is being used by police in some large cities.

sophisticated. Nevertheless, they are still grounded in the realities of day-to-day living. His latest creation, the Segway Human Transporter, allows people to move about more quickly and efficiently than they can by walking.

The machine is a simple battery-powered platform that has no brakes and no accelerator. Instead, a person standing on the vehicle leans forward to move the Segway forward, leans back to propel the Segway backward, and stands up straight to stop the Segway. Kamen made the Segway easy to operate so that people would be more likely to use it.

A Replacement for Cars?

In Kamen's ideal world, city streets would be quiet. Sidewalks would be filled with police officers, postal workers, and business people riding his two-wheeled invention. Companies such as GE Plastics and even the U.S. Postal Service are testing the Segway to improve navigation around their facilities and work areas.

Kamen hopes that the Segway will "be to the car what the car was to the horse and buggy." Half the world's population now lives in cities. So, if Kamen is right, the Segway could produce major changes to city transportation and to the environment.

Critical Thinking

1. **Making a Judgment** Kamen sees the Segway as a type of transportation between walking and driving. Does its limited purpose make the Segway more or less useful?

2. **Evaluating** People have criticized the Segway because it is expensive—more than $3,000. How important is cost as a factor when choosing new types of technology?

Key Ideas in Engineering and Technology

Benchmarks for Learning

- Every field of study has key ideas that distinguish it from other fields.
- In technology, key ideas relate to systems, resources, requirements, trade-offs and optimization, processes, and controls.
- Feedback is used to adjust a system if conditions around it change.

Reading Strategy

Listing As you read, list the different key ideas of technology. For each idea, write down a brief definition in your own words and provide one or more examples. For example,

1. **System**—group of parts that work together
   - CD player
   - air conditioner

Vocabulary

system	requirement	process
models and modeling	trade-off	control
resource	optimization	feedback

Systems

Every field of study includes a number of key ideas that set that field apart from others. In mathematics, key ideas relate to algebra, geometry, and measuring and analyzing data. In science, key ideas relate to the living environment and the physical setting. In technology, key ideas involve systems, resources, requirements, trade-offs and optimization, processes, and controls.

A **system** is a group of interrelated parts that work together to produce a desired result. A system can be any imaginable size, but its size does not affect its complexity. One example of a very small system is a computer chip, which includes millions of electronic parts (Figure 1.9). A system can also be as large as the global network that connects telephones all over the world.

Models and Modeling

A model is a representation of a system, process, product, structure, or environment. Models can be *graphic* (such as a drawing), *physical* (such as a scale model), virtual (such as a computer simulation), or *mathematical* (such as a graph, chart, or equation). Modeling alternative designs allows engineers and designers to think about advantages and

Figure 1.9 A computer chip is actually a small system.

Clarifying *In your own words, what is a system?*

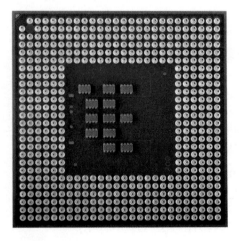

disadvantages of possible solutions before they invest time and money in producing their designs. In engineering and technology, using models and modeling techniques are important in helping to represent, analyze, predict, and explain the performance of a device or system.

Very often, physical models are built before the final product is produced. For example, a model is often created by architects and structural engineers before a building is built.

There are a variety of model types. Physical models, such as the model of the Freedom Tower in Figure 1.10A, model cars, and model rockets are one type. Drawings and diagrams such as the one shown in Figure 1.10B are also models.

A computer simulation that shows how a system works is another kind of model. Even a mathematical equation is a model. Mathematical models are used often by scientists and engineers to predict the behavior of a system. For example, a formula (known as Ohm's Law) that relates current flow (I) to the amount of voltage (E) and electrical resistance (R) in a circuit (I = E/R) allows a designer to predict how much current will flow through a light bulb when it is plugged into an electrical outlet that provides 120 volts.

Resources

All systems involve the use of resources to make them work properly. A **resource** is any type of supply, service, or

Figure 1.10A Model of Freedom Tower

Figure 1.10B Diagram of toilet tank flushing mechanism

Figure 1.10A

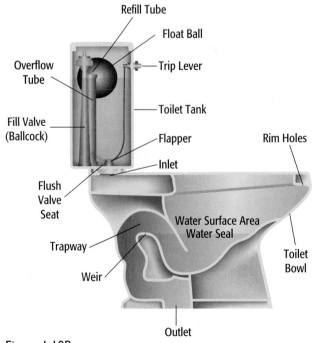

Figure 1.10B

support. Resources include the information used to design and test a new plane or the fiber used to make a new type of cloth. They also include the money, or capital, needed to pay people to make new products. The seven types of resources used in technology are people, capital, time, information, energy, materials, and tools and machines.

Requirements

Before systems can be designed, requirements need to be specified. **Requirements** are the features that define the performance of a system. Suppose that you are designing a new car. How quickly should the car accelerate? How fuel-efficient must it be? How should information be displayed to the driver? How much room is needed for seating and storage? These questions relate to the requirements of a system.

Trade-offs and Optimization

When technologists develop products, they often cannot include everything they want in their design. Sometimes parts are too costly, or certain desirable materials might be hard to find, or disposing of the product will harm the environment. Such realities require technologists to make trade-offs.

A **trade-off** is the giving up of one option in order to gain a better or more realistic option. For example, a manufacturer might be willing to pay more money to get a product that lasts longer. The process of choosing alternatives to make the solution work as well as it can is called **optimization.**

Processes

Systems often involve **processes**, or a series of steps by which resources are changed into desired products. Through specific processes, synthetic hormones are made into pills, plastic is molded into chairs, and moving water is converted to electricity. Systems may also use technological processes to organize, store, or move data and goods.

Controls

Systems need various controls to make sure that the entire operation functions properly. A **control** is a device that is used to adjust the operation of a system. A control can be

For: Car Construction Activity
Visit: www.mytechedkit.com

Build Your Own Car Online

Budget is a common limitation when making a large purchase. For example, when shopping for a car, consumers have many options to choose from, but each has a cost. Optimizing their wants versus their budget can be a challenge. Take on this challenge in an online activity.

All-terrain tires

Super Quadraphonic Stereo

8-cylinder engine

as simple as the handlebars on your bicycle, which you use to keep the bicycle moving in the correct direction, or as complex as the computer in a commercial airplane, which processes vast amounts of data and keeps the plane flying safely.

Controls are important for three reasons: (1) The goods, services, information, or energy produced should be exactly right. (2) A technological system should not harm people or the environment. (3) Resources should be used in the best way possible.

Many technological systems are controlled automatically by feedback. **Feedback** is information about the output of a system that can be used to adjust it. Because of feedback, systems adjust themselves as conditions around them change. Feedback can come from a human, as when your parents tell you to turn down the stereo, or from a mechanical device, as when a thermostat turns on the furnace because the room is too cold.

SECTION 3 Assessment

Recall and Comprehension

1. What are seven key ideas of engineering and technology? Make a list and provide a definition for each idea in your own words.
2. Give two examples of feedback that you have received today.
3. List the different types of resources used in systems.
4. What are three different types of models? Give an example of each.

Critical Thinking

1. **Applying Concepts** Your school functions as a type of system. Components of this system include teachers, students, and the school building. What are some of the controls that keep a school system operating correctly?
2. **Predicting** Suppose you are designing a new desk for your computer. What are your requirements? What resources will you need?

QUICK ACTIVITY

The drive system for a bicycle involves many different parts. Make a simplified diagram that shows how power is transmitted to a bicycle's wheels. Label the key parts of the system. What is the energy source that powers a bicycle?
For more related Design Activities, see pages 24–27.

Connecting to STEM

Picking the Best One

It is fairly easy to pick the best computer if the only thing you consider is how fast it processes data. But what happens when you want to consider other factors besides processing speed, such as cost, software, and monitors?

When you purchase a computer system, you often cannot afford to include every feature you want. You must make trade-offs, exchanging the benefits and disadvantages of one system for those of another. The process of making trade-offs to get the best possible solution is called optimization. Here, you will use mathematics to help you decide on trade-offs when buying a computer.

includes more software. The table below shows all the characteristics scored for Computer A and Computer B.

In the last two columns, each characteristic has been multiplied by its weighting factor. When these two columns are totaled, you get a surprising result: The faster computer, Computer A, is not the better choice. When all the important features are weighed carefully, Computer B is

Comparing Two Computers					
Features	Weighting Factors	A	B	A's Score × Weighting Factor	B's Score × Weighting Factor
Speed	10	10	7	100	70
Cost	10	6	10	60	100
Software	8	5	9	40	72
Size	7	7	8	49	56
Drives	5	9	6	45	30
				Total 294	Total 328

Which Features Are Important?

First, list the features that you think are important, such as processing speed, free and useful software, types of monitors, types of drives, physical size, brand, and cost.

Next, you need to establish a rating system. A rating of 10 means the feature is the most important to you, and 1 means that it is the least important. The value assigned to each choice is called a **weighting factor**. A weighting factor allows you to give more weight, or importance, to one choice than to another.

Comparing Two Computers

Now you can use these weighting factors to rate two computers, A and B. Computer A is 1.5 GHz faster than Computer B, but it costs $400 more. Computer B is more compact and

found to be the better product for your needs. A slower speed becomes a reasonable trade-off for better cost, better software, and smaller size.

Critical Thinking

1. **Extending** If all the features had been weighted equally, would Computer B still have been the better choice? Explain how you determined your answer.

2. **Applying** Select a product that interests you, such as a mountain bike, tennis racket, or running shoes. Perform the analysis you learned here to compare different versions of this new product.

Review and Assessment

Chapter Summary

- Our routines and the way we live are greatly affected by the devices, products, and services that technology provides. Today, technology is all around us and affects much of what we do.

- Technology uses knowledge to turn resources (such as materials and energy) into goods and services that people need.

- Science and technology have different aims. Scientists study the natural world and try to explain it based on evidence from their investigations. Technologists primarily design and maintain products and systems.

- People solve problems by using technology to develop new products and systems. Technology may also create new problems, however. The way technology is used determines whether it produces desirable or undesirable results. To use technology wisely, we must understand what it can and cannot do for us.

- Technology is based on a set of key ideas. These ideas involve systems, resources, requirements, optimization and trade-offs, processes, and controls.

Building Vocabulary

Your teacher may give you a crossword puzzle. Complete the puzzle using the following words from this chapter. Exchange puzzles with a classmate, and then check each other's answers.

1. control
2. designing
3. feedback
4. human-made world
5. innovation
6. invention
7. investigating
8. natural world
9. optimization
10. process
11. requirement
12. resource
13. scientific inquiry
14. system
15. technologically literate
16. technology
17. trade-off

See your teacher for the Crosstech puzzle.

Reviewing Content

1. Explain the different aims of science and technology.
2. How did technology affect your routine this morning? List at least three ways.
3. What are three benefits of technology? Three potential drawbacks?
4. Why is technological literacy important?
5. What are the key ideas of technology?

Applying Your Knowledge

1. Choose an ordinary device that you use in your home. Make a timeline of innovations and inventions for this product. What future changes, or innovations, might be made?

2. Research perpetual motion machines on the Web. Write a short argument either for or against the possibility of their existence.

3. Look in your newspaper for an example of a local, national, or global problem that uses technology as part of its solution. Explain how technological literacy can help you discuss the problem intelligently.

4. Describe a technological system that you have used recently. Identify the controls that are built into the system.

5. If you were designing a pen for yourself, what requirements would you specify? How might these requirements be different if you were designing pens to be used at a checkout counter?

Critical Thinking

1. **Evaluating** List five basic needs that all people have. How does technology address these needs?

2. **Making a Judgment** As a nation, we are producing an increasing amount of garbage in part because of inexpensive, disposable products. Should the technology that is used to make these products be discontinued? Why or why not?

3. **Extending** Factors that are considered in optimizing a new type of technology include the cost of materials, the cost of labor to make it, and the effects of the technology on the environment. What other factors might be considered? List as many as you can.

4. **Comparing** People in some parts of the world do not depend heavily on modern technology. These women are using ropes and pulleys to draw water from a well. What are the advantages and disadvantages of a life that uses simpler technologies?

5. **Applying** Your city is considering a law that will increase the speed limit on all local roads. Supporters of this law say that it will reduce gridlock and make local errands and deliveries more efficient. How might technological literacy affect your decision? What areas would you research if you were going to cast your vote?

 Connecting to STEM

science · technology · engineering · math

Science and Technology

The X-ray machine is an invention that is used in hospitals every day to pinpoint a disease or to check for broken bones. It is an example of how science and technology work together.

Using the Internet and resources in your library, research the history of the X-ray machine. Who invented it and when? What was its original purpose?

Design Activity 1
INNOVATION

Problem Situation

Changes and improvements are often made to a product to satisfy a need. People may ask that a device be made better in some way. Awareness of the needs of others can challenge us to improve products so more people can use them. For example, opening the screw top of a bottle is difficult for people who cannot grip objects tightly. Twist-off jar openers do exist, but they function primarily on jars and larger bottle tops. The tops on soda bottles and other beverage bottles are smaller. Your task is to improve on an existing jar opener so that it will work on beverage bottles.

Materials

You will need:
- bolts and nuts
- glue gun
- locking pliers
- metal strips
- rubber bands
- spring scale
- Velcro®
- wooden strips

Your Challenge

You and your teammates are to design and construct a bottle opener.

> Go to your **Student Activity Guide, Design Activity 1.** Complete the activity in the Guide, and state the design challenge in your own words.

1 Clarify the Design Specifications and Constraints

To solve the problem, your design must meet the following specifications and constraints:
- The bottle opener should accommodate a variety of top sizes.
- It should be easy for most people to use.
- It needs to have enough force to open most bottles.

> In your Guide, state the design specifications and constraints. Add any others that your team or your teacher included.

2 Research and Investigate

To better complete the design challenge, you need to first gather information to help you build a knowledge base.

> In your Guide, complete Knowledge and Skill Builder 1: Investigate Existing Products.

> In your Guide, complete Knowledge and Skill Builder 2: Size Requirements.

In your Guide, complete Knowledge and Skill Builder 3: Force Requirements.

3 Generate Alternative Designs

In your Guide, describe two possible solutions that your team has created for the problem. Your solutions should be based on the knowledge you have gathered so far.

4 Choose and Justify the Optimal Solution

Refer to your Guide. Explain why you selected the solution you did, and why it was the better choice.

5 Develop a Prototype

Construct your bottle opener. Include a drawing or a photograph of your final design in your Guide.

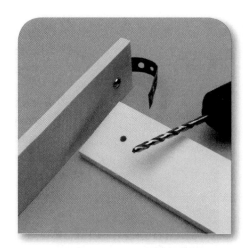

In any technological activity, you will use seven resources: people, capital, time, information, energy, materials, and tools and machines. In your Guide, indicate which resources were the most important in this activity, and how you made trade-offs between them.

6 Test and Evaluate

How will you test and evaluate your design? In your Guide, describe your testing procedure. Explain how the results will show that the design solves the problem and meets the specifications and constraints.

7 Redesign the Solution

Respond to the questions in your Guide about how you would redesign your solution. The redesign should be based on the knowledge and information that you have gained during this activity.

8 Communicate Your Achievements

In your Guide, describe the plan you will use to present your solution to your class. Include any handouts and/or PowerPoint slides that you will use.

Design Activity 2

THE GREAT SPINOFF

Problem Situation

Tops are one of the oldest known toys. In China during the 1800s, professional top spinners entertained large audiences by performing tricks with tops, such as making them jump up stairs and walk up and down an incline. Competitions with tops date to the 1700s. Contests were held to judge a top's decoration, speed, musical sound, and duration of spin. Today, tops are still popular all over the world. The most common top receives its motion from a string that is wrapped around it and then pulled. Your task is to design and create a top that would win a modern-day competition.

Your Challenge

You and your teammates are to build a top that will spin for the longest time in The Great Spinoff competition in your class.

> Go to your **Student Activity Guide, Design Activity 2.** Complete the activity in the Guide, and state the design challenge in your own words.

① Clarify the Design Specifications and Constraints

To solve the problem, your design must meet the following specifications and constraints:
- The top should spin longer than any other top made in your class.
- The top must be sturdy, attractive, and safe for younger children to use.
- The top must use the type of handle and pull string specified by your teacher.

> In your Guide, state the specifications and constraints. Write down any others that your team or your teacher would like to include.

② Research and Investigate

To better complete the design challenge, you need to first gather information to help you build a knowledge base.

> In your Guide, complete Knowledge and Skill Builder 1: Top Research.

> In your Guide, complete Knowledge and Skill Builder 2: Center of Mass Variation.

Materials

You will need:
- 1/16", 3/8", and 7/16" twist drills
- 3/8" dowel rod
- 6" squares of hardwood or softwood
- belt sander
- brass escutcheon pins or round head rivets
- drill press
- hole saw
- jigsaw or band saw
- paint and finishing supplies
- sandpaper
- stopwatch
- string
- wood glue

3 Generate Alternative Designs

In your Guide, describe two possible solutions that your team has created for the problem.

In each alternative design, be sure to note the variables that might affect the performance of the top. Variables are features of the top's construction that can be changed, such as its diameter and thickness.

4 Choose and Justify the Optimal Solution

Refer to your Guide. Explain why you selected the solution you did, and why it was the better choice.

5 Develop a Prototype

Now it is time to construct your top. Be sure to wear safety glasses and follow all the machine and hand-tool safety rules. Include a drawing or a photograph of your final design in your Guide.

In any technological activity, you will use seven resources: people, capital, time, information, energy, materials, and tools and machines. In your Guide, indicate which resources were the most important in this activity, and how you made trade-offs between them.

6 Test and Evaluate

How will you test and evaluate your design? In your Guide, describe your testing procedure. How will the design solve the problem and meet the specifications and constraints? Justify your reasoning.

7 Redesign the Solution

Respond to the questions in your Guide about how you would redesign your solution. Your redesign should be based on the knowledge and information that you have gained during this activity.

8 Communicate Your Achievements

In your Guide, describe the plan you will use to present your solution to your class. Show what handouts or overhead transparencies you will use.

Technology and Society

In this chapter, you will learn about the following important periods in the history of technology:

- Stone Age
- Bronze Age
- Iron Age
- Industrial Revolution
- Information Age

Technology has led to dramatic changes in the way we grow food, treat disease, manufacture complex machinery, commute to school and work, and even heat our homes. The pictures on this page tell the story of one of the most exciting changes in technology—how we communicate.

- The telegraph, invented in 1837, allowed people to send short written messages. Words were translated into a special code that was sent by electric pulses across a telegraph wire.

- In 1876, Alexander Graham Bell invented the telephone. For the first time ever, the sound of the human voice was transmitted by electrical signal.

- About 100 years later, the first cellular telephone was invented. Clunky and expensive, it was still innovative.

Today, telephone communication is inexpensive, and calls can be made from anywhere to anywhere. Wall jacks and traditional phone lines may soon be a thing of the past, as more devices are combining phones with computers and many people are replacing land lines with cell phones.

Telephones are only one example of the revolution in technology. In this chapter, you will learn about comparable changes in agriculture, manufacturing, energy, and transportation.

Technology and Early History

Benchmarks for Learning

- Technology has influenced history, and it continues to have a major impact on society.
- An invention or innovation was often developed without scientific knowledge.
- The design and construction of modern structures evolved from prior technologies and techniques.

Reading Strategy

Outlining   What were the important periods of early history? Which developments changed people's lives? Consider these questions and use an outline to organize the information you learn.

> I. Early Historical
>   Periods
>   A. Stone Age
>   1.
>   2.
>   B. Bronze Age
>   1.
>   2.

Vocabulary

Stone Age
Bronze Age
alloy

Iron Age
smelting

Agricultural Era
aqueduct

Figure 2.1   Early humans used stone, bone, and wood to make many simple tools.

Describing   *How did early humans use technology?*

Early Historical Periods

When did technology actually begin? People often think of our times as technological times and earlier eras as "before technology." In fact, people have always used technology. Recall that technology is simply using the knowledge we have to create things that people need. Because our knowledge has grown, technology has grown with it—from the simple spears and grinding stones of our earliest ancestors to the complex machinery and systems of today.

Much of our history has been defined by the materials that people used in a particular era. They used these materials to make tools that helped them carry out everyday tasks. The names of these materials have given us a way to classify historical periods. Three important periods of early history are the Stone Age, the Bronze Age, and the Iron Age.

Stone Age

The **Stone Age,** the period during which people used stone to make tools, lasted for almost 1 million years—from about 1 million B.C. to about 3000 B.C. (Other common materials used during this time were bones and wood.)

During this period, people lived very different lives from ours (Figure 2.1). Early humans used stone, bone, and wood to make axes, spears, scrapers, and even tools for starting a fire. They used the bow for hunting animals. They made needles from bone and used them to sew clothing from animal skins.

During the Stone Age, there were few villages. Most people lived nomadic lives, wandering from place to place. They hunted animals and gathered plants, fruits, seeds, and roots for food. They used fire for cooking meat and for protection from wild animals. When they had used up food sources in an area, they moved to a new location. The pace of technological change was very slow.

Bronze Age

During the **Bronze Age,** which lasted from about 3000 B.C. to about 1200 B.C., people began to craft tools and weapons from bronze rather than from stone. The Bronze Age actually began with the discovery of copper. Copper is a soft metal, so it was not practical for tools. During the Bronze Age, however, people learned that copper ore could be heated using charcoal to yield pure copper. They discovered that by melting other ores with copper, they could produce a stronger metal. Bronze is a mixture of the metals copper and tin. A mixture of two or more metals is called an **alloy.**

Bronze is lighter than stone, and bronze tools and weapons can be made sharper than those made of stone. Using bronze, early humans created new tools such as knives, hooks, and pins. They could also craft large urns and vessels to hold water and food (Figure 2.2).

Iron Age

The **Iron Age,** the era when iron came into common use, followed the Bronze Age at different times around the world. It began around 1200 B.C. in the Middle East and about 450 B.C. in Great Britain. It was shorter than either of the two previous periods, lasting only about 1,000 years.

Figure 2.2 Because bronze is lighter than stone, it was used to make large urns and vessels.

Hypothesizing *How might large containers make people's daily tasks more efficient?*

Figure 2.3 The iron-smelting furnace was used to soften iron. The blacksmith could then hammer the iron into different shapes.

Describing *What examples of technology can you find in this image?*

Iron was made from iron ore through a process called **smelting,** in which the ore was melted to take out the impurities. The early iron-smelting furnace was a clay-lined hole in the ground. Iron ore and charcoal were placed in the hole, and then air was pumped in with a bellows. Air from the bellows made the charcoal burn hot enough to soften the ore into a mass of iron. This mass was hammered into the desired shape while it was still red hot (Figure 2.3).

Iron eventually replaced bronze as the primary metal used in toolmaking because it was available in more areas and was cheaper to process. It could also be crafted into sharper, longer-wearing tools, such as chisels and saws. As with bronze, the techniques used to make iron varied in different parts of the world. People experimented with temperatures and with the amounts of the various ores they used. They often developed these early technologies by trial and error, without an understanding of the underlying scientific principles.

Key Inventions of Early History

During early history, people in different parts of the world used stone, bronze, and iron for many of their needs. Tools gradually became more sophisticated and easier to use. People in colder climates had a greater need for warm clothing and shelter. People who lived far from a river or lake wanted to be able to store water for later use. Over many thousands of years, important inventions were developed to meet these different needs.

Although many tools were used in similar ways throughout the world, the inventions in a particular area often reflected the specific needs of those people. For example, when horses were used as a form of transportation, the saddle was then invented so riders could ride more comfortably.

Agriculture

About 10,000 years ago—in 8000 B.C.—the nomadic, hunter-gatherer way of life changed. People learned how to grow their own food. They cultivated different types of grains for food, such as wheat and barley. They became farmers who grew flax, which provided fibers to make linen cloth. They invented the wheel and then vehicles with wheels. They built roads and stone houses.

The plow was first developed approximately 4000 B.C. This simple tool (Figure 2.4) allowed farmers to turn over soil for planting more easily. It made farming much more efficient. Soon, strong animals such as oxen were used to pull the plow. Farmers could now plow more land in less time.

Figure 2.4 The plow allowed most of the heavy work of turning the soil to be done by strong animals such as oxen.

Inferring *How did the use of strong animals change farming?*

Figure 2.5 The Pont du Gard is the largest and best preserved of the Roman aqueducts.

Hypothesizing *How do you think the use of aqueducts changed early cities?*

This period in history is described as the **Agricultural Era,** a time when most people lived off the land. Many tools and discoveries had to do with growing and harvesting crops.

Irrigation, Sewage Systems, and Aqueducts

With farming and the rise of towns and cities came the need for water systems. People developed irrigation systems to bring water to dry soil and increase plant production. Sewage systems helped safeguard health.

The ancient Romans built **aqueducts** to carry water from the nearby hills to the city. An **aqueduct** (Figure 2.5) is a channel built to carry water. The underground ones were channels carved into rock. Some were raised on arches to pass over rivers or valleys. All were built with a slight slope to allow for gravitational flow.

How Technology Works...

For: Water Wheel Activity
Visit: www.mytechedkit.com

See the Progress of Water Power Online

A desire for continuous improvement is the driving force behind innovation. For example, throughout early industrial history, people manipulated the design of water wheels to get as much energy from the water flow as possible. Learn about the development of water wheels in this online activity.

Overshot Water Wheels
Invented in about 1700 A.D., overshot water wheels use falling water to do their work.

Grist Mill
Grist is grain ready to be ground. Grist mills, such as the one above, have used water power to grind grains into flour for baking bread since at least the first century B.C.

Water Wheels and Machinery

An important early invention was the water wheel. In the earliest water-wheel systems, water from a river or stream turned a large wheel. This wheel turned a mill for grinding grain into flour. It replaced human muscle power with water power. Before the water wheel, grain was ground by hand using two large stones. This was such hard work that people were sometimes made to do it as punishment.

The water wheel introduced the machine age. It provided power not only for grinding grain but also for pumping water, making cloth, and heating iron ore for smelting. The use of machinery such as the water wheel was perhaps the most important technological development of the Middle Ages.

Writing and Printing

Writing enabled people to record their ideas and make them permanent. It increased the amount of knowledge that could be transmitted from one generation to the next and from one culture to another. People first wrote on stone or clay tablets. Years later, paper made from wood pulp was invented.

The printing press further transformed the way information was spread. Before the printing press, books were copied by hand, a process that often took weeks or even months. In the mid-1400s, a German goldsmith, Johann Gutenberg, devised a printing press that used small metal molds for all the letters of the alphabet. Once the letters were set on the press, text could be printed in a matter of minutes.

SECTION 1 Assessment

Recall and Comprehension

1. What are some of the earliest historical examples of technology?
2. Identify the first material that people used to create tools. For what purposes were these tools used?
3. Describe the advantages of iron over bronze.
4. Name the invention that made agriculture much more efficient.

Critical Thinking

1. **Analyzing** Explain how technologies such as writing and printing were responsible for other great advances in technology.
2. **Generalizing** Explain why the rate of technological change has increased.

QUICK ACTIVITY

Technology can be used to satisfy our desire for safety and security. Using a spring-type clothespin, string, a tongue depressor, insulated wire, aluminum foil, tape, a battery, and a buzzer, assemble a basic security system. The buzzer should sound when the tripwire string is pulled.
For more related Design Activities, see pages 52–55.

The Pace of Technological Change

Benchmarks for Learning

- Economic, political, and cultural issues are influenced by technology.
- Combining simple technologies can create more powerful technologies.
- Today, technology is changing at a faster rate than ever before.

Reading Strategy

Charting What effects were produced by the Industrial Revolution? Make a chart, and fill it in as you read.

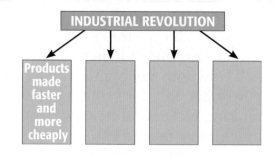

Vocabulary

Industrial Revolution
factory system

Information Age
exponential rate
of change

linear rate of change

Industrial Revolution

The **Industrial Revolution** was a period in which human and animal muscle power was replaced by machines. The Industrial Revolution began in Britain during the late 1700s and lasted until the mid-1900s. During this time, people invented machines that could spin wool, weave fabrics, and pump water.

An English metalworker named Abraham Darby was able to make iron that was very pure and extremely strong. He did this by heating it with coke. Coke is a carbon-like substance that comes from coal. Darby's process led to a much

Figure 2.6 During the Industrial Revolution, many machines, such as this early engine, were powered by steam.

Clarifying *What type of power did steam replace?*

better quality of iron. Iron could now be used to build large structures and new types of machines for industry. Many of these machines were powered by steam (Figure 2.6). This new technology provided a great leap forward in the manufacturing process.

A New Way of Making Things

With the invention of these new machines came the factory system of work. In the **factory system,** people made goods using machines rather than by hand. Before this innovation, craftspeople had used their own tools and workshops to make things. Shoes, cloth, and all sorts of tools were crafted individually. Each product was slightly different. With machines, however, products such as fabric were no longer created by an individual. Instead, massive looms manufactured yards of fabric at a time (Figure 2.7). Many products could be made faster and cheaper.

Richard Arkwright, a British inventor and cotton manufacturer, developed the first manufacturing system. He linked together machines of his own design. Using a sequence of steps, each machine carried out one job and prepared the material to be processed by the next machine. The cotton thread machine changed raw cotton into thread. Arkwright built many mills that produced cloth cheaper and faster than it could be produced by hand. Because of this new technology, fabric making was transformed from a cottage craft into a factory job.

In England and the United States, the assembly line became important for manufacturing. Assembly lines were based on a system of interchangeable parts. In a car, for example, identical copies of each part were assembled in the same way to build many identical cars.

Figure 2.7 Early machines allowed goods to be made more quickly than was possible by hand.

Hypothesizing *How did the use of machines change products that used to be handmade?*

Changes Produced by the Industrial Revolution

The factory system produced goods more cheaply and quickly than ever before. But it also created change and difficulty in people's lives. Craftspeople could not compete with machine manufacturing. Many had to give up their crafts and move to cities to find work. People became workers on factory assembly lines, using their employers' tools and machines instead of their own. They received wages for their work. Because parents were often gone from the home all day, schools were built to educate and care for children during the workday.

The technology of the Industrial Revolution also affected politics—how laws were made and civic affairs were governed. Men, women, and children often worked 14 hours a day in crowded and unsafe conditions. The mistreatment of employees led people to form labor unions. New laws addressed employees' rights, safety issues, working conditions, and fair wages.

Patent laws were also developed during the Industrial Revolution. People who invented a new device or improved on an old one could patent their idea. With patents, no one could legally use another's idea or invention without paying for it. In addition, patents gave official credit where and when it was due, because one's creativity was central to the idea. Because people could now make money from their inventions, there was an incentive to invent and improve products and systems. Existing ideas were used as building blocks for newer and more powerful technologies (Figure 2.8).

Figure 2.8 By combining two simple technologies, people were able to create newer and more powerful technologies.

Extending *Several important combinations are listed in the table below. Can you think of others?*

Combining Technologies

Existing Technologies		New Technology
Lenses	+ Chemistry	→ Photography
Shipbuilding	+ Steam engine	→ Steamship
Telegraph	+ Printing press	→ Newspaper
Internal combustion engine	+ Wagons & carriages	→ Automobile
Lightweight gasoline engine	+ Kite and glider technologies	→ Airplane
Materials technology (e.g., titanium)	+ Jet engine	→ Spy aircraft
Photography	+ Satellites	→ Exploration of the Earth

Figure 2.9

Figure 2.10

Information Age

The changes caused by machinery and factory production gradually gave way to a new type of technology: information. The last few decades—a tiny fraction of human history—have been marked by extremely rapid change. This period, which continues today, is called the **Information Age**. In 1947, the first transistor was invented using a piece of semi-conducting material and some wires. Transistors were soon combined with other components into integrated circuits. These were used in televisions and radios (Figure 2.9) and later in computers and electronic devices.

Inventions became increasingly focused on electronics and the computer. The first successful personal computer—the Apple II—was invented in 1978. Computers have since become much smaller, faster, and cheaper, and many of today's inventions are now based on computer technology. Computers also have changed how we work. Jobs depend on workers being well-educated and staying informed about changes in technology.

The Rate of Technological Change

The rate of technological change has increased over the course of history. During prehistoric times, it took many thousands of years to progress from stone tools to metal tools. As time passed, however, knowledge increased. People lived closer together in cities. They could publish and share their ideas in books and newspapers.

Technological change is faster now than it has ever been. Some people say that the world's body of knowledge is now doubling every four years. Today, technology undergoes an **exponential rate of change,** with extraordinarily rapid developments.

Information Age Inventions

Figure 2.9 The transistors used in these early radios were an important invention of the Information Age.

Figure 2.10 Another important invention of the Information Age has been the Internet.

Comparing *In terms of information, how are simple radios and the Internet similar?*

Figure 2.11 The pace of invention has continually increased over time.

Predicting *Based on this graph, what will the pace of technological change be for future generations?*

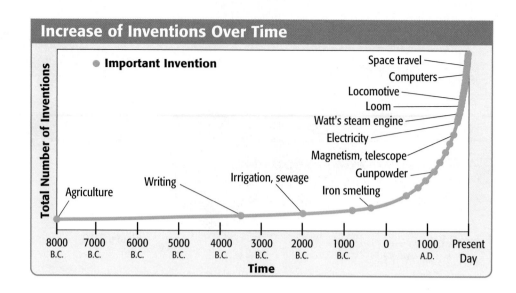

Increase of Inventions Over Time

● **Important Invention**

Space travel
Computers
Locomotive
Loom
Watt's steam engine
Electricity
Magnetism, telescope
Gunpowder
Irrigation, sewage
Iron smelting
Writing
Agriculture

Total Number of Inventions

8000 B.C.　7000 B.C.　6000 B.C.　5000 B.C.　4000 B.C.　3000 B.C.　2000 B.C.　1000 B.C.　0　1000 A.D.　Present Day

Time

When something changes at an increasing rate, we say that it changes exponentially. This is true of technological knowledge. It took about a million years to go from using stone tools to using tools made of bronze, but only about 5,000 years to go from using bronze tools to using machines.

Think about the changes that have taken place in the short time since the first personal computer was sold in 1981. Today's PCs have about 10,000 times as much memory and are a great deal faster than the early versions. The rate of technological change continues to increase (Figure 2.11).

SECTION 2 Assessment

Recall and Comprehension

1. What is the factory system? Give examples of the types of jobs it created.
2. Name the age we are living in. What types of jobs are important?
3. Has the rate of technological change increased or decreased over time? Describe the general pattern of change.

Critical Thinking

1. **Inferring** How did the formation of cities increase the rate of technological change?
2. **Summarizing** Explain why the assembly line made manufacturing faster.

QUICK ACTIVITY

Try to make some cotton thread from a ball of cotton. You will need to "comb" the cotton ball using a fine wire brush. Comb against a board to align the fibers, and then spin the fibers into a continuous thread. Who in your class can make the longest or strongest thread?
For more related Design Activities, see pages 52–55.

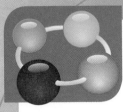

Connecting to STEM

Exponential Rate of Change

What does "exponential rate of change" mean? Let's look at an example. Sandra's mother agreed to let her download ten songs from iTunes a month. Her brother, Todd, wanted to download some, too. He got his mother to agree to this plan: He would download two songs the first month and then double the number he downloaded each month thereafter. Who do you think would have more songs after six months?

Sandra: Linear Rate of Change

At the end of the first month, Sandra had ten songs. Each month, she downloaded another ten. After six months, she had 60 songs, as shown in the graph. Because her supply increased by the same rate each month, the number of songs showed a **linear rate of change.** On a graph, this rate looks like a straight line.

Monthly Song Downloads		
Month	Sandra's Purchases (Number of CDs)	Todd's Purchases (Number of CDs)
1	10	2
2	10	4
3	10	8
4	10	16
5	10	32
6	10	64

Todd: Exponential Rate of Change

Todd started out by buying two songs. He doubled his purchases each month thereafter. In the second month, he bought four, and in the third month, he bought eight.

Todd's song supply increased faster each month. As a result, his supply increased exponentially. In a graph, an **exponential rate of change** shows a sharp curve. Their mother soon realized she would be spending all her money on songs for Todd, so she put an end to the arrangement. Use the graph of their purchases to answer the questions below.

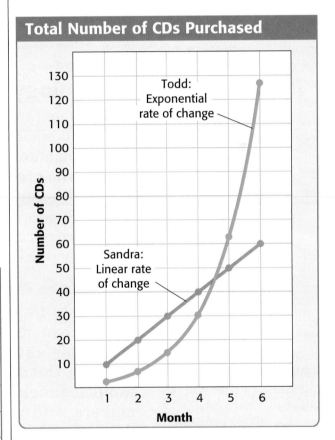

Total Number of CDs Purchased

Critical Thinking

1. **Interpreting** How many months did it take for Todd to have more CDs than Sandra?

2. **Extending** Can you think of another example of change at a linear rate? An exponential rate?

The Impact of Technology

Benchmarks for Learning

- People who live in different parts of the world have different technological choices and opportunities because of such factors as differences in economic resources, geographic location, and cultural values.

- The use of technology affects humans in various ways, including their safety, comfort, and attitudes.

- Technology must fit the needs of people, society, and the environment.

- The development and use of technology sometimes poses ethical issues.

- Economic, political, social, and cultural aspects of society drive improvements in technological products and systems.

- Some technological decisions involve tradeoffs between environmental and economic needs, whereas others have positive effects on both.

- Describe and analyze positive and negative impacts on society from the introduction of a new or improved technology, including both expected and unanticipated effects.

- Resources such as oceans, freshwater, and air are protected by regulating technologies in areas such as transportation, energy, and waste disposal.

Reading Strategy

Cause and Effect   As you read the selection, list the different impacts of technology, along with their causes.

Vocabulary

hybrid electric vehicle
ethical dilemma

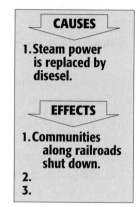

CAUSES

1. Steam power is replaced by diesel.

EFFECTS

1. Communities along railroads shut down.
2.
3.

Technology and Culture

Technology always causes change. Change can be good for some and bad for others. It can create jobs and restructure whole communities. It can also inflict economic hardship on people who are forced to move to find new jobs. People who live in different parts of the world have different technological choices and opportunities because of such factors as differences in economic resources, geographic location, and cultural values.

Technology Affects People's Jobs

In the late 1800s, the steam locomotive powered the westward expansion along the railroads. People moved west, and goods and services could now be exchanged easily between the east and west coasts. New towns were established along the rail lines, where machinists, boilermakers, and repair people maintained the equipment. Soon, diesel power replaced steam. Diesel engines were more fuel efficient, so they required fewer stops en route. The train communities withered as people moved in search of new jobs.

Similar changes have resulted from the new technologies of the Information Age. In the 1960s, transistors replaced vacuum tubes. Semiconductor manufacturers prospered and hired people, but vacuum tube manufacturers went out of business.

More recently, the Internet and the World Wide Web created a new industry almost overnight. The dot.com industry, made up of companies doing business only on the Web, created many new jobs, but even this high-tech industry has

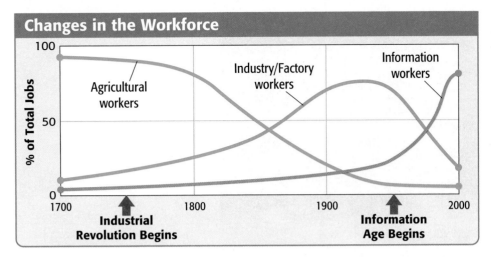

Changes in the Workforce

% of Total Jobs

Agricultural workers

Industry/Factory workers

Information workers

100

50

0

1700   1800   1900   2000

↑ Industrial Revolution Begins

↑ Information Age Begins

Figure 2.12 Changes in technology have led to the development of different types of jobs in different eras.

Predicting *What areas of technology might lead to new jobs in the future?*

had many businesses fail. This pattern of technological advancement, resulting in both winners and losers, is observed time and again (Figure 2.12).

Technological systems, processes, and products evolve and change over time in response to demands from people. These demands can be economically driven in the desire for less expensive goods and services; politically, culturally, or socially driven because of societal needs, such as clean drinking water; or technologically driven because of discoveries that allow new designs.

Technology Increases Comfort, Safety, and Accessibility

Using technology, products have been designed to make our lives more comfortable. Chairs are shaped for back support. Drinking glasses for small children are made so that they are easy to hold. Tables and desks are at a convenient height. Comfortable auto seats make driving more enjoyable. Designers match products to people's needs.

Products are also designed for safety and for accessibility to all types of people. The blades of kitchen machines such as food processors are protected to avoid the possibility of people accidentally cutting themselves. Automobiles are designed to be as safe as possible in the event of an accident. Special bicycles, cars, and skis are made so that people with disabilities can remain active and independent (Figure 2.13). Special buses allow those in wheelchairs to use mass transportation. Technology has done much to make public service accessible to all members of our society.

Technology Changes Our Landscape

Technology has also changed our surroundings. Since the Industrial Revolution, cities have grown in number and in

Figure 2.13 Created using technology, specially designed equipment allows people with all types of disabilities to enjoy an active lifestyle.

Extending *What other technological devices are used by people with disabilities?*

size. They have become more and more crowded. In the early twentieth century, the development of a mass transportation technology—the electric trolley car—allowed suburbs to flourish. The trolley was later replaced by the automobile. As roads become more congested, mass transportation becomes more important. Today, technological research is focused on exploring new types of transportation.

Technology and the Environment

Technology is often considered to have a negative effect on the environment. In fact, the conflict between technology and the environment has existed for thousands of years. In the 1600s, forests were cut down for fuel. Garbage was burned in open fires. Human waste was poured into rivers. However, all this occurred on a small scale.

Technology Has Changed Our Landscape

Today, technology has the potential to harm the environment on a worldwide scale. Pollution is a worldwide issue. Cars and trucks give off dangerous gases that pollute our air. Wastes from some factories pollute our rivers and oceans. We burn more than seven billion tons of coal worldwide each year to provide energy. Burning coal produces pollutants such as sulfur dioxide. In the United States, vehicles add huge amounts of pollutants to the air.

The chemicals given off by cars and industrial plants can produce acid rain. Acid rain has killed forests and fish in the United States and other countries (Figure 2.14). Chemicals are also used increasingly in farming to control pests and disease, but they can remain in the soil and air. They can get into the food we eat and the water we drink.

Figure 2.14 Pollution is a worldwide issue. Burning coal and other fuels can create acid rain, which destroyed these trees.

Extending *How does acid rain affect nonliving thingssuch as buildings or monuments?*

LIVING GREEN
Little Trips Add Up to Lots of Pollution

Did you know that the average family household with two cars can contribute up to 20,000 pounds of carbon dioxide emissions in just one year? In families that drive in urban areas, where traffic congestion is a problem, this number may be much higher. Automobiles emit carbon dioxide (CO_2). Carbon dioxide is a greenhouse gas, meaning it traps heat inside Earth's atmosphere and over time can contribute to global warming.

MAKE A CHANGE
Taking several small trips around town can really add to the amount of pollution your family car gives off. Encourage your family to combine errands, such as grocery shopping and trips to the post office, and to carpool to and from school events and sports practices.

TRY THIS
Make a list of all the activities you, your family, and your friends can do without the use of a car. Start simple, such as riding your bike or walking to school. Add up how much driving time (and CO_2 emissions) you can save in a week and how much good exercise you can gain.

Should We Go Back to Nature?

If we went back to a simpler life, would problems like pollution disappear? Probably not. Pollution has been a problem since people first began to live in cities. The rivers around Rome were so full of waste that people were forbidden to bathe there. One hundred years ago, before the automobile, horses produced tons of manure in the streets of New York and other cities. We need to create technologies that will protect the environment as well as serve our needs. Some technological solutions to environmental problems, such as dredging, involve trade-offs between cost and societal benefit. Others, such as clean drinking water, have positive economic and environmental effects.

Using Technology to Solve Environmental Problems

Advances in technology have helped solve pollution problems. Pollution control devices on cars and scrubbers on factory smokestacks have reduced pollution. Automobiles powered by nonpolluting energy sources are now available. **Hybrid electric vehicles** (HEVs) combine the engine of a conventional vehicle with the battery and electric motor of an electric vehicle (Figure 2.15). These cars achieve twice the fuel economy of conventional vehicles (about 50 miles per gallon).

Figure 2.15 This hybrid vehicle uses both electricity and gas. It captures energy from braking and uses it to recharge its batteries.

Taking a Position *Do you think the government should provide money to make HEVs less expensive than regular cars? Why or why not?*

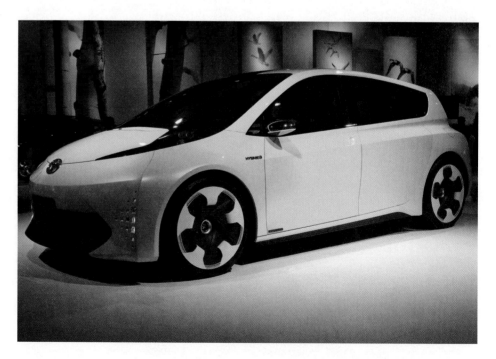

Advances in the technology of waste management have also helped protect the environment. Recycling items such as glass, paper, and aluminum has decreased the waste sent to landfills. In turn, this decreases the need for more land to be allotted to landfills. Wastewater treatment plants help keep our rivers and oceans clean by controlling the levels of pollutants released into them.

Many companies are trying to limit the harm caused by chemicals in the environment. They conduct technology assessment studies to determine the effects that chemicals will have on people. Such studies can save lives and prevent future health problems.

Our environment must be treated with care. We must use resources wisely and look for alternatives for scarce resources. We must develop technologies that do not cause environmental harm. For example, solar energy systems can provide for some of our energy needs without harming the environment.

Technology and Ethics

The development and use of technology can also pose a number of ethical questions. Sometimes, new technologies or new uses for a technology cause concern about whether using that technology is right or wrong. Society is often faced with these ethical dilemmas. The word *ethical* means "having to do with right and wrong" and the word *dilemma* means "a difficult decision." An **ethical dilemma** is a difficult decision that a person has to make about whether something is right or wrong.

The field of biotechnology, in particular, has created moral dilemmas for society. Scientists have successfully cloned sheep, mice, and cows. To clone the sheep, Dolly, cells were taken from an adult female sheep. Dolly became an exact duplicate of her biological mother. An ethical dilemma was posed: If cloning sheep and other animals is possible, what about cloning humans? Should society allow research into human cloning? Some say absolutely not because it is immoral. Others support cloning research because they believe we might learn how to avoid genetic diseases.

Another ethical dilemma concerns who will receive high-cost medical assistance. If a high-cost medical technology will keep a person alive, but with a low quality of life, should that technology be used? When is the use of certain technologies such as life support justified, given the financial and emotional costs? Should these technologies be available only to people who can afford to pay for them? All these questions must be considered as new technologies are developed.

Regulations

Government has an important role in ensuring the safety and well-being of its citizens, hence the need for laws and regulations. In democracies, the legislative branch of government passes laws; regulations are developed by the executive branch most often as a way to implement laws. This is also done internationally, with regulations affecting commerce between countries, by land, sea, and air and hence affecting technologies in these and other areas. Locally, there may be ordinances regulating proper waste disposal, whereas nationally there can regulations dealing with interstate transportation.

SECTION 3 Assessment

Recall and Comprehension

1. List three technologies that pose ethical dilemmas.
2. Give an example of one technology that is well matched to the needs of the environment, and one technology that is not.
3. In what ways are products designed to meet people's needs?

Critical Thinking

1. **Defending** Should scientists be allowed to research human cloning? Defend your answer.

2. **Generalizing** Cell phones have had major impacts on how we communicate with each other. However, there are negative consequences from the technology, such as texting during class or while driving. Analyze the positive and negative impacts on society and include anticipated and unanticipated effects.

3. **Summarizing** How did new types of transportation, such as the trolley and the car, change society?

QUICK ACTIVITY

Using a shallow aluminum pan containing 1/2" of water and a few drops of oil, simulate the cleanup of an oil spill. Use a variety of materials in your experiment, such as popsicle sticks, cotton, styrofoam, cloth, and liquid detergent. Keep a chart to record the results of each cleanup material. Which one worked best?
For more related Design Activities, see pages 52–55.

Technology in the Real World

Genetically Modified Foods

There is no question that genetic engineering lets us do amazing things. We can add vitamins to rice, producing better nutrition for millions of people. We can control the process of ripening in fruits—from tomatoes to coffee beans. We can even produce plants that "glow in the dark" when they need to be watered. All of these changes are produced by copying part of the genetic material from one organism to another.

Controlling a Sticky Pest

The bollworm, shown below, is a caterpillar that causes serious damage to crops such as cotton, corn, and potatoes. For decades, organic farmers have been treating their crops with a naturally occurring pesticide produced by the bacteria (small, single-celled organisms) *Bacillus thuringiensis,* or Bt for short. Bt is deadly to certain plant-eating insects, including the bollworm. Even better, it is harmless to most other insects.

Using genetic engineering, scientists made copies of the Bt gene that is harmful to the bollworm. They inserted this gene into common crop plants. The genetically modified crops remained pest free.

Genetic engineering is used widely to control the process of ripening in fruits, such as these tomatoes.

Some Unwanted Results

End of story? Not quite. The pests that were initially controlled by the Bt gene are starting to show signs of resistance to it. A greater percentage of each new generation of bollworms can tolerate the Bt gene. In addition, the Bt gene appears to be toxic to some beneficial insects, such as the monarch butterfly.

Scientists are quickly learning that genetic engineering can produce unanticipated and often unwanted results. Does this mean that we should stop using genetic engineering? Should we limit its use to certain types of organisms, such as plants and bacteria? Or should scientists be able to explore all the possibilities of genetic engineering, controlling and creating new life-forms at will? What do you think?

Critical Thinking

1. **Summarizing** What benefits did the new Bt crops produce for farmers?

2. **Extending** Given the problems that scientists encountered with Bt crops, what solutions might they explore?

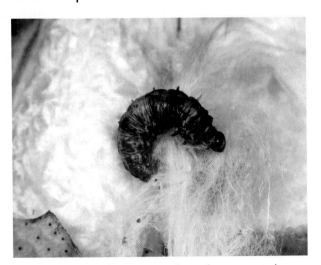

A bollworm munches its way through a cotton plant.

CHAPTER
2 Review and Assessment

Chapter Summary

- Technology is as old as the human race. Starting in prehistoric times, technology has developed to meet our needs for food, medical care, shelter, clothing, manufactured products, and communication.

- During the **Stone Age**, people used tools made of stone. During the **Bronze Age** and the **Iron Age,** tools and other items were made of metal. The **Industrial Revolution** began with the development of machines.

- In the present era, called the **Information Age,** most people use computers in their places of work.

- Technological change is increasing at an exponential rate. Each new technological development can give birth to new ideas, resulting in more inventions and more changes. Science has contributed to the pace of technological change.

- Technology affects our economy, our political systems, and our culture. Some results of technology are helpful and some are harmful, depending on how technology is used.

- The development and use of technology sometimes poses ethical issues. Advance planning can help limit the undesired effects of technology. Technology must fit the needs of people, and it must protect and work within the environment.

Building Vocabulary

Your teacher may give you a crossword puzzle. Complete the puzzle using the following words from this chapter. Exchange puzzles with a classmate and check each other's answers.

1. **Agricultural Era**
2. **alloy**
3. **aqueduct**
4. **Bronze Age**
5. **coke**
6. **ethical dilemma**
7. **exponential rate of change**
8. **factory system**
9. **hybrid electric vehicle**
10. **Industrial Revolution**
11. **Information Age**
12. **Iron Age**
13. **linear rate of change**
14. **smelting**
15. **Stone Age**

See your teacher for the Crosstech puzzle.

Reviewing Content

1. Define Agricultural Era, Industrial Revolution, and Information Age.

2. Bronze replaced stone as the primary material used to make implements. What properties of bronze made it more practical than stone?

3. Explain the purpose of an aqueduct. How did its introduction improve cities?

4. What are interchangeable parts? How did this innovation change manufacturing?

5. Give at least three examples of complex technologies that were developed from simpler technologies.

6. How can people exercise some control over the development of a technology that they believe may be harmful?

7. Describe a hybrid electric vehicle, shown below. Why is it considered environmentally friendly?

8. Provide one example of a problem that technology has caused and one of the solutions that technology has provided.

9. Give two examples showing that pollution is not just a recent problem.

10. What is cloning?

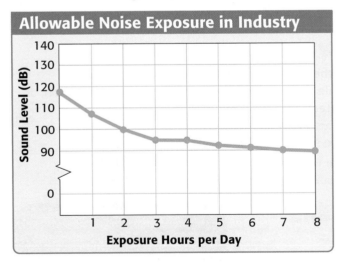

Allowable Noise Exposure in Industry

Sound Level (dB) vs. Exposure Hours per Day

Applying Your Knowledge

1. Draw a timeline that illustrates how technology is growing at an exponential rate.

2. Create a graph showing how the number of airplane miles traveled by U.S. residents has changed over the past ten years. You will have to do research. Look in an almanac, an encyclopedia, or on the World Wide Web.

3. Solar energy has been used to heat homes, but it has been less successful when applied to powering cars. Conduct research on solar power to find out why.

4. Assembly lines were used to make the first cars and other complex machines. Because interchangeable parts were used, each product was the same as the next. How many different parts would be needed to make the shoe you are wearing?

5. Explain why it is unethical to infect or damage other people's computers with viruses or to "hack" into other computer systems to gather or change information.

Critical Thinking

1. **Stating an Opinion** Do you think that technology is good, bad, or neutral? Explain your answer.

2. **Inferring** Why is technology more of a threat to the environment now than it was 5,000 years ago?

3. **Making Judgments** Do you think that cloning presents important ethical dilemmas? Why or why not?

 Connecting to STEM
science · technology · engineering · math

Noise in the System

Noise is a problem caused by some technologies. The higher the noise level, the less time people should be exposed to it. The amount of exposure should vary inversely to the noise level.

According to the graph, what is the allowable noise exposure measured in decibels (db) for one hour a day? Four hours a day? Is the relationship between allowable sound level and exposure direct or indirect? Explain your answer.

Design Activity 3

TECHNOLOGY MOBILE

Problem Situation

Today, we live in a modern, technologically sophisticated society because of the thousands of scientific and technological advances that have already taken place. The first inventions were ones that helped people survive and then live better. As societies became more complex, more sophisticated technologies were created. How can we illustrate these advances?

Materials

You will need:

- 1/8" and 1/4" dowel rods
- colored card stock
- colored pencils and/ or markers
- computer and printer
- fishing line or heavy- duty thread
- miter box
- paper cutter
- rubber cement or white glue
- scissors
- string

Your Challenge

You and your teammates are to choose an event, invention, or development from the Technological Timeline found in the back of your student text. Then, you will construct a hanging mobile that describes the technological development you selected. Hang your mobile in the proper place on the technological timeline strung in your classroom.

> Go to your **Student Activity Guide, Design Activity 3.** Complete this activity in your Guide, and state the design challenge in your own words.

① Clarify the Design Specifications and Constraints

Your design must meet the following specifications and constraints:
- The cards on the mobile should indicate clearly which event, invention, or development is illustrated.
- The mobile should be attractive as well as informative.
- The mobile must not be more than 3 feet long.

> In your Guide, state the specifications and constraints. Include any others that your team or your teacher included.

② Research and Investigate

To better complete the design challenge, you need to first gather information to help you build a knowledge base.

> In your Guide, complete Knowledge and Skill Builder 1: Developing a Database.

> In your Guide, complete Knowledge and Skill Builder 2: Constructing a Mobile.

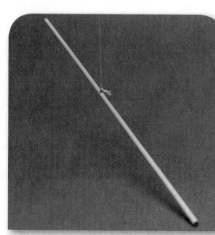

③ Generate Alternative Designs

In your Guide, describe two possible design solutions for your challenge. Examine all the information you have gathered about the technology, and then decide how you will make a mobile that clearly displays the information.

④ Choose and Justify the Optimal Solution

Refer to your Guide. Explain why you selected your solution, and why it is the better choice.

⑤ Develop a Prototype

Make the cards and construct your mobile. Include a drawing or a photograph of your final design in your Guide. In any technological activity, you will use seven resources: people, capital, time, information, energy, and tools and machines. In your Guide, indicate which resources were most important in this activity, and how you made trade-offs among them.

⑥ Test and Evaluate

How will you test and evaluate your design? In your Guide, describe the testing procedure. Explain how the results will show that the design solves the problem and meets the specifications and constraints.

⑦ Redesign the Solution

Respond to the questions in your Guide about how you might change your design. Your redesign should be based on the knowledge and information that you gained during the activity.

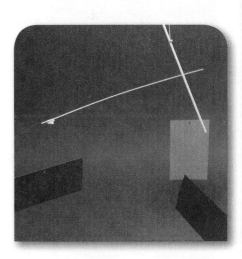

⑧ Communicate Your Achievements

In your Guide, describe the plan you will use to present your solution to your class. Include any handouts you will distribute on the inventions you researched.

Design Activity 4

IMPACTS AND OUTCOMES

Problem Situation

All of our actions have consequences, some that we anticipate and others that were not intended. Engineers and scientists invent something to solve a problem or satisfy a need. This is an intended consequence. But often there are unintended consequences. For example, improved burglar alarms on automobiles have caused an increase in noise pollution.

Materials

You will need:

- markers
- photographs or drawings of the invention
- poster board
- rubber cement, tape, or glue

Your Challenge

You and your teammates are to choose an invention and then create a poster board display that indicates the effects the invention had, both intended and unintended.

> Go to your **Student Activity Guide, Design Activity 4.** Complete this activity in your Guide, and state the design challenge in your own words.

① Clarify the Design Specifications and Constraints

Your design must meet the following specifications and constraints:
- Only one invention may be illustrated.
- The poster display must show both intended and unintended consequences.
- The poster display should be no larger than 24" by 30".

> In your Guide, record all the specifications and constraints. Include any others that your team or your teacher included.

② Research and Investigate

To better complete the design challenge, you need to first gather information to help you build a knowledge base.

> In your Guide, complete Knowledge and Skill Builder 1: Researching Your Invention.

> In your Guide, complete Knowledge and Skill Builder 2: Impact of the Invention.

③ Generate Alternative Designs

In your Guide, describe two of your possible designs for your challenge.

④ Choose and Justify the Optimal Solution

Refer to your Guide. Explain why you selected your design, and why it is the better choice.

⑤ Develop a Prototype

Make your poster board display of the invention and its consequences. Include a sketch or a photograph of your final design in your Guide.

In any technological activity, you will use seven resources: people, capital, time, information, energy, materials, and tools and machines. In your Guide, indicate which resources were most important in this activity, and how you made trade-offs among them.

⑥ Test and Evaluate

How will you test and evaluate your design? In your Guide, describe the testing procedure. Explain how the results will show that the design overcomes the problem and meets the specifications and constraints.

⑦ Redesign the Solution

Respond to the questions in your Guide about how you would change your design. The redesign should be based on the knowledge and information that you gained during the activity.

⑧ Communicate Your Achievements

In your Guide, describe the plan you will use to present your design to your class. Include any handouts you will distribute on the inventions you researched.

UNIT 2

Design for a Technological World

Unit Outline

"Design must play an important role in the development of a new product."

—Peter Lucas

Design and Problem Solving

In this chapter, you will learn about the process of informed design, including the following:

- clarify the problem
- research and investigate
- generate alternative designs
- choose an optimal solution
- develop a prototype
- test and evaluate
- redesign the solution and express your achievements

Designing is a process used to plan and produce a desired result. Very often, designing is done in teams. Collaboration between people and teams of people can involve working together in the same space at the same time, or at a distance at the same time or different times. Different communication technologies support this collaboration.

Designs come from original ideas. They serve different purposes, depending on the requirements of a particular task. The design process involved in creating the video game controllers on these pages indicates that designing is much more complex than sketching an idea on paper.

- One person or a team of people decides on the actions the characters will perform in the games.

- Another team studies how people play the games. It explores what people like and dislike about playing video games.

- Another team is then charged with building game consoles and controllers that produce the appropriate action and that work most comfortably for the user.

Their designs are produced using a combination of problem-solving and creative design techniques. This process is called *informed design*, and it is the process you have been using in your Guide.

The Process of Engineering Design

Benchmarks for Learning

- Technology addresses many types of problems and their solutions. Some problems require both science and technology to be resolved.

- Design is a creative, systematic, and iterative process for meeting human needs and wants, and leading to useful products and systems.

- Design steps can be performed in different sequences and repeated as needed.

- Design involves making a product or system and documenting the solution.

- Design requirements indicate criteria for success, specifications, constraints, or limits on the solution. There are often trade-offs between competing requirements.

Reading Strategy

Flowchart   Create your own flowchart showing the eight steps of informed design. Name each step, then draw a simple sketch to describe what takes place.

Vocabulary

engineering design	constraint	mathematical model
informed design	brainstorming	variable
specification	scale model	
design brief	prototype	

The Need for Problem Solving

Every day, we are faced with situations that require thoughtful solutions. Some of these situations, such as designing video games, are fairly short-term and involve a specific goal. Others are more general and long-term. Some problems, such as the following, involve society and the environment:

- Disposing of wastes without harming the environment.

- Producing enough energy to meet our increasing needs.

- Ensuring a continuing supply of clean, safe water.

Other problems have to do with people's needs or wants, as follows:

- Providing more independence for people with vision impairment or other disabilities (Figure 3.1).

- Improving the sound quality of recorded music.

- Allowing automobile drivers to communicate with each other to prevent accidents.

Each of these problems can be solved using science, technology, and design.

Figure 3.1 This blind woman is reading with the help of system that scans printed text and transfers it to Braille.

Extending *What other types of products might be useful to a vision-impaired person?*

Good Design in Problem Solving

Solving a problem is rarely quick and easy. It often takes time, thought, and creativity. We must understand exactly what is needed to solve the problem. More money, extra time, or other people's help may be needed. Solutions or products should be as inexpensive and easy to use as possible.

Problems in technology are seldom well-defined when we first encounter them. For example, if the problem is to design a house for a family, many questions need to be considered. How big will the house be? What styles and materials will be used? How will it be heated and cooled? How much money and time are available? The architect, builder, and customer must work together to clarify each problem and arrive at the best possible solution (Figure 3.2).

To solve any kind of problem, decisions and compromises need to be made. This is done through a process called informed design.

Figure 3.2 Before construction begins, many details and questions need to be discussed.

Hypothesizing *What problems might be encountered when building a house?*

Engineering Design

Engineering design is a systematic, creative, and iterative process for addressing challenges. It is a process for meeting human needs and wants and leads to useful products and systems. Engineering design is an **informed design** process. It is a method of making choices based on knowledge about the design problem, and then coming back to refine or revise those choices at any point in the process as often as necessary. Using the steps to make choices and decisions, more detail is added, resulting in a better defined problem, which leads to an acceptable solution.

An informed design process is used by engineers, architects, craftspeople, and artists. All of these people are designers. The choices that they make to create their designs often involve compromises. Cheaper materials may be used, or a simpler, less sophisticated design may be required to finish the project on time. As you learned in Chapter 1, compromises are also known as trade-offs. Designers often make trade-offs when selecting the best design from all the choices.

Clarify the Design Problem

In the first step of the informed engineering design process, shown in Figure 3.3, the designer describes the problem clearly and fully. Design requirements indicate criteria for success, specifications, constraints (or limits) on the solution. There are often trade-offs between competing requirements.

The **specifications** for a design problem are the performance requirements that the solution must fulfill. The **design challenge** states the problem and lists the specifications for a design. Specifications are different for each type of design project. The design specifications for toothpaste might be that it must clean plaque from teeth, taste good, and be squeezed easily out of a tube. A design specification for a certain type of bicycle might be that it is lightweight and can be used on paved roads as well as rocky trails.

Let's look at a specific example. You have decided that you need a new backpack to carry your things to and from school. The backpack needs to comfortably hold your school supplies. You also want separate compartments for your personal digital assistant (PDA) and for your CD player, which

Figure 3.3 Design process steps are usually completed in order, but sometimes certain phases need to be repeated.

Interpreting *You designed a new lamp but it doesn't turn on properly. Which stage of the design process should you repeat?*

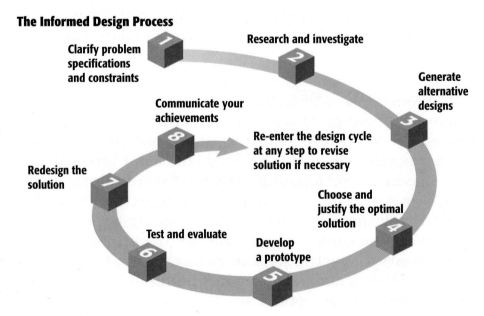

The Informed Design Process

1. Clarify problem specifications and constraints
2. Research and investigate
3. Generate alternative designs
4. Choose and justify the optimal solution
5. Develop a prototype
6. Test and evaluate
7. Redesign the solution
8. Communicate your achievements

Re-enter the design cycle at any step to revise solution if necessary

Design Brief

Design Problem: Design a durable, comfortable, and moderately priced backpack to carry materials for school and other activities.

Requirements
- Backpack should be able to hold at least 3 textbooks, a CD player, a thin jacket or fleece, a palm pilot, and a cell phone.
- Backpack should have several smaller pockets for money, keys, pens, and pencils.
- Material should be brightly colored, waterproof, and durable.

Constraints
- Backpack cannot cost more than $35 retail.
- Backpack must be adjustable so one size can be used by 5th–8th graders.

Figure 3.4 This figure describes the specifications for a new backpack.

Analyzing *If different people design a backpack, why might each of their design briefs be different?*

you listen to as you walk home from school. Figure 3.4 shows a design for your backpack.

Design specifications may also include safety considerations. A passenger elevator, for example, might require a safety factor ten times greater than the load it is expected to carry, or the front of an automobile may need to withstand a collision at 5 miles per hour without damage.

Constraints are the limits imposed on a design solution. Constraints are often related to resources, such as what kind of materials the designer can use, how much money a finished product can cost, or how much time is available to produce it. Other limitations may involve employing only workers with special skills or limiting negative effects on the environment.

Research and Investigate the Problem

Search for and discuss existing solutions to solve the problem at hand or similar problems. Through this process, you will learn a lot about good and bad solutions.

Through your research, you can identify problems, issues, and questions that relate to the design challenge. You can gather information from other people, from research in the library or on the Internet, and from visits to stores that sell products similar to the one you are designing.

You will also need to take measurements and collect data about different needs the solution must fulfill. The **design challenge** states the problem and lists the specifications for a design. Specifications state the types of materials and

Figure 3.5 Market research helps designers make a product that appeals to the greatest number of people.

Extending *What questions might help you to improve your backpack design?*

the performance of those materials. Using mathematics and other skills, you will need to rate different alternatives to compare them. As you research, always keep the design criteria in mind. They will help you identify the important questions that you need to answer.

To learn more about a product and how it will be used, many companies perform market research or scientific investigations (Figure 3.5). For example, a company that is developing a new toothpaste might ask potential buyers to try the toothpaste and then fill out a questionnaire. Do they like the taste? What kind of dispenser do they prefer to use—a tube or a pump? A more scientific investigation could be used to determine what levels of fluoride and tartar control are most effective in stopping cavities.

For your backpack, you will need to research several options. How large should the backpack be? How many pockets should it have? What types of materials should be used? Should it have zippers, Velcro®, or some other type of fastener? You may want to collect data about people your age and how much weight they can carry. You may also want to look at existing products to get new ideas.

Generate Alternative Designs

Don't stop at the first solution that might work. Continue to approach the challenge in new ways. As you continue your research, you should develop one or more potential solutions to your problem (Figure 3.6). The ideas can be totally different, or they may be improvements on your first idea. These different ideas are called alternatives—two or more possibilities from which choices can be made. The process of generating different designs is not the same as solving a mathematics problem in your textbook. Usually, mathematics problems are well-defined, with only one correct answer.

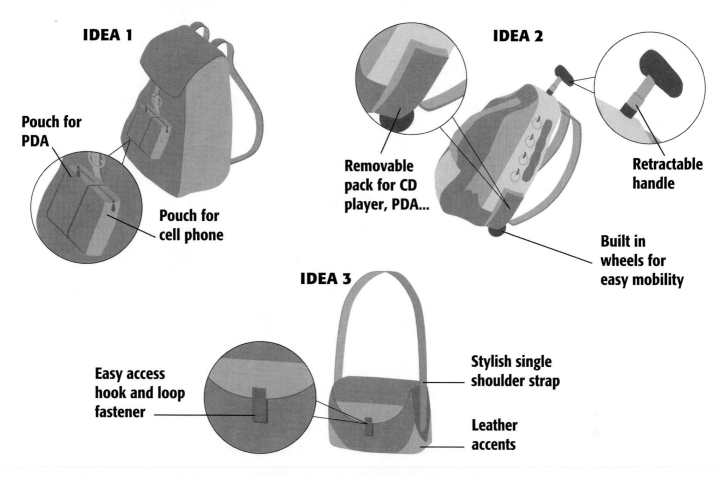

IDEA 1

Pouch for PDA

Pouch for cell phone

IDEA 2

Removable pack for CD player, PDA...

Retractable handle

Built in wheels for easy mobility

IDEA 3

Easy access hook and loop fastener

Stylish single shoulder strap

Leather accents

Because there are many possible design solutions, good designers are rarely satisfied with the first idea that pops into their minds. **Brainstorming** is a method by which a person or group of people can come up with alternative solutions. During brainstorming, each person in a group suggests ideas. One person writes down all the ideas, and no one is allowed to laugh at or criticize them—no matter how foolish or unusual they might seem. After many ideas have been proposed, the group reviews them all. The best ideas are developed further.

Figure 3.6 Each of these sketches represents a possible design solution.

Comparing and Contrasting
How are the different backpack designs all similar? How are they different?

Choose and Justify the Optimal Design

Decide on a design that best meets the specifications, fits within the constraints, and has the least number of negative characteristics. Good design solutions are those that work well, are inexpensive, and cause little or no harm to the environment or to people. They meet all of the design requirements within all of the limitations (such as cost and time) that have been imposed.

For each of your design alternatives, you will need to list its strengths and weaknesses in relation to the original design criteria. Each of the alternatives must be examined to see if it meets the original design criteria, specifications, and

Figure 3.7 The preliminary design represents the best fit for the product's specifications and constraints.

Evaluating *Does this design meet all the original specifications?*

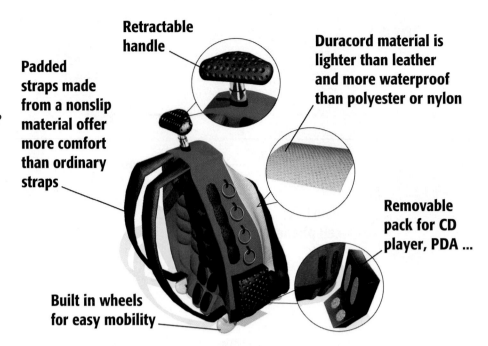

Retractable handle

Padded straps made from a nonslip material offer more comfort than ordinary straps

Duracord material is lighter than leather and more waterproof than polyester or nylon

Removable pack for CD player, PDA ...

Built in wheels for easy mobility

constraints. The alternative you choose will be the basis of your preliminary design (Figure 3.7).

You may find that you have to conduct more research or gather more data to examine each alternative completely. Improvements to a design can lead to better performance, increased safety, and lower cost. The process of improving each alternative is called *optimization*. When people choose the best solution, they normally make trade-offs, giving up one desirable thing for another.

Once you have chosen a design, be prepared to justify your choice. Why is it the optimal choice? How does it meet your design specifications? Use all the information you have gathered to defend your decision.

Develop a Prototype

Make either a model or a prototype of the design solution. A **scale model,** which may be larger or smaller than the final product, is a model of an object with all parts in correct proportion to those of the object. For example, on a scale model, 1 inch may represent 1 foot. A **prototype** is a full-scale, fully operational version of the solution (Figure 3.8).

During the design process, drawings and models of the proposed solution are made in order to work out any design problems before the product is put into production. If the final product is very small, such as an integrated circuit, a large-scale version might be built so that all the parts and connections can be seen clearly. If the final product is large, such as a jet plane or a museum exhibit, a smaller-scale version may be used.

Figure 3.8 Skilled craftspeople and technicians often make prototypes before full-scale production is started.

Summarizing *Why is prototyping a necessary part of the design process?*

Models are important for several reasons. The solution or product may be very large and costly. Alternatively, the individual product might be inexpensive, but plans to make several thousand would involve a large expense. Sometimes, the proposed solution might need evaluation of its possible risks to people or the environment. For example, a model of a nuclear power plant would be built and debugged before construction of an actual power plant. In all of these instances, a scale model or prototype makes it easier for a designer to modify and refine the design.

Another kind of model is a mathematical model. In a **mathematical model,** equations describe how a product will function or perform. For an electrical device, a mathematical model might show how current in a circuit is directly related to the amount of voltage and inversely related to the amount of resistance.

Test and Evaluate the Design Solution

Develop one or more tests to assess the performance of the design solution. Test the design solution, collect performance data, and analyze the data to see how well the design satisfies the original constraints and specifications (Figure 3.9).

Observing, or monitoring, the results of tests may suggest improvements to a design or its construction. You should note each different factor that affects the performance of your design. These are called **variables.**

Sometimes, tests are run on a computer using computer simulations, in which a computer tests different models to predict how a product will perform. Computer simulations

Figure 3.9 Before it is manufactured, this automobile is tested many times.

Hypothesizing *What types of features might be most important to test in this product?*

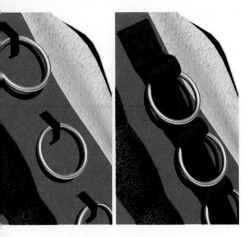

Figure 3.10 Slight modifications can improve a design and facilitate its production.

Summarizing *Why is a redesign phase necessary for making a good product?*

are most useful when a large number of calculations must be carried out.

Redesign the Solution with Modifications

Examine your design critically. Note how other students' designs perform to see where improvements might be made to yours. Even during this phase, you should be willing to consider new ideas.

During the redesign phase, you will need to identify and change any variables that affect the performance of your design. Think about any science or math concepts that underlie these variables.

In the backpack design, for example, you may have to modify a feature in order to reduce the cost or make the backpack lighter. In Figure 3.10 the rings now have stronger attachments.

Communicate Your Achievements

Present your final design, along with a summary of how it meets each of your specifications. This is your opportunity to "show off" your final design. Keep in mind that your presentation is as important as the design itself. Describe the final product and how it will be used. You may also want to describe phases of the design process that were important for creating the final design. You will read more about presenting design solutions in Chapter 4.

SECTION 1 Assessment

Recall and Comprehension

1. What is brainstorming? Why is it helpful in coming up with new ideas?
2. How are optimization and trade-offs used in the design process?
3. What is a prototype? Why is it used in the design process?
4. How does data from testing affect a product's final design?

Critical Thinking

1. **Contrasting** What is the difference between specifications and constraints? Use examples to explain the difference.
2. **Applying** Describe an improvement you would like to make in an existing technology. For example, you might want to change the size of a telephone receiver to make it easier for your 5-year-old sister to use.

QUICK ACTIVITY

Trade-offs, compromises, or even mistakes made during the design of a product can sometimes come back to haunt a manufacturer. Using newspapers, magazines, and the Internet, write a report about a product that was recalled. Include the reason for the recall, and explain what could have been done to prevent the problem.
For more related Design Activities, see pages 80–83.

Connecting to STEM

Independent and Dependent Variables

On a class trip to a science museum, Susan observed a large pendulum moving back and forth. Susan's teacher challenged her to determine how the pendulum worked and what made its swing shorter or longer. With the help of her friend Jason, Susan set about planning the experiment.

Researching the Problem

First, she did some research. She found that pendulums have various lengths and weights. They all move in a back-and-forth motion, and they can be released at different pullback distances. With this information, Susan built a pendulum using items in her classroom.

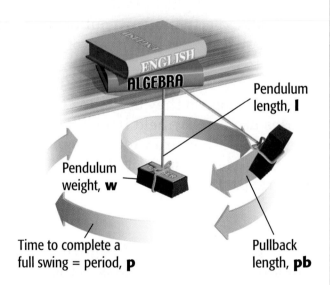

Pendulum length, **l**

Pendulum weight, **w**

Time to complete a full swing = period, **p**

Pullback length, **pb**

The time for the pendulum to complete a full swing is called its *period*, **p**. Susan knew from her research that the period is a *dependent variable*—it depends on the values of other variables, which are called *independent variables*. Susan and Jason hypothesized that the period would depend on three independent variables: the pendulum's length (**l**), its weight (**w**), and its pullback length (**pb**).

Testing Different Variables

To test their hypothesis, Susan and Jason built another pendulum. They decided to test different values for the pendulum's weight (**w**). They attached a single washer to the string, measured the length of the string, pulled the washer back 10 cm, and then released it. Jason recorded the time for 10 swings and then divided it by 10 to get the period. They performed many different experiments and recorded their results in the table.

Pendulum Weight, w	Period (p)
w = 1 washer	1.1 seconds
w = 2 washers	1.0 seconds
w = 3 washers	1.1 seconds
Pullback Length, pb	
Pb = 10 cm	0.9 seconds
Pb = 20 cm	1.1 seconds
Pb = 30 cm	1.0 seconds
Pendulum Length, l	
l = 10 cm	0.9 seconds
l = 20 cm	1.6 seconds
l = 30 cm	2.1 seconds

Critical Thinking

1. **Interpreting** Did the period change when Susan and Jason varied the pendulum's weight? How do you know?

2. **Extending** Are length, weight, and pullback independent variables? Keep in mind that if changing something has no effect on the dependent variable (the period), it is not an independent variable.

Principles of Engineering Design

Benchmarks for Learning

- Ergonomic design is concerned with how machines, tools, and the workplace fit with the human body.
- The use of technology affects humans in various ways, including safety and comfort.
- Designs are created that are pleasing to the eye.

Reading Strategy

Outlining   Create an outline of the principles of design. Next to each principle, write a definition in your own words. Provide one example of a product that illustrates each principle.

Vocabulary

functionality   anthropometry
quality   aesthetics
ergonomics

> I. Principles of Design
>   A. Functionality: product does what it's supposed to do
>   1. Toaster toasts bread the same way every time

Functionality

Functionality refers to the capability of a product, system, or process to fulfill its intended purpose over the course of its desired life span. For example, a light bulb designed to give 1,000 hours of service is not expected to burn out before that amount of time, and a software program should complete all the tasks it was designed to perform.

The form of an object is often determined by its function. Dining tables hold plates of food, so they are made to be flat. Because drinking glasses are usually held in one hand, they are designed to be higher than they are wide. It would make no sense to design dining tables with slanted tops or drinking glasses too wide to grasp.

Figure 3.11   These kitchen products were designed for use by many different people.

Interpreting   *What makes these products ergonomic?*

Quality

The product, system, or process must be designed to meet certain minimum standards of quality. **Quality** is the degree of excellence with which a product is made. For example, it is essential that medicines use the same formulation in every batch produced. Products made in different locations all need to meet the same standard for quality.

Ergonomics

Ergonomics is the science of adapting the work environment to people. Also called human factors engineering, it deals with designing products so they can be used easily and comfortably (Figure 3.11). Ergonomics combines an understanding of the human body with the techniques of design and engineering. A good design fits the user's size and capabilities. A desk chair, for example, must be designed for human comfort. The design solution would differ depending on the specifications—whether it's for a teenager, a man, a woman, a child, or all four. Another example of ergonomic design is an automobile's dashboard. All the information is displayed so that the driver can read it in a glance.

When designing for people, the designer must keep in mind that people come in many sizes and shapes. There really is no average person. To determine appropriate sizes, designers depend on data from the field of anthropometry.

3D VIEW
SEE IT LIVE ON THE WEB

How Technology Works...

For: Ergonomics Activity
Visit: www.mytechedkit.com

Design a Space Travel Chair Online

Ergonomics is the study of people in their working environment. It determines how new technologies are applied to everyday products. When combined with advances in technology, ergonomics allows us to improve products. Go online to see how it is used to develop a chair for space travel.

Supports forearms

Supports lower back

Anthropometry is the science of measuring people. It provides information about the average size and shape of people's bodies. Most often, designers accommodate the middle range of body sizes. They ignore the bottom 5 percent and the top 5 percent of dimensions. For this reason, a pocket calculator may have buttons too small for people with very large fingers.

Safety

A product, system, or process must be designed so it is safe for consumers to use. For example, cooking utensils that are made of heat-conducting materials should have handles made of materials that do not conduct heat well. Safety codes and regulations must be met. These are especially important for children's products and large equipment, such as automobiles or amusement rides (Figure 3.12).

Figure 3.12 Safety features are an important part of good design.

Hypothesizing *What principles other than safety were probably important in designing this roller coaster?*

LIVING GREEN
Greening Up Your Design Process

Many engineering and design companies are beginning to incorporate environmentally responsible decisions into their design process. This can mean using sustainable resources, such as wood made from renewable lumber, and manufacturing products in factories that use solar or wind energy. Several colleges and universities are beginning to offer 'green' design and engineering certifications as part of their programs. With limited nonrenewable resources and the concern for the effects of technology on the health of our environment, this trend is sure to continue.

MAKE A CHANGE
Take a look at products around your house, at school, or the next time you're in a home improvement or electronics store. See if you can find labels that inform you that the product was made from recyclable materials or in a factory that uses green energy. Make an attempt to purchase products that consider the environment in their design.

TRY THIS
In your next technology design activity, include an environmentally responsible principle into your design. This might mean choosing to use recycled materials in your design, or perhaps considering how easily your final product can be recycled.

All things have a life span. Living organisms die and physical devices fail to work. One important responsibility of a designer is to ensure that a product's failure does not endanger people. When failure occurs, a pump should not explode and an electric motor should not burst into flame. Ensuring people's safety is a critical part of a designer's job.

Aesthetics

Designers know that if a product is going to sell well, people must like the way it looks. **Aesthetics** is the study of the way something looks and how that affects people's feelings about it. For some products, aesthetics is used as the overriding principle in its design. In other cases, greater quality or functionality also makes a product more aesthetically pleasing.

Maintainability

Equipment and structures need to be maintained. A machine part needs periodic lubrication. Measurements (such as electrical resistance) need to be taken. A design has to allow for access to certain parts for maintenance or repair and to others—the filter on a clothes dryer, for example—for regular cleaning.

SECTION 2 Assessment

Recall and Comprehension

1. What is anthropometry? How is it related to product design?
2. What is quality? Give an example of what you might check to determine the quality of a new sofa.
3. How is maintenance of machine parts related to accessibility?

Critical Thinking

1. **Inferring** Why are crayons for young children larger than crayons for older children?
2. **Contrasting** Discuss two styles of jackets. Contrast their functionality with their aesthetic appeal.
3. **Interpreting** Examine a safety top on a medicine container. Why has it been designed in this way? In your own words, explain how it works.

QUICK ACTIVITY

Sculpt modeling clay around an inexpensive plastic pen to make it more comfortable for your writing grip. Why would an ergonomically designed pen be slightly different for each individual? What problems would you need to address if you wanted to mass-produce and market your new pen design? **For more related Design Activities, see pages 80–83.**

People in Technology

Peter Lucas and MAYA Design, Inc.
Creating User-Friendly Designs

"No one wants a remote control with a zillion buttons. It just won't sell."

—Peter Lucas

Peter Lucas is the co-founder of a company called the MAYA Design in Pittsburgh, Pennsylvania. MAYA Design, Inc. creates user-friendly designs for very complex products—items such as DVD players, remote controls, medical instruments, and computer software. According to Peter, design that is based on a product's function or technology is "hot, and will remain hot. . . . No one wants a remote control with a zillion buttons. It just won't sell."

The Importance of Design
MAYA stands for **M**ost **A**dvanced **Y**et **A**cceptable design. It is a phrase coined by a famous designer named Raymond Loewy. He used it to describe an ideally designed product—one that is advanced enough to take advantage of new technology but is still user-friendly.

Peter Lucas

One of Peter's central beliefs is that design must play an important role in the development of a new product. Rather than letting a product's design evolve slowly, Peter believes that companies need to treat design more seriously. Because the impact of design on a final product is so great, companies need to consider it as an important part of the way industry does business.

Creativity through Brainstorming
At MAYA Design, Inc., designers work in areas as diverse as consumer electronics, multimedia, and product design. They are encouraged to be as creative as possible.

To brainstorm new ideas, they gather in a circular room with white, washable walls to write on. The room is called a kiva. It gets its name from the circular underground rooms that the Pueblo Indians built for their religious ceremonies. "You can work your way around the erasable white board. When you get back to where you started, you are done with your idea!" jokes Peter. Or, more likely, they start erasing and keep going with even better ideas.

The MAYA Design, Inc. uses a whiteboard to brainstorm design ideas for new projects.

Critical Thinking

1. **Applying** How could you simplify the design of a remote control for a DVD player?

2. **Extending** Sketch possible ideas for a new design for a whiteboard. Explain your sketches in writing so others will understand your design.

Solving Real-World Problems

Reading Strategy

Listing Make a list of products or solutions that might be affected by real-world problems. Start your list by writing down each of the five headings in the section.

Social and Environmental Concerns
- Building a new airport
- Enlarging a shopping mall

Vocabulary

values   risk/benefit trade-off   probability

Benchmarks for Learning

- Sometimes, there are trade-offs between social and personal desires.
- Design solutions should benefit society but have minimal impact on the environment.
- Values and beliefs shape people's attitudes toward technology.

Social and Environmental Concerns

Most problems that can be solved in your technology class do not involve issues of the real world. Often, they are a bit simpler, such as designing a backpack or making a product more ergonomic. As you learn to use problem solving, you will see that solutions to real-world problems often involve more limitations. Engineers and designers must keep in mind how their solutions will affect society and the environment. If an airport is being planned, it cannot be built too near a residential area because the noise from arriving and departing planes would be annoying (Figure 3.13).

In real-world problem solving, the needs of society or the community must be considered. When a technological solution has potentially negative effects, people must decide whether the solution's benefits outweigh its risks. A good technological solution meets the needs of people and has little adverse effect on the environment and creates a win-win situation for the economy and the environment.

Politics

Often, solutions to real-world problems are affected by politics (Figure 3.14). Groups of people join together to promote a special interest. For example, some groups oppose the building of nuclear power plants. They believe that if an accident were to happen, radioactive material could poison the environment. They point out that, for now, there is no safe way to get rid of radioactive waste.

Figure 3.13 Design can affect society and the environment.

Extending *How might problems of noise and pollution be lessened for this community?*

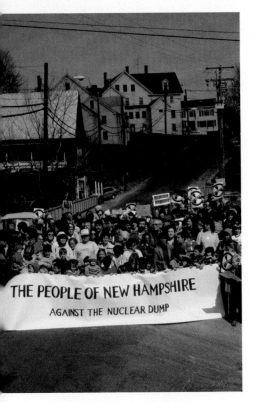

Figure 3.14 Design solutions affect local politics. Here, students are protesting the location of a new site for dumping radioactive waste.

Extending *What constraints might exist for the design and location of a new superstore?*

Other groups favor nuclear power plants. They believe that nuclear power is an important source of energy. They think nuclear power will make us more self-sufficient and less dependent on other countries for oil to fuel our power plants. They are convinced that the risk of an accident is small, and that the benefits of a reliable source of energy are worth the risk.

Values

Our **values,** or what we think is important, influence all of our decisions. They are a factor in our decisions for or against something. For example, if we look at cars only as transportation, we might decide to buy a basic vehicle that gets good gas mileage. If we think that cars are neat and that driving is fun, we might decide to buy a sports model.

Risk/Benefit Trade-off

A common trade-off made in solving complex problems is a **risk/benefit trade-off.** To obtain desired benefits, designers must accept some risk, but they try to keep the risk as low as possible. They may not implement a solution if the risks are too high.

For example, when we travel by car, we accept a very low risk of being hurt in an accident. In exchange for this risk, we receive the benefit of being able to travel quickly and comfortably. It is never possible for products and projects to be 100 percent risk free, but a well-designed product presents the minimum possible risk to people as well as to the environment.

Determining risk requires a combination of human knowledge and mathematics. The goal is to determine the probability of some negative event (the risk). **Probability** is the likelihood that something will take place (Figure 3.15). Determining probability requires the use of mathematics as well as a good sense of judgment.

Continued Monitoring

Solutions to real-world problems must be monitored and maintained to make sure there are no undesirable outcomes. A material may not be as durable as expected, or a product may be used in ways that were not anticipated. Information from monitoring should be obtained and evaluated over the life of the product (or solution) to ensure that it continues to meet the design criteria.

Figure 3.15

Figure 3.16

This information may indicate that changes need to be made to a product's design. When this occurs, the product is recalled. For example, the windshield wiper motors on a car might have to be redesigned because information from customers indicates that the original motor presents a safety hazard. A child may use a toy in ways that a designer did not predict (Figure 3.16).

Other types of recall are more serious. Certain medicines, for example, have been found to produce serious and often harmful side effects. Proper testing and continuous monitoring reduce the likelihood of negative effects.

Figure 3.15 Wise investing involves an understanding of the potential risks and benefits of a particular decision.

Inferring *How might probability relate to the amount of money a person chooses to invest?*

Figure 3.16 Even after extensive research, products may be used in unintended or unexpected ways.

Explain *Why do you think manufacturers put age limits on their products?*

SECTION 3 Assessment

Recall and Comprehension

1. What is probability? How is it related to design?
2. Why is it necessary to monitor how a product is used?
3. How do real-world problems differ from most that you solve in technology class?

Critical Thinking

1. **Hypothesizing** Imagine that your neighborhood needs low-income housing because many people cannot afford existing housing. Discuss the challenges in developing a plan to build low-income housing.
2. **Making a Judgment** Discuss the problems in developing a medicine that, though helping seriously ill people, may have long-term side effects.

QUICK ACTIVITY

A local government is considering a change to its building code, requiring that all new driveways and parking lots use green technology, such as crushed stone, rather than asphalt. Discuss the benefits and liabilities of each technology, taking into account environmental impacts.
For more related Design Activities, see pages 80–83.

3 Review and Assessment

Chapter Summary

- People have always been problem solvers. We use technology to make life easier, and to solve problems involving society, the environment, and the individual.

- Design is a process used to create new processes, products, and systems. Different solutions are carefully thought out and refined. There is usually more than one good solution for any given problem.

- Informed design assists people in solving problems. It allows them to begin decision making without complete knowledge. Later in the process, they can return and refine decisions when more information is available.

- Alternative designs must be compared. Designers must choose the one that works best, is the most economical, and has minimal negative impact on people and the environment.

- When technologists design, they consider principles such as functionality, quality, ergonomics, safety, aesthetics, and maintainability.

- Most of the technological problems presented in school are less complex than real-world problems. Real-world problems involve political and environmental issues, as well as those relating to values.

Building Vocabulary

Use the chapter vocabulary words listed below to complete a crossword puzzle. Exchange puzzles with a classmate. Complete the puzzle and then check each other's answers.

1. aesthetics
2. anthropometry
3. brainstorming
4. constraints
5. design brief
6. ergonomics
7. functionality
8. informed design
9. mathematical model
10. probability
11. prototype
12. quality
13. risk/benefit trade-off
14. scale model
15. specifications
16. values
17. variables

 See your teacher for the Crosstech puzzle.

Reviewing Content

1. Give three examples of problems: one involving society; another the environment; and the last, the individual.

2. What are the eight steps of the informed design process?

3. What is ergonomics? Give three examples of products that use ergonomic design.

4. What are the six main principles of design? Select a piece of sports equipment that you have used, such as a tennis racket, and describe each of the principles in terms of that product.

5. What is market research? When is it used in the design process?

Applying Your Knowledge

1. Describe a problem in society or the environment that needs to be corrected.

2. Give an example of how a person's values might affect a decision about the kind of car to buy.

3. Suppose that you are an architect hired to design a house for this neighborhood. Planes fly overhead approximately every 7 to 10 minutes. What types of solutions might you explore?

4. You are in charge of market research for a new backpack. Design a questionnaire you will use to gather information from your classmates.

5. Draw simple sketches of three different designs for an ergonomic bottle. How is each one ergonomic?

6. Participate in and assume a leadership role in a design or science competition (e.g., A Technology Student Association, Lego League, First Robotics, Science Olympiad, or Science fair competition).

Critical Thinking

1. **Applying** Think of a problem that involves a personal issue. Propose a workable and economical solution to the problem. Include a sketch with explanations.

2. **Analyzing** Designers create a prototype or a smaller model after they have completed the preliminary design. Why is a prototype especially important when many final products will be made?

3. **Inferring** Your city has run out of land for a landfill (refuse disposal). The city government has chosen to build an expensive incinerator to handle the garbage problem. What might be some of the trade-offs that were made in reaching this decision?

4. **Analyzing Cause and Effect** Why are products sometimes recalled? How might designers help to prevent product recalls?

5. **Extending** Advertisers spend a lot of time and money trying to appeal to people's values. What types of values might be important when advertising high-quality, expensive sports gear?

Connecting to STEM
science • technology • engineering • math

Determining Risk

Risk is a part of life. Designers must keep this in mind as they create new products, such as a car. We know that the risk of a fatal accident is very low—about 24 deaths per billion miles, or 0.000000024. If you were to take a trip of 100 miles, what is the probability of your being involved in a fatal accident? How might designers reduce this number as they design new cars?

Design Activity 5

TEAM LOGO

Problem Situation

School has just started, and many of your classmates don't know each other. The teacher is organizing the class into teams so students can work together on technology projects. What should your team be named? How can you display your new name?

Your Challenge

You and your teammates are to create a name and logo that indicate to the class the common interests of your team.

> Go to your **Student Activity Guide, Design Activity 5.** Complete the activity in your Guide, and state what the design challenge is in your own words.

① Clarify the Design Specifications and Constraints

To solve the problem, your design must meet the following specifications and constraints:
- The logo must indicate the names or numbers of members of the team.
- The logo must indicate common interests shared by the team.
- The logo must be readable by the teacher at a distance of 6 feet.
- The logo must stand by itself using a 24" wooden stick as a support.

> In your Guide, state what the specifications and constraints are. Include any others that your team or your teacher provided.

② Research and Investigate

To better complete the design challenge, you need to first gather information to help you build a knowledge base.

> In your Guide, complete Knowledge and Skill Builder 1: Logo Investigation.

> In your Guide, complete Knowledge and Skill Builder 2: Common Interests and Names.

> In your Guide, complete Knowledge and Skill Builder 3: Eye Test.

> In your Guide, complete Knowledge and Skill Builder 4: Stands.

Materials

You will need:
- 8" x 20" foam board
- 24" wooden stick

③ Generate Alternative Designs

In your Guide, describe two possible solutions that your team has created for the problem. The solutions should be based on the knowledge you have gathered so far.

④ Choose and Justify the Optimal Solution

Refer to your Guide. Explain why you selected your solution and why it is the better choice.

⑤ Develop a Prototype

Create your team's logo. Include a drawing or a photograph of your final design in your Guide.

In any technological activity, you will use seven resources: people, capital, time, information, energy, materials, and tools and machines. In your Guide, indicate which resources were most important in this activity, and how you made trade-offs among them.

⑥ Test and Evaluate

How will you test and evaluate your design? In your Guide, describe the testing procedure. Explain how the results will show that the design solves the problem and meets the specifications and constraints.

⑦ Redesign the Solution

Respond to the questions in your Guide about how you would redesign your solution. The redesign should be based on the knowledge and information that you gained during the activity.

⑧ Communicate Your Achievements

In your Guide describe the plan you will use to present your design to your class. Include any handouts that will explain the logo you have designed.

Design Activity 6
FOLDING CHAIR

Problem Situation

Whether at the beach, at home, or watching your football team play a game, you never know when you could use an extra chair. When you attend a picnic with your family, you might have to sit on the ground because there is a shortage of chairs. When you have company for dinner, you may need an extra chair for an unexpected visitor. A portable chair would be useful for either of these situations.

Materials

You will need:

- 1-inch cardboard strips, 12 inches long
- 1-inch foam sheets (for padding)
- 1-inch PVC tubing
- 3-mil plastic sheeting
- chair webbing (can be denim or canvas)
- dowels
- fasteners (screws, nuts and bolts, washers, nails)
- glue
- hinges
- metal strips
- paint
- paper fasteners
- paper punch
- string
- twine
- wooden slats

Your Challenge

You and your teammates are to construct a portable folding chair.

> Go to your **Student Activity Guide, Design Activity 6.** Complete the activity in your Guide, and state the design challenge in your own words.

① Clarify the Design Specifications and Constraints

To solve the problem, your design must meet the following specifications and constraints:

- You must determine where the folding chair will be used.
- The folding chair should be ergonomically appropriate for a typical student in your class.
- The folding chair should be easy to open and close.
- The folding chair should be safe to use.
- The folding chair should be lightweight but durable.

> In your Guide, record each specification and constraint. Include any others that your team or your teacher provided.

② Research and Investigate

To better complete the design challenge, you need to first gather information to help you build a knowledge base.

> In your Guide, complete Knowledge and Skill Builder 1: Aesthetics.
>
> In your Guide, complete Knowledge and Skill Builder 2: Pivot Points and Folding Devices.
>
> In your Guide, complete Knowledge and Skill Builder 3: Ergonomics.

③ Generate Alternative Designs

In your Guide, describe two possible solutions your team has created for the problem. The solutions should be based on the knowledge you have gathered so far. In your description, indicate what you consider to be each solution's strengths and weaknesses.

④ Choose and Justify the Optimal Solution

Refer to your Guide. Explain why you selected your solution, and why it is the better choice.

⑤ Develop a Prototype

Construct a working model of your chair. Before using the actual materials, you may want to make some of the elements out of cardboard strips first to make sure the design works. Include a drawing or a photograph of your final design in your Guide.

In any technological activity, you will use seven resources: people, capital, time, information, energy, materials, and tools and machines. In your Guide, indicate which resources were most important in this activity, and how you made trade-offs among them.

⑥ Test and Evaluate

How will you test and evaluate your design? In your Guide, describe the testing procedure. Explain how the results will show that the design solves the problem and meets the specifications and constraints.

⑦ Redesign the Solution

Respond to the questions in your Guide about how you would redesign your solution. The redesign should be based on the knowledge and information that you gained during the activity.

⑧ Communicate Your Achievements

In your Guide, describe the plan you will use to present your solution to your class. Include any handouts or PowerPoint slides that you plan to use.

Communicating Design Solutions

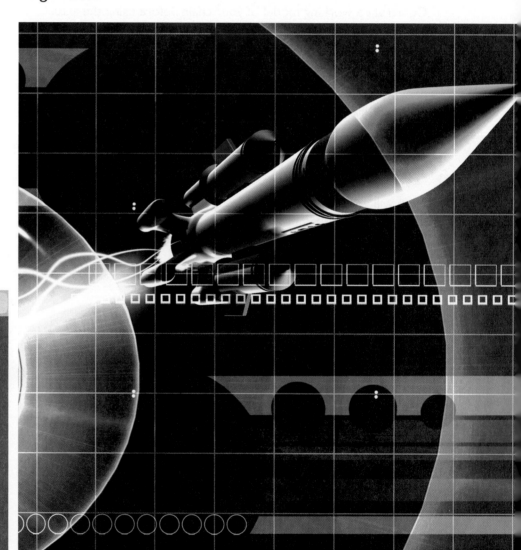

In this chapter, you will learn the following different ways of presenting design ideas:

- ideas
- sketching
- freehand drawing
- technical drawing
- computer-aided drawing [CAD]
- visual aids
- oral presentation

The saying that "a picture is worth a thousand words" is especially true in technology. At each stage of design, architects, engineers, and all types of designers use pictures to communicate ideas.

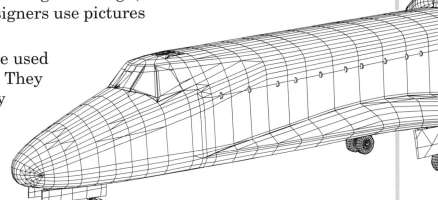

- Sketches and simple drawings are used during the early stages of design. They help clarify the problem, and they are used to brainstorm ideas with other people.

- During the later stages of design, more detailed technical drawings are used to present a variety of ideas and solutions. These drawings may use color and texture.

- Three-dimensional models are often used to show the final design idea. They can range from a simple cardboard model to a computer simulation.

- Many design ideas are also illustrated using tables, graphs, and charts.

Drawing Techniques

Benchmarks for Learning

- The use of symbols and drawings promotes clear communication.
- Objects can be shown using two- or three-dimensional representations.
- Drawings and sketches are used to organize, record, and communicate ideas.

Reading Strategy

Mapping Create a concept map that shows the different types of drawing techniques described in this section.

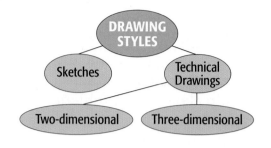

Vocabulary

crating
orthographic drawing
pictorial drawing

oblique drawing
isometric drawing

perspective drawing
rendering

Symbols and drawings

Figure 4.1 Effective graphic messages use carefully selected words, symbols, and fonts.

Figure 4.2 Crating helps create sketches from simpler geometric shapes.

Interpreting *Name several occupations where drafting is used.*

Graphic Techniques

Graphic techniques are methods used to communicate ideas visually. They include the use of letters, symbols, sketching, and drawing, among others. Each technique can be used to communicate a specific design or visual idea.

Using Text Symbols

One way of communicating a graphic idea is to combine words, symbols, and fonts. A font is a particular size and style of type. The words you are reading were set in a font that is clear and easy to read. Other fonts are used to convey a mood or to get the reader's attention (Figure 4.1). The

Figure 4.1

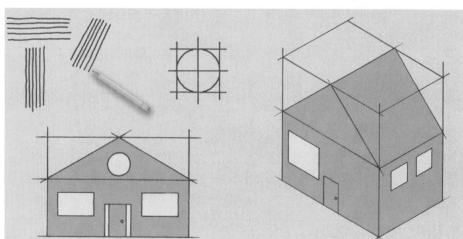

Figure 4.2

use of symbols can further promote clear communication by providing a common language through which ideas are expressed. Type sizes and fonts can underscore the message.

Sketching and Freehand Drawing

Sketching is a simple, visual way of communicating a design. A sketch is a rough drawing of an object. It is done quickly, giving only an outline and a few details. Sketching is usually done freehand, without drawing tools or instruments.

Try some different exercises to help you learn to sketch (Figure 4.2). Draw about 20 parallel horizontal lines on a piece of paper. Then, draw some vertical lines and some diagonal lines. Try to move your whole arm when you draw. Now, draw geometric shapes such as squares, rectangles, triangles, and circles. Most objects can be drawn by combining these shapes. This technique of working from boxlike geometric shapes is called **crating.**

Technical Drawing

An artist can illustrate things fairly accurately using freehand drawings. For more precise drawings, however, technical drawing is used. Technical drawing is a method of producing highly accurate drawings using special instruments and tools. Artists, engineers, architects, designers, and drafters use it to communicate an object's true size and shape.

Technical drawing can be done by hand or by computer. To make a drawing by hand requires a drawing board, a T square, a plastic triangle, paper, and a pencil (Figure 4.3).

Figure 4.3 A drafting table, T square, and plastic triangle are basic tools used to create technical drawings by hand.

Inferring *When drawing lines, why is a T square used instead of a simple ruler?*

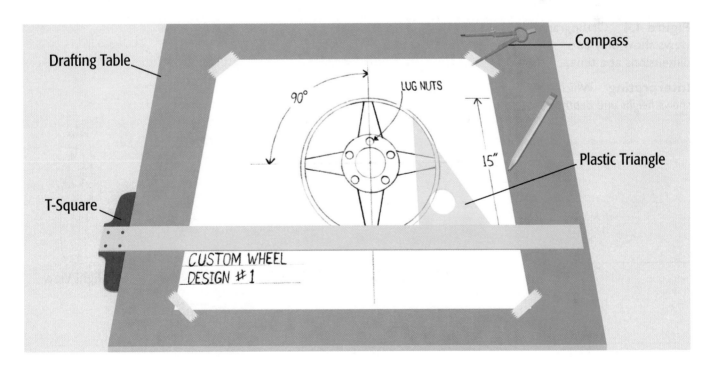

Drafting Table

Compass

T-Square

Plastic Triangle

90°

LUG NUTS

15"

CUSTOM WHEEL
DESIGN #1

A T square is a long straightedge with a crosspiece at one end. It is held against the edge of the drawing board and then moved up or down. In this way, horizontal lines can be drawn exactly parallel to each other. The plastic triangle is placed against the long edge of the T square and used as a guide to draw vertical lines or lines at an angle.

Two-Dimensional Drawing

When we draw flat shapes, we draw in two dimensions. A two-dimensional, or 2-D, drawing is also called an **orthographic drawing.** Orthographic drawings let us look at the front, side, or top of an object (Figure 4.4). Sometimes, other views, called auxiliary views, are shown.

Orthographic drawings are used to illustrate details of an object in a certain view. The front view of a stereo receiver, for example, can be used to show the different user controls. However, orthographic drawings do not give a realistic view of an object. They show only two of the three dimensions (height, width, and depth). Depending on the view, the two dimensions could be width and depth (top view), width and height (front view), or depth and height (side view).

Three-Dimensional Drawing

A three-dimensional, or 3-D, drawing is more realistic than a 2-D drawing. Because it gives a clear picture of the object, it is called a **pictorial drawing.** The three dimensions—height, width, and depth—are all visible in a single 3-D drawing.

Figure 4.4 Orthographic views show only two dimensions at a time.

Interpreting *Which view shows height and depth?*

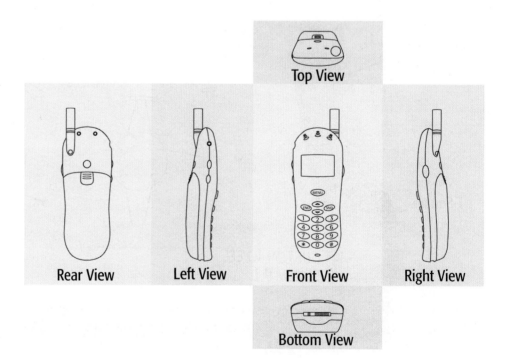

Top View

Rear View　　Left View　　Front View　　Right View

Bottom View

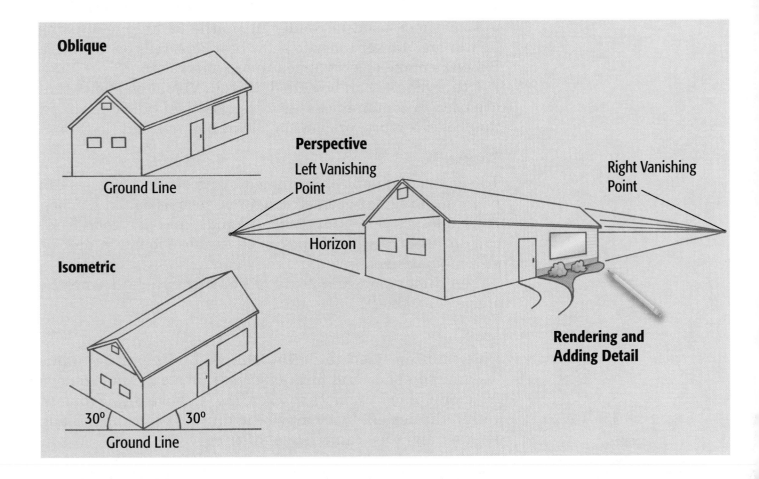

Oblique

Ground Line

Perspective

Left Vanishing Point

Right Vanishing Point

Horizon

Isometric

30° 30°

Ground Line

Rendering and Adding Detail

There are three kinds of pictorial drawings: oblique, isometric, and perspective. An **oblique drawing** shows one surface of an object with a straight-on view. The other two surfaces are shown at an angle (Figure 4.5). An oblique drawing is probably the easiest type of pictorial drawing to make, but it also tends to distort the look of the subject.

An **isometric drawing** is drawn within a framework of three lines, or an isometric axis. These three lines are like the edges of a cube. The two base lines of the axis are drawn at an angle of 30 degrees to the horizontal. An isometric view is fairly quick to draw and presents a less distorted view than that in oblique drawings.

A perspective drawing is the most realistic type of pictorial drawing. In a **perspective drawing,** parts of the object that are farther away appear smaller. These types of drawings can show one or two vanishing points.

Elements of Effective Drawing

Once a 2-D or 3-D drawing is made, more detail can be added to make it more realistic. **Rendering** is a process of making perspective drawings look more realistic. Shading, which

Figure 4.5 3-D drawings provide more detail than 2-D drawings.

Interpreting *What effects are produced by the use of color and texture?*

is sometimes used in rendering, is the technique of using lighter and darker tones to show how light falls on an object. Shadows make objects appear more natural.

Other details can be added as well. Visual elements that are used to communicate designs effectively include shape, line, color, texture, proportion, balance, unity, and rhythm.

Shape

The shape of an object is defined by its outline. Shapes can be those of regular geometric figures, such as squares, cubes, rectangles, circles, triangles, pyramids, and polygons. They can also be irregular shapes, such as those found in nature, or combinations of regular and irregular shapes.

The dimension and scale of your drawings define the shape of an object.

Line

Line is often used to define shape. Lines put boundaries around space and form objects. Lines can lead the eye from one place to another, creating a sense of movement within the design. Lines may be of different thicknesses and lengths, and they can be straight or curved.

3D VIEW
SEE IT LIVE ON THE WEB

How Technology Works...

For: Design Activity
Visit: www.mytechedkit.com

Communication of a Design Idea Online

Have you ever tried explaining how an object looks to someone and end up sketching it out for them? When you do so, you are communicating information using dimension and scale, not words. Go online to see how the elements of design are used to express an idea.

Color

Color lends excitement and interest to a drawing. It can also affect how we react to or feel about it (Figure 4.6). For example, blue colors make an object appear colder. Red colors make an object seem warmer. Colors can relax or upset us. They can make objects appear larger or smaller, closer or farther away, lighter or heavier.

Texture

Texture refers to the way the surface of an object looks and feels. Different materials have different textures, and they can be illustrated with drawing techniques. Texture is often used as a surface finish to improve the way something works. For example, a rough leather finish on a car's steering wheel provides a good grip for the driver and makes driving safer. A smooth finish on a kitchen or bathroom countertop makes cleaning easier.

Proportion

Proportion involves the relationship among the different sizes and dimensions within a drawing. Certain proportions are pleasing to the human eye. The Golden Rectangle, discovered by the Greeks, has a pleasing shape. People use golden rectangles in art and architecture because the width—five units—and the height—three units—are in perfect proportion. This ratio has been noted in the proportions of both human and animal bodies and most notably in the famous ancient Greek temple—the Parthenon.

Balance

Balance refers to the way the various parts of the drawing relate to one another. Some drawings have symmetry, in which one side is a mirror image of the other side (Figure 4.7). Drawings that have asymmetry have two sides that are different from each other (Figure 4.8). Many designers seek symmetry in their work, as that is usually more pleasing to the eye. However, asymmetry is also eye-catching when it does not interfere with the overall harmony of an image.

Unity

Unity refers to the way in which all parts of a drawing come together to produce a single general effect. Rhythm, as in music, has to do with movement. In a drawing, rhythm relates to the way the eye of the viewer moves around the entire work.

Figure 4.6 The use of line, shape, and color strongly affects how we feel about a drawing.

Evaluating *What design elements did the artist use to make this poster?*

Figure 4.7

Figure 4.8

Symmetry in Architecture

Figure 4.7 The Lincoln Memorial is a symmetrical building. The left side is a mirror image of the right side.

Figure 4.8 The Guggenheim Museum in Bilbao, Spain, is made up of unusual, asymmetrical shapes.

Interpreting *What sense of rhythm is created by the asymmetrical building?*

All the elements you just learned are taken into account by designers and product developers alike. The appropriate combination of these elements as well as an understanding of the intended audience and purpose of the drawing can promote clear communication of many types of messages.

For example, many people find it easier to follow illustrated instructions when putting together something like a swing set or bicycle. Images can provide a common language that everyone can understand. Technologists and scientists often use specialized symbols or technical illustrations to convey ideas.

SECTION 1 Assessment

Recall and Comprehension

1. How does freehand drawing differ from technical drawing?
2. Explain the difference between orthographic and pictorial drawings. When might a designer choose to use each type of drawing?
3. Design a poster that points out three possible safety hazards to watch out for in your technology lab.

Critical Thinking

1. **Making a Judgment** Draw rectangles with different proportions (width to height), including one with the proportions of the Golden Rectangle. Which do you find most visually appealing? Why?
2. **Applying** Draw sketches that describe the following words: tree, shut, sharp. Would other people understand your sketches? Why or why not?

QUICK ACTIVITY

Internationally recognized symbols are used at airports and other public locations. The "no smoking" symbol and the "handicapped" symbol are two such examples. Do some research to find some other international symbols. Make sketches of them and present your findings to the class. **For more related Design Activities, see pages 104–107.**

SECTION 2
Computer-Aided Design (CAD)

Reading Strategy

Analyzing Cause and Effect Show the different effects that CAD may have on every process of design, from drafting, revising, and manufacturing to production.

Vocabulary

computer-aided design (CAD)

drafting tablet
plotter

USING CAD

1. Drafting
2. Revising
3. Manufacturing/ Production

EFFECTS

1. Drawing can be made faster and more accurately
2.
3.

Benchmarks for Learning

- Computers are used to improve productivity in various applications.
- A computer-aided design system uses a computer to produce designs and drawings quickly and accurately.

Using CAD

The computer has brought about great changes in drawing and design. It has made drafting much easier. **Computer-aided design (CAD)** is an important software application that allows people to produce sophisticated drawings and designs using a computer instead of traditional technical drawing tools. CAD has become an important tool used by engineers, drafters, and designers.

The CAD Workstation

A CAD workstation has many parts (Figure 4.9). It includes a computer, keyboard, display screen, drawing tablet, and plotter. CAD software offers a variety of drawing tools that can be accessed using either the keyboard and mouse or the drafting tablet. With a keyboard and mouse, the drafter chooses a drawing tool from the assortment shown on the screen. Different tools may produce heavy or thin lines as well as different shapes. A **drafting tablet** is a special surface on which the drafter uses an electronic pen to make lines and drawings that appear directly on the computer screen.

Producing Final Drawings

Once the drawings have been made, they can be changed using keyboard commands or the drawing tablet. Programs can be used to draw lines and arcs, add symbols to a drawing, turn the drawing around or upside down, or zoom in for a closer look at a particular detail. CAD software can produce 3-D models that look solid and realistic. The size and scale of

Figure 4.9 A basic CAD workstation.

Identifying *What are other advantages of using CAD?*

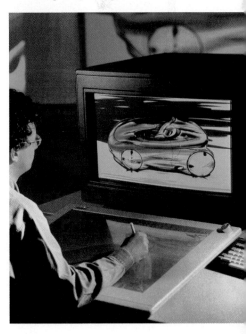

CHAPTER 4 Communicating Design Solutions **93**

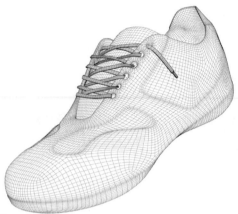

Figure 4.10

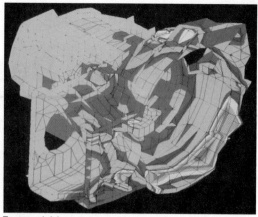

Figure 4.11

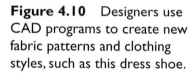

Figure 4.10 Designers use CAD programs to create new fabric patterns and clothing styles, such as this dress shoe.

Figure 4.11 CAD mechanical drawings often include a list of parts needed as well as their cost.

Comparing *What is similar about these two drawings?*

the final drawings can also be changed. These drawings can be stored in a computer's memory for later review and revision.

When the final design is ready, the drawing is printed on a large sheet of paper. A wide, large-scale printer, called a **plotter,** is used to print mechanical and architectural drawings.

Applications of CAD

CAD computer programs make it possible to produce clear drawings of almost any designed product. CAD drawings may show mechanical parts, illustrate entire buildings, or detail tiny electronic circuits. Clothing designers use CAD software to create patterns and images for new fabrics and clothing (Figure 4.10). Engineers often use CAD-generated drawings to label the parts of a machine or indicate how it is assembled using a 3-D exploded view (Figure 4.11). Landscape architects use CAD to show the arrangement of trees, shrubs, walkways, and lighting in an outdoor area. Urban planners show how new streets and traffic lights will affect the flow of traffic through a city. Other CAD-designed products include cars, aircraft, furniture, rugs and other textiles, and electronic games.

CAD is used to create many different views for a design presentation. Before building begins, an architect may want to have an overall 3-D image that shows the client how the final building and landscaping will appear. Other designers can create physical 3-D scale models from the CAD images. This allows clients to see a physical representation of the design. Once construction is ready to begin, the architect uses CAD to generate 2-D layouts for others on the project.

Why Use CAD?

Once an architect or engineer has mastered the use of the software, a CAD system can have many advantages over a hand-drawn design.

- CAD saves time by combining design and drafting into a single process. Architectural or mechanical drawings can be created much faster with CAD than by hand.

- Information stored in the computer can help prevent errors in the design.

- Because designs can be revised on screen, much less time is needed to change or improve existing designs. All one has to do is create an object only once, and then save it as a file.

- Designs are more accurate. Changes can be made more consistently from one design to the next. In addition, each change can be saved as it own version, in case one wants to refer back to a previous design.

- CAD systems can provide a list of parts needed to make a product, as well as the cost of each.

- A CAD system can help identify a design's strengths and weaknesses. It can show how an electronic circuit will work, or whether a mechanical part will hold up under use.

Because less time is needed to produce layouts and drawings and to determine budgets, CAD systems free designers, drafters, and engineers to spend more time creating and thinking about design.

SECTION 2 Assessment

Recall and Comprehension

1. What does the acronym CAD stand for? Describe a CAD system in your own words.
2. Discuss the advantages of CAD drawings compared to those that are hand-drawn.
3. How does CAD help with budgeting and manufacturing a product?

Critical Thinking

1. **Extending** What are the disadvantages of using CAD? When might it help to have hand sketches or a physical 3-D model?
2. **Applying** You are designing a new series of snowboards, but your boss does not want to set up a CAD system for the office. Convince her that CAD would improve design and manufacturing. Write an outline of the points you will make.

QUICK ACTIVITY

Use CAD software to draw an American flag. Draw only one star and use the Duplicate or Copy-and-Paste function to draw and properly situate all 50 stars in the blue field. Refer to the flag hanging in your classroom if necessary. Be sure to save your flag to a disk, as you may need it for another activity. **For more related Design Activities, see pages 104–107.**

Technology in the Real World

Virtual Imaging and Real Estate

You have read about how architects and engineers use CAD to design new buildings and machines. InterArc is an example of a company that is building on this existing technology and creating exciting new applications in the real world. Many architectural companies use 3-D technology today as a means to visualize what a shopping center or office park is going to look like. But when viewers look at these designs, usually they can only see the sites from preprogrammed angles or along preset pathways.

Condominium Design Meets Gaming Technology

The designers at InterArc were tired of only being able to see these views provided by the programmers. Because they are gamers, InterArc's designers thought, "Why not use game technology to put the camera in the viewers' hands?" Just like a first person perspective game, they wanted to let the viewer "walk around" in the 3-D space. In InterArc's virtual environments, viewers can go anywhere they like, walk through rooms in any order, and look in any direction. They can even redecorate, changing the colors of floors and walls or selecting different furnishings.

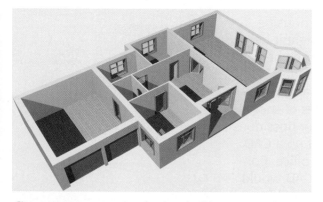

Clients can compare and select building materials as they walk through the designs.

InterArc uses game technology to enable clients to walk through building designs.

InterArc initially designed their application to be a sales tool for condominium developers that allowed prospective buyers to "walk around" in a new condo before the building was even completed. Since then, the application has developed into much more. Developers can now visualize their properties in far greater detail, experiencing things the blueprints just cannot tell you. With these virtual environments, you can tell right away if a first-time visitor can easily see the restaurant from the entrance of a building's lobby. You can see whether a 300-gallon aquarium will be big enough in a 7,000 square foot lobby, or if you really need a 600-gallon one. And so much more. This cuts down on building expenses and streamlines the entire process from blueprint to sold units. Imagine being able to walk around in a condo or office that has not even been built yet, pick out every detail of how you want it to be finished, and see it right then!

Going Worldwide

Currently, InterArc is bringing this technology online, so a businessperson in Japan can explore office space in New York, Las Vegas, or Dubai, and see how it will look with different styles of office furniture, without ever leaving Japan. You can learn more about InterArc at www.InterArc.us.com.

Presenting Your Design Ideas

Reading Strategy

Listing   Make a list of the different visual aids that can be used in a presentation. Next to each item on your list, write a quick description to identify the visual aid.

> **Types of Visual Aids**
> **1.** Outline—list of headings or topics
> **2.** Table

Vocabulary

circle graph   bar graph

Benchmarks for Learning

- The processes and procedures of design need to be documented and communicated to audiences.
- Common symbols are used to communicate key ideas.
- A variety of presentation techniques can be used to improve communication.

Effective Presentations with Visual Aids

A good presentation is an important part of the design process. Often, presentations are made to convince clients that a design is worth producing. They are also made to market the product to potential customers. Without the support and enthusiasm of these people, the design would remain a good idea instead of being developed into a successful product.

Visual aids are an excellent help when making a presentation (Figure 4.12). They provide something for the audience to focus on as they listen. They also make a presentation clearer and more interesting. Effective visual aids include outlines, tables, graphs, photographs, transparencies, and audio and video clips (Figure 4.13).

Outlines and Tables

An outline is a simple list of topics or concepts that helps an audience follow a presentation. An outline communicates key ideas. Each topic listed in the outline can be expanded on during the presentation.

Tables are used to organize and display information in a clear and concise way. Tables can be used to present numbers, text, or even small pictures and photos. The headings at the top of a table indicate what type of information is being presented.

Graphs

Graphs provide other ways of presenting information visually. They should make the presentation interesting but should be kept as simple as possible so they do not distract the audience.

Figure 4.12   Effective presentations of design ideas are just as important as the ideas themselves.

Identifying   *How do visual aids help hold an audience's attention?*

Figure 4.13 Many visual aids can be prepared and even presented using a computer.

Assessing *During a presentation, do you think people prefer to see photos, text, or a mix of both? Why?*

A graph is a visual representation of information and relationships. Graphs are usually used to show how something changes over a period of time. For example, you might use a graph if you need to illustrate how many people will use your product during the next five years. Different types of graphs are used to represent different types of relationships.

Circle graphs (also called pie charts) illustrate the fraction or percentage of the total for different products or categories. They allow people to compare things relative to each other and to the whole. A bar graph also lets people compare different categories. A **bar graph** indicates different quantities by the variations in the lengths of the bars.

Other Visual Aids

When presenting designs to audiences, the use of different types of media can lend impact. In addition to using outlines, tables, and graphs, the inclusion of drawings, photographs, slides, overhead transparencies, and video or audio clips can enhance a presentation.

Traditional media for visual aids include paper charts (often 2" x 3" flip charts) and 8" x 11" transparencies. In transparencies, handwritten notes or drawings are enlarged.

A popular visual-aid tool is Microsoft's PowerPoint graphics software. PowerPoint® presentations offer the advantage of integrating information and animation. Using Power-Point, you can readily include photographs, slides, and short movie clips in your presentation.

Preparing the Oral Presentation

Now that you have prepared your visual aids, how do you get ready for the oral part of a presentation? Few people can stand up and talk about a subject without preparing what to say. Just as in writing, when speaking you must have a clear

Figure 4.14 Research and preparation are especially important for a group presentation. Everyone needs to practice presenting their part.

Hypothesizing *Why might students benefit from doing research as a group, rather than individually?*

idea of what you want to express. You will need to spend some time beforehand preparing an outline of the key topics you want to cover, and then identifying the theme that connects them. Use an outline to help you organize your thoughts.

If you are nervous, it may help to write out exactly what you want to say under each key topic in your outline. Go over your presentation several times. If you are presenting with a team, make sure each person knows the topic he or she will cover (Figure 4.14).

LIVING GREEN
Earth-Friendly Presentations

Many times, materials that we use for visual presentations include poster boards, paper, inks and some type of adhesive. Of course paper and poster boards are made from trees, and although trees are a renewable resource, it takes a great deal of time for them to grow. Some glues and inks used in markers, some pens, and even laser printers are manufactured using petroleum (oil). These chemicals are not only nonrenewable resources, but can also make recycling the paper they are used on difficult.

MAKE A CHANGE
What do you do with your presentation boards once you have finished with them? If you toss them, at least make sure you toss them in the recycling bin. But maybe you can reuse them! Make sure to use both sides of your poster board (you're bound to have more than one presentation in school this year). When shopping for supplies, look for papers and posters that are made from post-consumer recycled content, and look for markers and inks that are made from soy or are marked as having "low VOCs." VOCs are <u>v</u>olatile <u>o</u>rganic <u>c</u>ompounds. These can affect the environment and human health adversely.

TRY THIS
Often, people shy away from purchasing recycled or renewable brands of products because they think they are too expensive. Is this really true? Ask your instructor for a copy of the latest technology laboratory supply catalog and make a chart (from recycled or scrap paper) to compare the prices of the most commonly used presentation supplies. Present your research to your instructor or person responsible for purchasing your lab supplies at your school and see if it makes *cents* to make a switch.

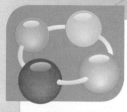

Connecting to STEM

Bar and Circle Graphs

Imagine that you are writing a report on bicycles, and you want to show which models are most popular. You have compiled your information in a table, shown below, but you want to present it visually as well. This can be done using different types of graphs.

Bicycle Sales by Model

Model	Percentage of Total (%)
All-terrain	25
Mountain	37.5
Racing	25
Other	12.5

Creating a Circle Graph

A circle graph is a useful way of presenting information that is part of a whole. A circle, which consists of 360 degrees, is divided to show the sizes of different categories. For example, sales of all-terrain bicycles make up 25%, or one quarter, of the total. In the circle graph shown, therefore, this segment is shown as one quarter of the total area of the circle.

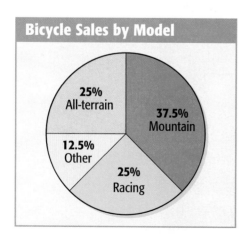

Bicycle Sales by Model

Creating a Bar Graph

A bar graph of the same data shows percentages along a vertical axis, called an x-axis. The length of each bar indicates the percentage of that model of bicycle. The horizontal y-axis indicates the different bicycle models.

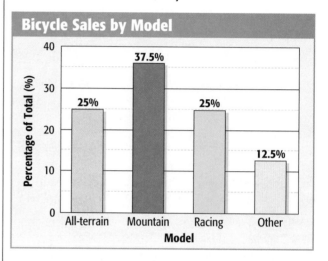

Bicycle Sales by Model

Using Computers to Generate Graphs

Computers can be used to make many different kinds of graphs. If you create a spreadsheet in Excel, it can generate graphs from the data.

To create a graph using Excel, first enter the data for bicycle sales on a spreadsheet. Create a table like the one shown. Highlight both columns, then click on the chart (graph) symbol. Click on each chart option to see which one will best present the information you have.

Critical Thinking

1. **Making a Judgment** Does one of the graphs shown provide the information about bicycle sales more clearly? Explain your answer.

2. **Making Graphs** Using Excel, create a graph that shows the following information: In a class of 24 students, 6 students received As, 8 received Bs, 8 received Cs, and 2 received Ds.

Communicating Effectively

Just as the appearance of a written report affects people's view of its content, your personal appearance affects the audience's reaction to your oral presentation. When you give your talk, you should look presentable, neat, and well dressed. You are presenting not only your material but also yourself as the speaker (Figure 4.15).

Remember that communication is a two-way street. Information is transmitted, or sent, by you and received by the audience. It is important to remember the following points:

- Establish eye contact with different people. Look at individuals directly, and shift your attention from one person to another.

- Speak clearly and loudly enough for the audience to hear you.

- Allow time for listeners to hear and then grasp the ideas you are presenting.

By focusing on your audience in this way, you will keep them involved and can see if your message is getting through.

In most circumstances, plan on 5 to 10 minutes for your presentation. If it takes longer, the audience may lose interest. After you give your presentation, you may be asked questions. Answering questions is often the most informative and important part of a presentation. Be prepared to expand on your topics, and be open to new ideas from your audience.

Figure 4.15 Appearance and body language are an important part of giving a presentation.

Generalizing *What body language is this speaker using?*

SECTION 3 Assessment

Recall and Comprehension

1. What are some ways in which visual aids can improve a presentation?
2. How does appearance affect what is said to an audience?
3. What is a circle graph? How does it differ from a line graph?

Critical Thinking

1. **Making a Judgment** The new gas grill that you designed is tested by five different groups of potential users. Each group has given overall scores for appearance, usability, and seven other factors. What type of visual aid would you use to present these numbers to your audience?

QUICK ACTIVITY

The use of nuclear reactors to produce electricity has always been a subject of controversy. Using charts, diagrams, PowerPoint software, or other visual aids, convince your classmates to support or oppose the continued operation of nuclear power plants. Your teacher may assign you to debate someone who holds opposing views.
For more related Design Activities, see pages 104–107.

Chapter Summary

- Graphic techniques are used to communicate design solutions. These techniques include **sketching,** freehand drawing, and computer-aided technical drawing.

- Sketches and **freehand drawings** are simple, give few details, and are done quickly. **Technical drawings** are more detailed. Instruments and tools are used to make technical drawings. Rendering can also be used to make drawings more realistic. Rendering techniques include shading, adding shadows, and using texture.

- Drawings can be two- or three-dimensional. Two-dimensional drawings provide views of the front, side, top, or bottom of an object. Three-dimensional drawings are more realistic.

- Eight elements of effective design are used to develop appealing designs. These elements are shape, line, color, texture, proportion, balance, unity, and rhythm.

- **Computer-aided design (CAD)** is designing or drawing on the computer screen. A CAD system uses people, software, and hardware. CAD saves time, improves accuracy, and makes changing a design faster and easier.

- **Visual aids** can enhance a presentation. They include outlines, tables, charts, graphs, drawings, photographs, slides, overhead transparencies, and audio and video clips.

- The purpose of communication is the transmission and reception of information, of ideas and results. When giving an **oral presentation,** you need to speak clearly and loudly enough for the audience to hear you, slowly enough for them to follow.

Building Vocabulary

Your teacher may give you a crossword puzzle. Complete the puzzle using the following words from this chapter. Exchange puzzles with a classmate to check each other's answers.

1. asymmetrical
2. balance
3. bar graph
4. circle graph
5. computer-aided design (CAD)
6. crating
7. drafting tablet
8. freehand
9. graph
10. graphic techniques
11. isometric drawing
12. oblique drawing
13. orthographic drawing
14. perspective drawing
15. pictorial drawing
16. pie chart
17. plotter
18. proportion
19. rendering
20. rhythm
21. shading
22. sketch
24. T square
25. technical drawing
26. texture
27. unity

 See your teacher for the Crosstech puzzle.

Reviewing Content

1. Explain how texture and color can make designs more effective.

2. List the following types of drawings from the simplest to the most detailed: orthographic drawing, sketch, perspective drawing, isometric drawing.

3. What types of visual aids can be used during a presentation? List them and then draw a sketch of what each one looks like.

4. What equipment or tools does a designer need to create a drawing using CAD?

5. Do these images show a 2-D or 3-D view? What is the specific type of view shown? Be able to explain your answer.

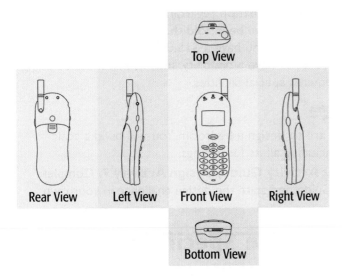

Top View

Rear View   Left View   Front View   Right View

Bottom View

Applying Your Knowledge

1. Write separate outlines for brief presentations explaining how to shoot a basketball, how to make brownies, and how to locate a file on your computer. For each presentation, write a list of the visual aids you will use.

2. Make a technical drawing of a cube that is three inches on a side. Include the measurements in your drawing.

3. Trace the outline of a car or an article of clothing in a magazine photograph. Render it to make it appear more realistic, using shading, shadows, and texture. Use color as well if you wish.

4. Place a bottle of water in front of you on a table. Using the techniques you learned in this chapter, make a simple, freehand sketch of the bottle.

5. Using the Internet and other resources from your library, do some research on well-known public speakers. These might include politicians, teachers, or actors.

Critical Thinking

1. **Generalizing** How has the introduction of CAD affected the fields of drawing and design?

2. **Extending** Investigate the designs of Native Americans. Many of them use symmetry as an important style element. Replicate a design used by a Native American people that exhibits symmetry, and point out the lines of symmetry in the design.

3. **Comparing** Using a painting or drawing program on your computer, make several geometric shapes, such as a circle, square, ellipse, and rectangle. On a piece of paper, draw the same forms by hand. Compare the process and the results. Which drawings are easier to change? Which are more accurate?

4. **Making Judgments** CAD systems can be expensive to set up and use, so they are not used for every type of project. In which project might it be better to use CAD: the design of a new 50-story building, or the layout of new furniture for a loft? Why?

 Connecting to STEM
science · technology · engineering · math

Graphing Data

A recent survey indicates that during a typical day, students divide their time among the following activities: sleep, 8 hours; school, 7.5 hours; watching TV, 1 hour; eating, 2.5 hours; homework, 1.5 hours; other, 3.5 hours. Decide on the best type of graph for presenting this information, then create the graph by hand or on a computer.

Design Activity 7

BEDROOM DESIGN

Problem Situation

You are moving to a new house that is being built for your family. The architect designing the house needs information regarding your family's living style to determine the best design. You have been told that you can design your own bedroom, but there is a catch: You have a $15,000 budget. This amount can be increased by $2,000 if you have a disability, such as blindness or being wheelchair-bound, that requires special facilities.

Materials

You will need:

- card stock or index cards
- colored paper
- foam board or cardboard
- markers

Your Challenge

You and your teammates are to design a bedroom. You will build a scale model of the bedroom, including all its furnishings.

> Go to your **Student Activity Guide, Design Activity 7.** Complete the activity in your Guide, and state the design challenge in your own words.

① Clarify the Design Specifications and Constraints

To solve the problem, your design must meet the following specifications and constraints:
- The window area must be equal to at least 20% of the floor area.
- The minimum room size is 120 square feet, and the minimum closet size is 8 square feet. The minimum height of all ceilings is 8 feet.
- The room has two outside walls and two interior walls.
- The bedroom budget is $15,000.
- The cost of basic construction is estimated at $75 per square foot of floor area.

> In your Guide, state the design specifications and constraints. Add any others that your team or your teacher included.

② Research and Investigate

To better complete the design challenge, you need to first gather information to help you build a knowledge base.

In your Guide, complete

- Knowledge and Skill Builder 1: Geometric Shapes

- Knowledge and Skill Builder 2: Ratio and Proportion
- Knowledge and Skill Builder 3: Sketching
- Knowledge and Skill Builder 4: Aesthetics
- Knowledge and Skill Builder 5: Pricing Information

③ Generate Alternative Designs

In your Guide, describe two possible solutions that your team has created for the problem. Your solutions should be based on the knowledge you have gained so far.

④ Choose and Justify the Optimal Solution

Refer to your Guide. Explain why you selected the solution you did, and why it was the better choice.

⑤ Develop a Prototype

Construct a scale model of your bedroom, including the furnishings. Include a drawing or a photograph of your final design in your Guide.

In any technological activity, you will use seven resources: people, capital, time, information, energy, materials, and tools and machines. In your Guide, indicate which resources were most important in this activity, and how you made trade-offs among them.

⑥ Test and Evaluate

How will you test and evaluate your design? In your Guide, describe the testing procedure. Explain how the results show that the design solves the problem and meets the specifications and constraints. Describe the calculations you used.

⑦ Redesign the Solution

Respond to the questions in your Guide about how you would redesign your solution. The redesign should be based on the knowledge and information that you gained during the activity.

⑧ Communicate Your Achievements

In your Guide, describe the plan you will use to present your solution to your class. Show any handouts or overheads that you will use.

Design Activity 8

WHIRLIGIG

Problem Situation

Whirligigs are simple kinetic, or moving, sculptures that have been designed and created for hundreds of years. Your school's principal wants your class to create a dynamic whirligig display to put in front of the building—something that will hold visitors' attention and move with a breeze. Your class has decided that whirligigs created around a common theme would provide an interesting display. You will need to decide on a theme, perhaps one related to your town, your school, or even a magical land with unusual animals and plants.

Materials

You will need:

- 1" x 6" boards for making whirligig body
- coping saws
- drill
- files
- glue
- 16d and 20d nails
- sandpaper
- screws and washers
- string
- 3/16" diameter tubing
- wooden base, 4" square or larger
- wooden rods or dowels, 12" long
- wooden slats 1/8" thick x 1" wide

Your Challenge

You and your teammates are to design and construct a whirligig.

> Go to your **Student Activity Guide, Design Activity 8.** Complete the activity in your Guide, and state the design challenge in your own words.

① Clarify the Design Specifications and Constraints

To solve the problem, your design must meet the following specifications and constraints:
- The whirligig must relate to the common theme.
- The whirligig's parts must move in a breeze.
- The whirligig must be safe, with no sharp edges.
- The whirligig should be eye-catching.

> In your Guide, state the specifications and constraints. Add any others that your team or your teacher included.

② Research and Investigate

To better complete the design challenge, you need to first gather information to help you build a knowledge base.

> In your Guide, complete Knowledge and Skill Builder 1: What Are Whirligigs?

> In your Guide, complete Knowledge and Skill Builder 2: The Center of Gravity.

In your Guide, complete Knowledge and Skill Builder 3: Constructing Simple Pivot Points.

3 Generate Alternative Designs

In your Guide, describe two possible solutions that your team has created for the problem. Your solutions should be based on the knowledge you have gained so far.

4 Choose and Justify the Optimal Solution

Refer to your Guide. Explain why you selected the solution you did, and why it was the better choice.

5 Develop a Prototype

Make a full-scale drawing of the whirligig, and then construct it. Include the full-scale drawing in your Guide.

In any technological activity, you will use seven resources: people, capital, time, information, energy, materials, and tools and machines, In your Guide, indicate which resources were most important in this activity, and how you made trade-offs among them.

6 Test and Evaluate

How will you test and evaluate your design? In your Guide, describe the testing procedure you will use. Explain how the results show that the design solves the problem and meets the specifications and constraints.

7 Redesign the Solution

Respond to the questions in your Guide about how you would redesign your solution. The redesign should be based on the knowledge and information that you gained during the activity.

8 Communicate Your Achievements

In your Guide, describe the plan you will use to present your solution to your class. Show any handouts or overheads that you will use.

Resources for Technology

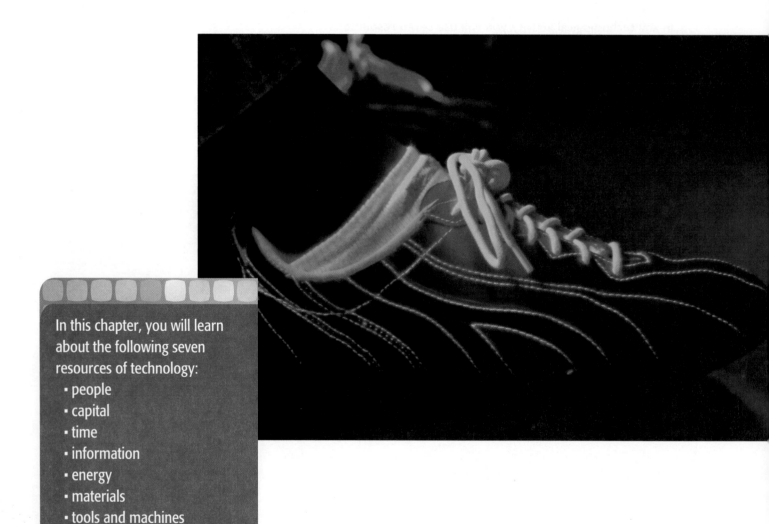

In this chapter, you will learn about the following seven resources of technology:
- people
- capital
- time
- information
- energy
- materials
- tools and machines

Resources are the things we need to get a job done. Think about running shoes, for example. In the United States, more than 370 million pairs of shoes are sold each year. Making these shoes requires the following resources:

- **People** to design the shoes, manufacture the shoes' components, and assemble the shoes.

- **Capital,** or money, to buy the raw materials and pay the people who make the shoes.

- **Time** to complete the shoes' design and manufacture.

- **Information** that will ensure the shoes are well suited to people's needs and are attractive to consumers.

- **Energy** to transport raw materials to manufacturing plants, process the materials into shoes, and transport the finished product.

- **Materials** to make the running shoes durable and comfortable.

- **Tools,** such as **machines** for molding, stitching, and gluing the shoes.

Every technological system involves the use of these seven kinds of resources.

People, Capital, Time, and Information

Benchmarks for Learning

- Every system involves the use of seven basic resources.
- Capital is needed to create products and systems.
- Time is an increasingly important resource in our information age.
- Processing data into information is key to the operation of systems.

Reading Strategy

Outlining How are people, capital, time, and information used in technology? Consider this question as you read through the section, and outline your response.

Vocabulary

consumers	dividends	data
capital	interest	nanoseconds
stock		

I. Resources for Technology
A. People
  1. Technology is a response to the needs and wants of people.
  2.

Figure 5.1 Apollo 11 astronaut Edwin Aldrin stands facing the U.S. flag on the moon.

Summarizing *Why did the United States feel the need to send astronauts into space?*

People

People are at the very heart of technology, which can be defined as the application of human knowledge to solve practical problems. For this reason, technology should be considered a response to the needs and wants of people. Companies use people as a resource to design and create products that maximize resources for our use. People are also the **consumers** (those who buy products or services) of technology.

Advances in technology are generated by people's needs, as well. For example, the Soviet Union sent the first satellite, Sputnik I, into space in 1957. The United States government decided to match this achievement. In 1958, NASA (National Aeronautics and Space Administration) was created to direct the space program. In 1969, astronauts from the United States landed on the moon (Figure 5.1).

Technologists are constantly designing and developing new technology to meet ever-changing needs. For instance, for the Apollo 15 mission in 1971, NASA scientists had to combine their knowledge from earlier moon landings with new ideas to develop a lunar roving vehicle (LRV), as well as a way to get it to the moon and back safely.

People are hired by companies as a resource for their ideas and expertise. Of course, people provide the labor on which technology depends. Many workers are needed to create the products and services we use every day.

All of the seven resources you will learn about are important, but it is human knowledge that turns resources into goods and services.

Capital

Capital is another of the seven technological resources. To build houses or factories, to make toasters or automobiles, to move people or goods, capital is needed. Any form of wealth is capital. Cash, stock, buildings, machinery, and land are all forms of capital (Figure 5.2).

A company needs capital to operate. To raise capital, a company may sell **stock,** which allows people to own a small piece of that company. Each share of stock has a certain value. When people buy stock in a company, their money is then used to operate or expand the business. These investors become part owners (shareholders) in the company. Shareholders hope that the company will do well and that their stock will become more valuable. When a company has done well, it may turn back some of its profits to investors in the form of payments called **dividends.**

Companies also borrow money from banks, for which the banks charge a fee, or **interest.** This means that the amount of money that will have to be paid back is more than the amount borrowed. A company borrows money with the hope that profits will pay for both the loan and the interest.

Another reason that capital is such an important resource is that companies use it to pay their employees, the people who not only come up with new ideas but also produce and manufacture the products that are based on those ideas. Capital is often directed to the development of new technology in order to help a company increase its business.

Figure 5.2 Capital resources, such as cash, are necessary for any technological project.

Applying *How might insufficient capital affect the schedule for a project?*

Figure 5.3

Figure 5.4

Measuring Time

Figure 5.3 Stonehenge, in England, is believed to have been built in ancient times to be used as a calendar.

Figure 5.4 Satellites use atomic clocks that are precise to within one billionth of a second.

Summarizing *How has the concept of time changed throughout history?*

Time

Time has become an increasingly important resource in our information age. It takes time to plan, design, and create products. People need and use time as a resource to complete these tasks. How much time is used can often determine how other resources, such as capital, are used.

Early people measured time by the rising and setting of the sun and by the change of seasons (Figure 5.3). When people began farming, time was measured in days. Later, clocks were used to measure time periods shorter than a day. People began to measure time in hours, minutes, and seconds. Measuring and using time in a more efficient manner became a driving force behind improving technology.

In today's information age, time is broken down into fractions of a second. The most precise timekeeper is the atomic clock, a timepiece that is regulated according to the very precise and unchanging vibration frequency of atoms or molecules. Because atomic clocks keep time better than any other clock, they are used to calibrate other clocks, especially those used in scientific research (Figure 5.4).

Computers can process **data** (raw facts and figures) in **nanoseconds** (billionths of a second). This technology saves huge amounts of time, energy, and labor, maximizing the resource of time by storing and calculating data quickly.

Technology has also greatly influenced the time it takes to distribute information. Once, it took weeks to send a message across the country. Today, email is delivered around the world in a matter of minutes. Time saved in this way is a valuable resource in our everyday lives.

Information

Technology requires information. We need to know what to do and how to do it. A surgeon must know not only how to perform a delicate operation but also which tools to use. A farmer must know what type of corn will grow best in local soils. A factory worker must know how to operate a machine safely and correctly.

Technology has grown quickly during the last few decades because of an explosion of information. This information can be shared throughout the world as a result of new and better ways of communicating. Everyone in today's technological world uses information, which begins as data. Data processing is the act of turning data into useful information: data is collected, recorded, classified, calculated, stored, and retrieved. All of the facts and figures included in this data have now become information that people can use.

Information can be found in many places, such as computer files, books, films, and museums, to name a few. The Internet has created a vast, worldwide information-sharing network. Search engines, such as Google or Yahoo, allow us to answer specific questions in a matter of seconds.

Information is not valuable until we make use of it. It can help advance technology to better suit people's wants and needs. Information can also help us determine the best way to use available resources. For example, sometimes one type of material can do the same job as another for a lot less money. Information gathered from the past can be used in the future to help us make decisions about these issues.

SECTION 1 Assessment

Recall and Comprehension

1. What are the seven resources of technology?
2. Explain the difference between a stock and a dividend.
3. In what ways are people a technological resource?
4. What are the components of data? How is data processed?

Critical Thinking

1. **Relating** How are time and capital related?
2. **Applying** Give examples of the seven resources that are used in your school.
3. **Summarizing** Why is information so important to technology?

QUICK ACTIVITY

Since ancient times, people have used the "barter" system: trading skills or property to get another person's goods or services. For example, a carpenter may build a table for a dentist in exchange for dental care. List your own skills or abilities that could be used to barter for something you want or need. **For more related Design Activities, see pages 132–135.**

People in Technology

Nearly a Century of Inventing
Harvey Severson

"I want to encourage the readers of your book to be studious and to stay away from alcohol and drugs. It's a much bigger adventure to get into technology."

—Harvey Severson

Harvey Severson was 82 when his concern for the environment led him to invent a machine that recycles waste wood. Wood blocks that normally would be disposed of in landfills are processed to create curled wood shavings that replace the Styrofoam "peanuts" used for packaging. A rotating cutter shaves off thin slices and forms the shavings into round curls that are hollow and won't get crushed easily.

Hard Work

Mr. Severson was born on a farm in Iowa in 1912. "I lived in the horse and buggy days. I went to town in a horse and carriage. There was no electricity, only kerosene lamps. I have seen quite a transition." Mr. Severson grew up during the Great Depression in the early 1930s. "There was no work during the Depression. I hitchhiked to New York City and landed there with two dollars in my pocket. For 12 years, I worked for an oil company to save enough money to get an education. Then, I went to college in Lincoln, Nebraska, to study aeronautics."

Designing Pays Off

For a while, Mr. Severson worked for the Wright Brothers in Paterson, New Jersey. He worked on the engine of the plane Charles Lindbergh piloted on the first solo flight across the Atlantic, and he was the chief mechanic for the first American bombers to raid Tokyo during World War II. In 1994, he received the National Inventor of the Year Award for his wood-recycling machine.

Mr. Severson says we all should try to recycle products to protect the environment: "I've got plenty of ideas, but I won't spend a nickel on anything unless it has market potential and will benefit people."

Critical Thinking

1. **Extending** Visit **www.uspto.gov**—the Web site of the U.S. Government's Patent and Trademark Office—and research Severson's patent application. It is number 05427162.

2. **Summarizing** What are the main benefits of Mr. Severson's invention?

Energy, Materials, Tools and Machines

Reading Strategy

Outlining   How are energy, materials, and tools and machines used in technology? As you read the section, continue the outline you began in Section 1.

Vocabulary

renewable energy
  sources
nonrenewable energy
  sources

raw materials
renewable raw
  materials

nonrenewable raw
  materials
synthetic materials

Benchmarks for Learning

- Systems require energy to be converted from one form to another.
- Most energy is derived from sunlight.
- Tools, materials, and skills help people make things and carry out tasks.
- A system remains balanced when one action is counterbalanced by another.

Energy

All systems in the natural and synthetic worlds require energy to function properly. Energy is used to make products; to move goods and people; and to heat, cool, and light the places where people work and live.

Like materials, energy sources are either renewable or nonrenewable. **Renewable energy sources** are those that can be replaced (Figure 5.5). Human and animal muscle power are renewable. So is the energy we get from burning wood. Solar, wind, gravitational, tidal, geothermal, nuclear fusion energies are other renewable types of energy. **Nonrenewable energy sources** are those that cannot be replaced once we use them up. These include coal, oil, natural gas, and nuclear fission (atomic energy).

The original source for nearly all forms of energy on Earth is the sun. Plants receive the power to grow from the sun. People and other animals get energy from the plants and other foods they eat. Coal, oil, and gas come from decayed plant and animal matter.

The following are the seven different kinds of energy sources:

- human and animal muscle power—from food

- solar energy—from the sun

- wind energy—from the airflow across Earth's surface

- gravitational energy—from tides and falling water

Figure 5.5   These solar panels convert energy from the sun.

Clarifying   *Is energy from the sun considered renewable or nonrenewable?*

- geothermal energy—from heat inside the Earth
- chemical energy—from fossil fuels and wood
- nuclear energy—from the conversion of radioactive matter into energy

We can use these sources of energy directly. We also can convert them into other forms of energy, such as mechanical, electrical, and light energy. For example, a dam can convert gravitational energy into electrical energy.

LIVING GREEN
Check the Status

Most energy used in technology has traditionally come from non-renewable energy sources. Coal, oil, and natural gas are burned to create electricity and other types of energy. In the United States, coal was the source that accounted for 45 percent of all energy produced in 2009. By comparison, all renewable energy sources accounted for just over 10 percent produced in this country. This percentage has remained almost unchanged for at least the past 10 years.

MAKE A CHANGE
Americans rely a great deal on nonrenewable energy sources but, as the name suggests, they cannot last forever. Without knowledge, people cannot change their habits. Take the time to learn as much as you can about the renewable energy sources mentioned here through your technology and science classes. Once you're armed with knowledge, start talking! Tell your family and friends and anyone who will listen about the benefits of using renewable energy as a source for electricity.

TRY THIS
Find out if your school and community use any renewable energy sources and compare their renewable energy use (in percentage terms) to the country's production of renewable energy. Start by contacting your school's chief engineer or head custodian and ask what kinds of energy are used to generate the school's electricity, heat and air conditioning. Then take it a step further and contact your town or city's major utility companies and find out what kinds of resources they use to generate electricity for homes and other buildings.

Materials

When most people hear the word *resources*, they think first about materials. Products are made from a wide variety of materials that have many different properties. Materials are sometimes tested to determine their characteristics before they are used to make products.

Materials are an important resource for technology. Some are natural, whereas others are created by people. Natural resources are materials that are found in nature. These include air, water, land, timber, minerals, plants, and animals. Natural resources that are used to make finished products are called **raw materials.**

Countries that are rich in natural resources have lots of raw materials. The United States is rich in some natural resources, such as timber, oil, coal, iron, and natural gas. However, we must import other materials such as chromium, platinum, and industrial diamonds.

Raw Materials

There are two kinds of raw materials: renewable and nonrenewable (Figure 5.6). **Renewable raw materials** are those that can be grown and therefore replaced. Wood, which is produced by trees, is a renewable raw material. **Nonrenewable raw materials** cannot be grown or replaced. Oil, gas, coal, and minerals are nonrenewable. Once the supplies of these resources have been used up, there will be no more.

Some resources, such as sand, iron ore, and clay, are available in great amounts. Others are in short supply. If possible,

Figure 5.6 Renewable and nonrenewable raw materials.

Contrasting *How do renewable and nonrenewable materials differ?*

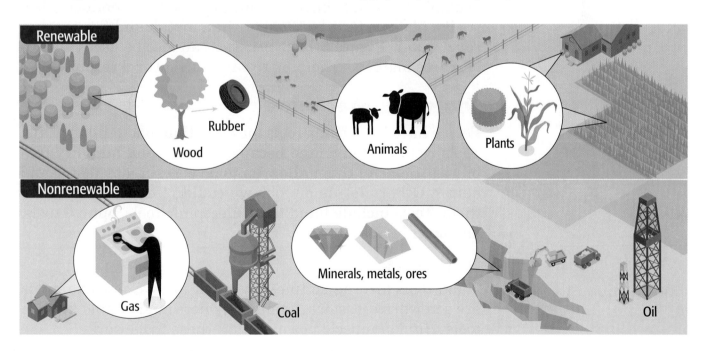

Figure 5.7 At Denver's airport, this bold use of synthetic materials echoes the shape of the nearby Rockies.

Interpreting *What do you think inspired the architects of this airport?*

plentiful materials should be used instead of scarce ones. Freshwater is a resource that is scarce in some places. Many people think that there will be a shortage of clean, freshwater in the future.

Synthetic Materials

People have long used technology to make substitutes for some resources. Materials made in the factories are called **synthetic materials.** Many everyday materials are synthetics. Plastics such as acrylic, nylon, and Teflon are not found in nature. Instead, they are made from chemicals.

Many synthetics are less costly and more useful than natural materials. They can also be made stronger, lighter, and more long-lasting than the materials they replace. For example, scientists have developed glass that conducts electricity, plastics that last longer than metal, and fabrics that repel water (Figure 5.7). Synthetics can also be used in place of scarce materials, helping to save our natural resources.

Tools and Machines

People have been using tools for more than one million years. Tools were first invented because they extend human capabilities. Some of them allow us to do certain jobs faster and better. Others let us do jobs we couldn't do at all without them. Tools include hand tools, electronic tools, optical tools, and machines.

Hand Tools

Hand tools are the simplest types of tools. Some examples are screwdrivers, saws, hammers, pliers, and simple kitchen tools. Human muscle power makes these tools work.

Figure 5.8

Figure 5.9

Machines

Machines, such as the ones in Figures 5.8 and 5.9, change the amount, speed, or direction of a force. Early machines were mechanical devices that used human, animal, or water power. They used the principles of the following six simple machines:

- lever
- wheel and axle
- pulley
- screw
- wedge
- inclined plane

Many modern machines have moving parts that are based on these simple machines. Other machines use electricity to move mechanical parts, as in machines that have electric motors. These machines are called electromechanical devices. A robot is such a device. Automatic machines do not need people to operate them. They must only be started and maintained by workers to make sure they are working properly. Maintenance is important for safe operation.

Electronic Tools and Machines

The computer is an electronic tool. Computers are used to process information and can be used to run factory machinery. Some electronic tools, such as meters, are used to test electrical circuits.

People and Tools

Figure 5.8 We use kitchen tools to prepare food.

Figure 5.9 Portable machines are often powered by electricity.

Comparing and Contrasting
How are tools and machines similar? How are they different?

For: Simple Machine Activity
Visit: www.mytechedkit.com

Identify Simple Machines Online

Simple machines provide a mechanical advantage. They multiply the force or distance that results from our work. Go online to find everyday examples of simple machines in a kitchen.

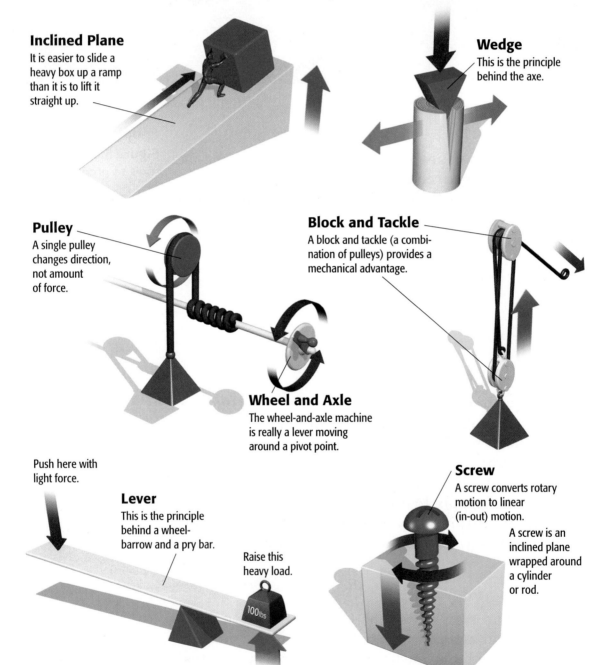

Inclined Plane
It is easier to slide a heavy box up a ramp than it is to lift it straight up.

Wedge
This is the principle behind the axe.

Pulley
A single pulley changes direction, not amount of force.

Block and Tackle
A block and tackle (a combination of pulleys) provides a mechanical advantage.

Wheel and Axle
The wheel-and-axle machine is really a lever moving around a pivot point.

Push here with light force.

Lever
This is the principle behind a wheelbarrow and a pry bar.

Raise this heavy load.

100lbs

Screw
A screw converts rotary motion to linear (in-out) motion.

A screw is an inclined plane wrapped around a cylinder or rod.

Figure 5.10 This welding robot can form a seam between two pieces of metal.

Hypothesizing *What are some advantages of using robots? Disadvantages?*

Optical Tools

Some optical tools extend the power of the human eye. Lenses magnify objects, making them easy to see and study. Eyeglasses and contact lenses are common examples of optical tools. Microscopes and telescopes are tools that have given people the ability to see the smallest cells and farthest planets.

Another optical tool is the laser. The term laser is short for "Light Amplification by Simulated Emission of Radiation." Lasers produce very strong bursts of light energy. People have developed laser technology for use in many different industries. Lasers are used to cut and weld materials ranging from sheet metal (Figure 5.10) to the delicate tissues of the eye.

SECTION 2 Assessment

Recall and Comprehension

1. Name three renewable and three nonrenewable sources of energy.
2. Which tools are considered obsolete today?
3. Give five examples of synthetic materials.

Critical Thinking

1. **Describing** Explain how information is used in making running shoes.
2. **Distinguishing** Compile a list of all the tools and machines you use in a typical day from morning to night. Note which energy sources are being used to make the tools or machines perform.

QUICK ACTIVITY

A wheel and axle is classified as a simple machine. Using cardboard, dowels, tape, and glue, make two sets of wheels and axles. One set should have 1" diameter wheels and the other, 2" diameter wheels. Which set would probably be better for a rough road? Demonstrate your reasoning using a diagram or model. **For more related Design Activities, see pages 132–135.**

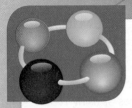

Connecting to STEM

An Elephant Seesaw

A seesaw is a lever that is used on many playgrounds. We know that if two people weigh the same, the seesaw will be balanced. If people don't weigh the same, then the lighter person needs to move farther from the fulcrum for the seesaw to balance.

On the lever below, ten 100-pound people balance one 1,000-pound elephant.

This is represented as

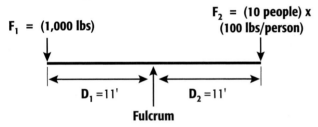

$F_1 = (1{,}000\ \text{lbs})$

$F_2 = (10\ \text{people}) \times (100\ \text{lbs/person})$

$D_1 = 11'$ $D_2 = 11'$

Fulcrum

The forces F_1 and F_2 represent weight, or force. If these weights are equal, then the lever bar is balanced and does not tip. The equation representing this is

$$\left(\begin{array}{c}\text{Distance from}\\ \text{Fulcrum}\end{array}\right) \times \left(\text{Force}\right) = \left(\begin{array}{c}\text{Distance from}\\ \text{Fulcrum}\end{array}\right) \times \left(\text{Force}\right)$$

$\quad\quad D_1 \quad\quad\quad\quad F_1 \quad\quad\quad\quad\quad D_2 \quad\quad\quad\quad F_2$

$$11 \times 1{,}000 = 11 \times 1{,}000$$

It is difficult to get 10 people to stand or sit together on the lever. Where should the fulcrum be located so that one person can balance the elephant? If we shift the fulcrum to the left, as shown in the sketch at the top of the next column, the elephant's weight is balanced by only one person.

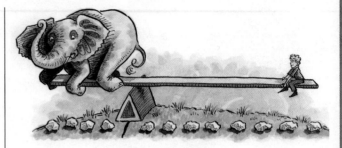

To balance the lever bar,

$F_1 = (1{,}000\ \text{lbs})$ $F_2 = (100\ \text{lbs})$

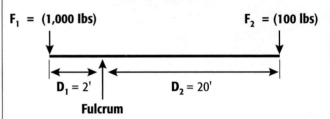

$D_1 = 2'$ $D_2 = 20'$

Fulcrum

$$\left(\begin{array}{c}\text{Distance from}\\ \text{Fulcrum}\end{array}\right) \times \left(\text{Force}\right) = \left(\begin{array}{c}\text{Distance from}\\ \text{Fulcrum}\end{array}\right) \times \left(\text{Force}\right)$$

$\quad\quad D_1 \quad\quad\quad\quad F_1 \quad\quad\quad\quad\quad D_2 \quad\quad\quad\quad F_2$

$$2 \times 1{,}000 = 20 \times 100$$

By shifting the position of the fulcrum, 100 pounds (one-tenth the previous force) balances 1,000 pounds. As Archimedes, the famous Greek scientist who discovered this principle, once said, "Give me a place to stand and a lever long enough and I can move the world."

Critical Thinking

1. **Calculating** The elephant and his friends want to continue with their balancing game, but the only plank available is 20 feet long. If the elephant and the 10 people stand on opposite ends of the plank, where does the fulcrum need to be so that the two loads are balanced?

2. **Solving** The elephant's 500-pound younger sister wants to play, too. How many people will be needed to balance her if she takes her brother's place on the seesaw?

Choosing Resources

Reading Strategy

Listing   Make a list of the various ways in which manufacturers choose resources. After reading this section, add to the list using your own examples.

Vocabulary

labor costs   composite

- To choose wisely, people must understand the uses and limitations of each resource.
- Important factors in choosing resources are cost, availability, and appropriateness.
- We must sometimes make trade-offs to reach the best possible solution.

Selection of Resources

Resources must be selected very carefully. One resource may be better than another for an intended purpose. Others may need to be substituted if they are harmful to people or the environment. Sometimes, one resource can do the same job as another but costs a lot less.

When choosing resources to solve technological problems, we have to consider and weigh various factors. Three important factors are

- cost,
- availability, and
- appropriateness.

We might decide that the best location for a new factory is overseas because workers' salaries would be lower. In this case, lower **labor costs** is a more important consideration than a readily available supply of local materials.

We must also determine which resources are best suited for the task at hand. For example, if a corporation in the United States wanted to set up a company in France, the company would look for qualified people who also speak French and English. A person who did not speak both languages well would not be an appropriate resource.

Cost of Resources

Because companies are in business to make a profit, the cost of resources used is very important. It is important to consider not only the cost of a location but also the costs of energy, materials, and the available workforce.

In some businesses, such as the automotive industry, decisions must be made about whether to hire human workers or to purchase robots. Robots never complain about getting tired, never call in sick, and produce high-quality work. However, they cost a lot of money to buy and require maintenance. The company must decide whether human workers or robots will be more cost effective in the long run.

A company's location is another factor that affects costs. When building a new factory, should a company build the plant in the United States, where workers earn high salaries and materials can be obtained easily, or should it build the plant in a foreign country, where labor costs will be cheaper but materials may be more difficult to obtain?

Availability of Resources

We also make certain choices based on the availability of material resources. For example, if we are producing jet aircraft for the armed forces, we need a special kind of metal called titanium. The United States does not produce much titanium. We obtain it from other countries, primarily Australia and India, so the cost of titanium is very high.

People use technology to create new materials that can be substituted for existing materials. Car bodies can be made from plastic instead of metal. Airplane parts can be made of special materials called composites (Figure 5.11). A **composite** combines several materials in order to improve their properties. One composite is plywood, which is created from several crisscrossed layers of wood, making it much stronger than a single layer of the same thickness.

Appropriateness of Resources

Figure 5.11 The B-2 bomber uses high-tech composites to achieve its stealth capabilities.

Extending *What other types of composites are in common use today?*

When choosing tools and machines, we need to consider how appropriate they are for the setting in which they will be used. For example, if we were hired to do farming in an area without technically trained people, would we want to use modern tractors? Modern tractors are complicated pieces of equipment. They require trained service people to keep them in good running order. In some countries, there are many unskilled workers and relatively few technicians. Although tractors can plow much faster than people and animals, human and animal labor would be a much more appropriate resource for such areas.

We must also consider the appropriateness of energy resources. In a country such as Israel, oil prices are very high,

Figure 5.12

Figure 5.13

but there is abundant sunshine and high temperatures. Solar energy is therefore used a great deal. Almost every house has a solar water heater on the roof. Other countries find it more cost-effective to build small, local energy-generating plants rather than large central facilities, which require high maintenance.

Trade-offs in Choosing Resources

When we choose resources, we make trade-offs, or give up some first choices to reach an acceptable solution. Corporations make decisions like this all the time to come up with the optimal solution (the best solution, all things considered). For example, a car manufacturer may use plastic instead of metal due to its lower cost and improved quality.

Normally, there are two kinds of trade-offs to consider when we choose resources.

1. We can trade off one type of resource for another within the *same* resource category—for example, we can substitute plastic for metal when manufacturing a garbage bin (Figure 5.12).
2. We can trade off a resource in one category for a resource in *another* category—for example, we can substitute human labor for machines (Figure 5.13).

Maintenance of Resources

We must consider how well the resources we choose will stand up over time. Photocopy machines are used in most offices, but they require service and parts replacement to keep them in good working order. Service and repair costs can be very expensive. A supply of trained service people must be available, and the budget must include the replacement of costly parts.

Trade-offs in Choosing Resources

Figure 5.12 Same-category trade-off: The material used is cheaper, more suitable, or more available than another.

Figure 5.13 Cross-category trade-off: Human labor can be replaced by a machine.

Applying *Can you think of a product that was made from one material but that is now made of another?*

Figure 5.14 Medical research often involves decisions made according to people's values.

Applying *Give some examples of how cultural values influence the adoption or rejection of technology.*

Cultural Values

Our choice of resources is also influenced by our cultural values (Figure 5.14). Some people believe animals should not be used in medical research. Others support this research because they believe it will benefit human lives. Similar issues exist with energy resources. Although nuclear energy is renewable and efficient, many people are opposed to using it because it produces hazardous waste and possible damage to the environment and our health.

SECTION 3 Assessment

Recall and Comprehension

1. What are the three considerations when choosing resources?
2. In choosing resources, we must consider how well they will stand up over time. What is an advantage and a disadvantage of choosing resources that will last a long time?

Critical Thinking

1. **Analyzing** Explain why an available resource, such as nuclear fuel, may not be an appropriate resource to use in generating power.
2. **Speculating** How do our cultural values influence the way we look at solar energy versus nuclear energy?

QUICK ACTIVITY

Magnetic Levitation (MagLev) trains "float" on powerful magnetic fields, substantially cutting the friction found in traditional trains. This allows MagLev trains to reach very high speeds. Aluminum is a metal that is used extensively on MagLev trains. What other materials could probably be used in MagLev trains? Why? **For more related Design Activities, see pages 132–135.**

Maintenance and Troubleshooting

Benchmarks for Learning

- Many different kinds of products and processes require regular maintenance and replacement of parts to ensure proper functioning.
- Troubleshooting uses logical decision making in determining the cause of a problem.
- Condition monitoring enables us to monitor the condition, or health, of equipment.

Preventive Maintenance

Maintenance has often been thought of as how to prevent something from wearing out. However, it is much more than that. In general, it relates to activities necessary for maintaining proper technological systems and product functioning or restoring the system/product in case it ceases to function properly. Many different kinds of products and processes require regular maintenance and replacement of parts to ensure proper functioning. We maintain good health by eating a proper diet, exercising, and brushing our teeth, to ensure that our body's system functions well. If we become sick, or have an accident, we need to undertake additional activities to bring us back to full health. So it is with technological systems and products.

There are several strategies for proper maintenance. One involves periodic activities, such as replacing the oil in a car every 5,000 miles. Changing air filters in a home heating and/or air conditioning system is done after so many weeks of operation. There is no inspection or knowledge about the system performance needed before the maintenance is performed. Typically, preventive maintenance focuses on cleaning, lubrication, and correcting deficiencies found through testing and inspections.

If we consider a large facility, be it a school building, a factory, or a power plant, there are advantages to having a sound preventive maintenance program, which includes the following:

- Reduced production downtime, resulting in fewer machine breakdowns.

- Better conservation of assets and increased life expectancy of assets, thereby eliminating premature replacement of machinery and equipment.

- Reduced overtime costs and more economical use of maintenance workers due to working on a scheduled basis instead of a crash basis to repair breakdowns.

- Timely, routine repairs to circumvent fewer large-scale repairs.

- Reduced cost of repairs by reducing secondary failures. When parts fail in service, they usually damage other parts.

- Reduced product rejects, rework, and scrap due to better overall equipment condition.

- Identification of equipment with excessive maintenance costs, indicating the need for corrective maintenance, operator training, or replacement of obsolete equipment.

- Improved safety and quality conditions.

Condition Monitoring

More intelligence can be applied to maintenance programs where the "condition" or health of the equipment is monitored. The essential idea is that by using information about the condition of the system, one will be able to determine when to perform maintenance before failure occurs, but not too soon, thus minimizing expense and time when the equipment is not functioning because it is not running for maintenance operations. Of course this can be a little tricky, because how do you tell when the equipment will fail? Figure A illustrates a schematic showing how the vibration level of a rotating compressor might change with time. The vibration will increase with time as parts become worn and do not fit smoothly together. At some point if the parts are not replaced, the compressor will fail and not be able to run any more. The idea is to stop and replace the parts before failure, but not too far before so you are able to use the equipment for as long as possible before spending money for renewal.

The challenge occurs in figuring out which line the compressor vibration is following. This is where expertise and experience are needed. Line A is linear; however, line B is nonlinear and changes quickly so it is more difficult to predict when to stop and replace equipment parts.

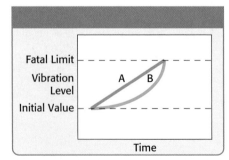

Figure A

Troubleshooting

Troubleshooting is a problem-solving technique used to solve equipment operational malfunctions. Consider a car's engine. Perhaps it will not start, what could be the cause of the problem? If the engine does run, but not well, perhaps with a smoky exhaust, why is that happening? At other times while performing routine maintenance, a problem may be found, such as extra wear on a bearing. The process of troubleshooting begins by gathering all information observed and information supplied by the person operating the equipment, in this case the car's engine. Based on this information, the troubleshooter proceeds in a logical fashion through the various symptoms. Often a problem does not create one symptom, but several, all of which must be analyzed to lead to the correct problem and its solution.

The service manual that accompanies equipment is very valuable. Once all information has been gathered, the best approach is to check the troubleshooting charts in the manual and proceed by doing the easiest one first. Along with doing the easiest task first, do not leap to conclusions when starting the troubleshooting task. Gather all the information, take your time, think about what has happened, and then proceed in a logical fashion, one step at a time, eliminating various symptoms and finally finding the cause of the problem.

SECTION 4 Assessment

Recall and Comprehension

1. Explain in your own words four advantages of a preventive maintenance program.
2. Discuss the strategy for troubleshooting a problem, such as a door that will not stay closed.

Critical Thinking

1. **Relating** How are condition monitoring and maintenance related?
2. **Applying** List five examples of maintenance that needs to be done in your school.
3. **Summarizing** Why is preventive maintenance important?

QUICK ACTIVITY

Construct a tower platform by attaching four 10-12" long straws to a rectangular platform made from a 6" square of cardboard with craft sticks glued to the perimeter as a place to attach the straws. Test the stability of the tower by placing books on it. Enhance the stability by adding horizontal and/or cross bracing of the tower straws. Test the stability again. What did you find? Is the tower system less likely to fail? Why?

5 Review and Assessment

Chapter Summary

Every technological activity involves the use of seven resources: people, capital, time, information, energy, materials, and tools and machines. We must make informed choices about which resource we use.

- **People's** needs drive technology. People not only create but also use the products and services of technology.

- **Capital** is any form of wealth. Cash, shares of stock, buildings, machinery, and land are all forms of capital.

- **Time** is needed to make products. It has become an increasingly important resource in the information age.

- **Information** is needed to solve problems and to create new knowledge.

- **Energy** sources are either renewable or nonrenewable. Renewable energy sources include human and animal muscle power, water, wind, geothermal energy, and solar energy. Nonrenewable energy sources include oil, gas, coal, and nuclear fission.

- **Materials** found in nature are called raw materials. Renewable raw materials are those that can be replaced. Nonrenewable raw materials cannot be replaced once they have been used up. Synthetic materials are human-made materials.

- **Tools** extend the capabilities of people. Hand tools extend the power of human muscles. **Machines** are tools that change the amount, speed, or direction of a force. Machines that use electrical energy to move mechanical parts are called electromechanical devices.

When we choose resources, we must sometimes make trade-offs to reach the best possible solution.

Building Vocabulary

Your teacher may give you a crossword puzzle. Complete the puzzle using the following words from this chapter. Exchange puzzles with a classmate to check each other's answers.

1. availability
2. appropriateness
3. capital
4. coal
5. cost
6. energy
7. finite
8. gas
9. geothermal
10. hydroelectricity
11. inclined plane
12. information
13. laser
14. lever
15. machines
16. material
17. nuclear energy
18. oil
19. people
20. resources
21. solar
22. synthetic
23. time
24. tools
25. trade-offs

See your teacher for the Crosstech puzzle.

Reviewing Content

1. What are the seven main sources of energy?

2. A kayak is made from synthetic materials. What are synthetic materials? What are some advantages of using them?

3. Define a machine in your own words. Then, use your definition to determine whether the following items are machines:
   (a) a baseball bat
   (b) software for a video game
   (c) a radio
   (d) a hand-operated drill
   (e) a wrench

4. What are two advantages that synthetic materials have over natural materials?

5. Which factors should be considered when choosing a resource?

Applying Your Knowledge

1. This man is using a machine to sand wood. How does a machine differ from a hand tool? What are some of the advantages of using tools and machines?

2. If a person wanted to start a company, how might he or she arrange to get capital? Describe three different ways to get capital.

3. Suppose we choose to locate a toy-making factory in Korea instead of the United States because Korea's labor is cheaper. What are the trade-offs?

4. List five different sources of information that you use. Which of these is most useful to you? Least useful?

5. Suppose you need to lift the front end of a car to tow it to a different location. Read an automobile owner's manual and describe the towing procedure suggested by the auto manufacturer. Which simple machine might make this job easier? Explain your answer.

6. Suggest how you might redesign an existing tool to make it easier for a person with limited strength in his or her hands, to accomplish a task.

7. How are bicycle tires manufactured, how are they disposed of when their usable life is over? Examine the life cycle of a bicycle tire and discuss the tire's environmental impact.

Critical Thinking

1. **Analyzing** When building a house, a contractor chooses to invest extra money to install solar panels rather than cheaper gas heaters. In terms of energy, why might this be a good decision?

2. **Inferring** What type of resource is provided by the Internet and the World Wide Web? How has technology been changed by the Internet?

3. **Taking a Position** Wood is a renewable resource. Does that mean that we can cut down all the trees we need? Explain your answer.

4. **Analyzing** Your family needs to buy a new lawnmower. There are many different types to choose from—hand mowers, electric mowers, gas-powered mowers, and riding mowers. Describe how your values might influence this decision.

5. **Predicting** Of the seven different kinds of energy sources, name some that we might be using more of in the future. Why?

Connecting to STEM
science · technology · engineering · math

Budgeting for New Materials

You have been assigned the task of budgeting the cost of materials for a new type of cafeteria tray. Each tray will be 18 inches long and 12 inches wide. Your school needs to know the cost of materials for 300 new trays. How many square feet of materials will you need? If two different types of plastics cost 23 cents and 29 cents per square foot, what will be the overall difference in cost?

Design Activity 9

PINBALL MACHINE

Problem Situation

It is a rainy day and you have to stay home, but you would rather go to the arcade and play pinball. You have just been discussing how wonderful it is to play pinball. There is actually a lot of science involved in the game. Your mom gives you a challenge: to make a pinball machine of your own using materials found in your home.

Materials

You will need:

- 6–centimeter sticks
- 24" x 24" piece of foam board
- 24" x 24" piece of plywood
- ball bearings with 1/2" diameters
- glue
- Plexiglas strips, metal strips
- push pins, nails
- rubber bands of different widths (1/4", 3/8", and 1/2")

Your Challenge

You and your team members are to create a mechanical pinball machine—with no electronics—that is fun to play.

> Go to your **Student Activity Guide, Design Activity 9.** Complete this activity in the Guide, and state the design challenge in your own words.

1 Clarify the Design Specifications and Constraints

To solve the problem, your design must meet the following specifications and constraints:

- The board must be tilted at a minimum of 10 degrees to the horizontal.
- The ends of the flippers must be no closer to each other than one inch (twice the ball diameter).
- The ball must travel at least twice the board length in a game.

> In your Guide, state the design specifications and constraints for your project. Include any others that your team or your teacher listed.

2 Research and Investigate

To complete the design challenge, you need to first gather information to help you build a knowledge base.

> In your Guide, complete the Knowledge and Skills Builder 1: Visit to a Toy Store.

> In your Guide, complete the Knowledge and Skills Builder 2: Angles, Bumpers, and Balls.

> In your Guide, complete the Knowledge and Skills Builder 3: Creating a Maze.

In your Guide, complete the Knowledge and Skills Builder 4: Design Considerations in Making Flippers.

③ Generate Alternative Designs

In your Guide, describe two possible solutions that your team has created for your problem.

④ Choose and Justify the Optimal Solution

In your Guide, explain why you selected the solution you did, and why it was the better choice.

⑤ Develop a Prototype

Construct your solution. Put a photograph or sketch of your final design in your Guide.

In any technological activity, you will use seven resources: people, capital, time, information, energy, materials, and tools and machines. In your Guide, indicate which resources were most important in this activity, and how you made trade-offs between these resources when you made the pinball machine.

⑥ Test and Evaluate

How will you test and evaluate your design? In your Guide, describe the testing procedure. Justify how the results will show that the design solves the problem and meets the specifications and constraints of this project.

⑦ Redesign the Solution

Respond to the questions in your Guide about how you would redesign your solution. The changes you make should be based on the knowledge and information that you gained during the activity.

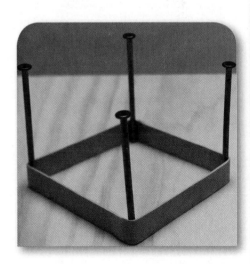

⑧ Communicate Your Achievements

In your Guide, describe the plan that you will use to present your solution to your class. Show the handouts and/or PowerPoint slides that you will use.

Design Activity 10

SEVEN RESOURCES GAME

Problem Situation

Imagine that you have to take care of a younger brother while your mom goes shopping. He asks you about your technology class, and you mention that seven resources are used in all technological processes—even something as simple as making a hamburger. Now, he wants to know what the seven resources are. There must be a fun way to describe them. Wait, you have an idea!

Materials

You will need:
- cardboard
- dice
- foam board
- index cards
- markers
- plastic strips

Your Challenge

Design and create a board game that requires knowledge of the seven resources used in all technological processes: people, capital, time, information, energy, materials, and tools and machines.

> Go to your **Student Activity Guide, Design Activity 10.** Complete this activity in the Guide, and state the design challenge in your own words.

① Clarify the Design Specifications and Constraints

To solve the problem, your design must meet the following specifications and constraints:
- The game should be fun for a third- or fourth-grade student to play.
- The game should include all seven resources and should take at least 10 minutes for two people to play.

> In your Guide, state the specifications and constraints of your project. Include any that your team or your teacher listed.

② Research and Investigate

To complete the design challenge, you need to first gather information to help you build a knowledge base.

> In your Guide, complete the Knowledge and Skills Builder 1: Board Game Investigation.

> In your Guide, complete the Knowledge and Skills Builder 2: Seven Resources.

> In your Guide, complete the Knowledge and Skills Builder 3: Getting the Game Rolling.

③ Generate Alternative Designs

In your Guide, describe two possible solutions to the problem. Include such considerations as the objective of the game (collect points, move along path), preliminary rules of the game, and a sketch of the game board.

④ Choose and Justify the Optimal Solution

In your Guide, explain why you selected the solution you did, and why it was the better choice.

⑤ Develop a Prototype

Construct your board game. Include a drawing of the board or a photograph of your model in step 6 of your Guide.

In your Guide, indicate which resources were most important in this activity, and how you made trade-offs between these resources when you created the game.

⑥ Test and Evaluate

How will you test and evaluate your design? In your Guide, describe the testing procedure you will use. Justify how the results will show that the design solves the problem and meets the specifications and constraints of this project.

⑦ Redesign the Solution

In your Guide, describe how you would redesign your solution based on the knowledge and information that you gained during the activity.

⑧ Communicate Your Achievements

In your Guide, describe the plan that you will use to present your solution to your class. Show the handouts and/or PowerPoint slides that you will use.

Technological Systems

In this chapter, you will learn about the following different parts of a system:
- input
- process
- output
- feedback
- sensor
- comparator
- controller

A bicycle is a type of system that allows us to travel from one place to another more quickly than on foot. It can be designed and made for different purposes—dirt riding, street and ramp riding, long-distance racing, or all-around use. It has numerous components, many of which are used to make the bicycle go faster or slower.

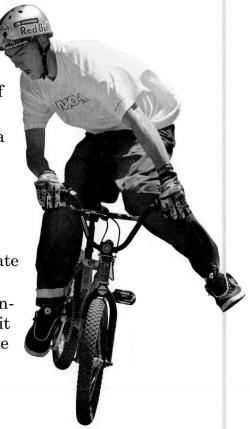

- Gears determine how fast you need to pedal to achieve a certain speed.

- Brakes let you slow down gradually or stop quickly.

- Shock absorbers can make the ride more comfortable, especially on bumpy roads or trails.

- Wider tires provide more support, but they can also create a slower ride.

And you—the rider—are also part of the system. You control how fast the bicycle goes, where it goes, and how well it is maintained. In this chapter, you will learn how a bicycle and many other systems work.

Parts of a System

Benchmarks for Learning

- All systems include inputs, processes, and outputs.

- In a system, processes are used to combine resources. The processes provide an output in response to an input command.

- Subsystems can be connected to one another to produce more powerful systems.

- Technological systems are designed to achieve goals. They incorporate various processes that transform inputs into outputs. They all use energy in some form.

- It is important for citizens to reduce the negative impacts and increase the positive impacts of their technologies on the lives of people in another area or on future generations.

- Technological systems can interact with each other to perform more complex functions and tasks than the individual system could by itself.

Reading Strategy

Outlining Using the headings as your guide, make an outline of the important points in this section. Next to each term in your outline, jot a quick definition of that term.

> I. Basic System Model
>   1. Inputs
>   2. Processes
>   3.

Vocabulary

system	process	subsystem
input	output	

The Universal System Model

A **system** is any group of interrelated parts designed collectively to achieve a desired goal. The desired goal can be anything from driving an automobile to making toast to calling a friend (Figure 6.1). Different parts of a system work together to produce a specific result.

A technological system produces results through the use of technology. A car, for example, is a technological system used for traveling from one place to another. A telephone is an electronic or cellular system used for speaking to someone in a different location. A system may be huge, such as the space shuttle, or it may be small, such as a pocket calculator. All systems include an input, a process, and an output. The universal system model, shown in Figure 6.2, describes any technological system. It shows how the parts of a system work together.

Input

An **input** is the command given to a system. It is also the desired result. When we turn on a television set, for example, we are giving it a command: "Give us a picture and sound." With a car, shown in Figure 6.3, we provide the input by driving. A car moves when we "tell it to" by stepping on the gas. The input command (or desired result) might be: "Go 30 miles an hour." Other inputs to the system include resources such as time, energy, and materials.

Process

The **process** is the action part of a system. It combines resources and produces a result, which is the output.

The process is a response to the input command. During the process, the seven technological resources are used.

In the driving process, both the car (machine) and the driver (people) are essential. The driver gives the input command ("Go 30 miles an hour") and then steps on the accelerator pedal. Gas is converted to energy to make the engine run and move the car forward. The car and the gasoline cost money, or capital. The driver uses information to make the car produce the desired result.

Figure 6.1

What Is a System?

Figure 6.1 A microwave is a technological system. The desired result is a hot meal.

Figure 6.2 All systems include an input, a process, and an output. Each part of a system affects the other.

Extending *What is the input in preparing microwave pizza?*

Figure 6.3 In response to the input command, a car uses resources to produce an output.

Interpreting *What is the output of this system?*

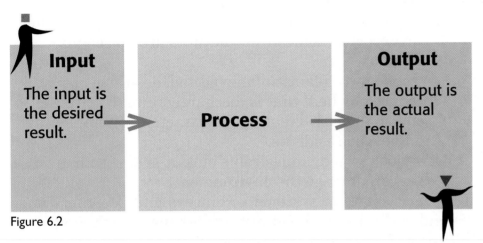

Input
The input is the desired result.

Process

Output
The output is the actual result.

Figure 6.2

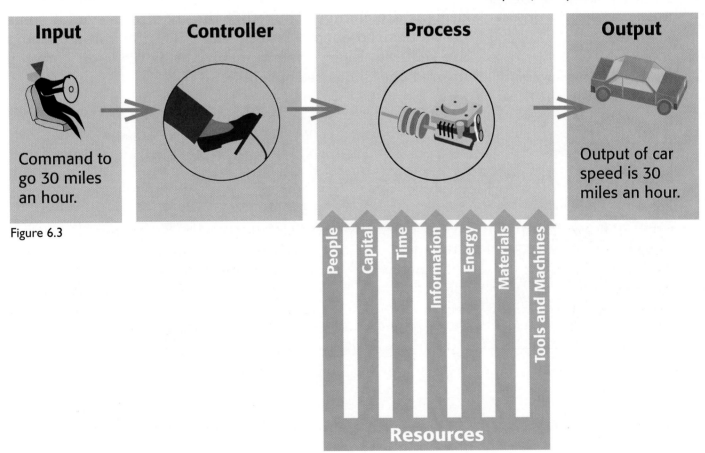

Input
Command to go 30 miles an hour.

Figure 6.3

Controller

Process

Output
Output of car speed is 30 miles an hour.

People · Capital · Time · Information · Energy · Materials · Tools and Machines

Resources

Output

The **output** is the product of a system. It is the actual result. If a system works correctly, the output matches the input, or desired result. With the car system example, we wanted the car to go 30 miles an hour. A properly functioning car will produce this output in response to the input command.

Most systems have more than one output. Outputs can be desirable and expected (for example, the car moves), or they can be undesirable and unexpected (the car produces exhaust and heat) (Figure 6.4). Let's take a look at the outputs of a power plant that produces electricity (Figure 6.5).

1. The expected, desirable outputs are electricity and heat.
2. The expected, undesirable outputs are noise and smoke.
3. At one power plant, an unexpected, desirable output is produced by heat that is discharged into the river. Tropical fish flourish, creating an attraction for employees and nearby residents.
4. An unexpected, undesirable output is acid rain. Acid rain can damage the environment.

When designing systems, engineers and designers must consider the unexpected outputs as well as the expected outputs. They must understand how the different parts of a system can produce unexpected results. Sometimes, designs are changed to modify these outputs even if some of the desired output is lost when the unwanted outputs are reduced. Every system involves this trade-off between desired and undesired outputs.

Figure 6.4 Although the desired result of a car system is transportation, its outputs also include exhaust and heat.

Extending *What other outputs are produced by a car?*

Figure 6.5 Outputs from a system can include expected or unexpected, desirable or undesirable results.

Analyzing *What are the undesirable outputs of this system?*

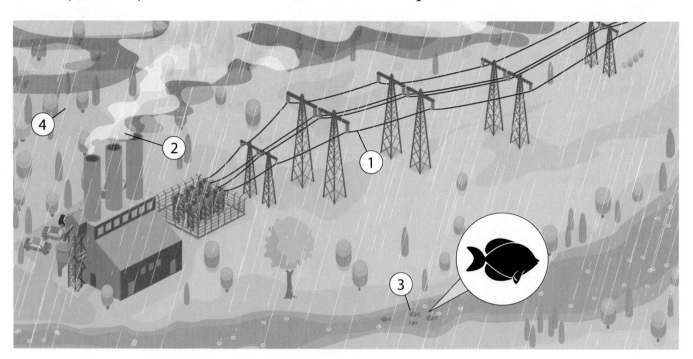

Home Entertainment System

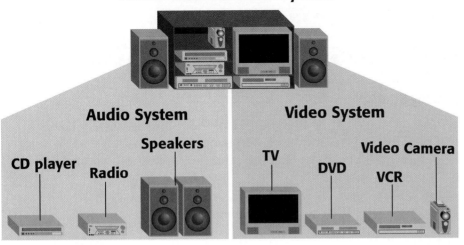

Figure 6.6 Even a home entertainment system is a complex system that includes many different subsystems.

Interpreting *What components make up the audio system?*

Subsystems

Systems are often made up of many smaller systems. Each of these smaller systems is called a **subsystem**. To understand a large system, it might be helpful to break it into subsystems. Then, you can study each one separately.

Suppose that you want to look at a transportation system that carries goods by truck from Los Angeles to New York City. You could break down the large system into smaller subsystems, such as the vehicle system, the management system, and the communication system. On a smaller scale, a home entertainment center is another example of a complex system (Figure 6.6). Combining different subsystems produces more powerful and complex systems that can perform tasks that independently operating subsystems could not accomplish.

SECTION 1 Assessment

Recall and Comprehension

1. What is a system input?
2. Which part of a system combines resources? What types of resources are used?
3. What is a subsystem? How does it relate to a system?
4. Describe the four outputs that might be produced when you turn on the television.

Critical Thinking

1. **Applying Concepts** Think of a system to squeeze orange juice. Identify the input, process, and output.
2. **Interpreting** In our car system example, suppose that we want to come to a complete stop. What is the input command? What type of process would produce the desired output (stopping)?

QUICK ACTIVITY

Using a variety of materials of your own selection, design a device that will always point into the wind. Identify the input and the output. Demonstrate your model in class, and describe whether the output matched the input. Explain any problems you encountered during the process of construction. **For more related Design Activities, see pages 158–161.**

Open-Loop and Closed-Loop Systems

Benchmarks for Learning

- An open-loop system has no feedback.
- Systems may require the use of feedback and control subsystems.
- A closed-loop system uses feedback to adjust the output to match the desired result.
- Systems thinking involves considering how every part of a system relates to the other parts.

Reading Strategy

Listing   Make a list of all the types of feedback described in this section. Then, make another list to show the output for each type of feedback. After you have finished reading the section, add several examples of your own.

Feedback
- No traffic—car is going too fast

Output
- Traffic light changes pattern—car slows down

Vocabulary

open-loop system   closed-loop system   feedback

Open-Loop Systems

If you have ever sat in a car waiting for the traffic light to turn green, even when there are no other cars around, you have used an open-loop system.

A system that has no way of monitoring or adjusting itself is called an **open-loop system**. An open-loop system cannot alter its output based on changing conditions. A simple traffic light is set to stay red for a specific period of time, regardless of whether or not there are cars present.

A model railroad with an on/off switch is another example of an open-loop system (Figure 6.7). The train runs at a fixed speed on a level track. Traveling uphill, however, the train slows down. Downhill, it speeds up. The speed is not maintained at a constant level. In such an open-loop system, the output (speed) changes when conditions change. The changing conditions are the different slopes of the track. Another example of a technological open-loop system is a home heating system that has no thermostat. In such a system, the heater is turned on and continues to run. The house gets warmer and warmer, but nothing tells it to turn off. Without a thermostat to detect the increase in room temperature, the system has no way to respond to changes that have taken place.

Open-loop systems can be controlled to some degree. Timers and other kinds of programmed controls can modify an operation based on expected conditions. For example, an oven can be set to turn off after a certain period of time,

Uphill: Train goes more slowly.   Downhill: Train goes faster.

Figure 6.7 An open-loop system, such as this train, has no way to regulate its output.

Analyzing *How does the train's output change?*

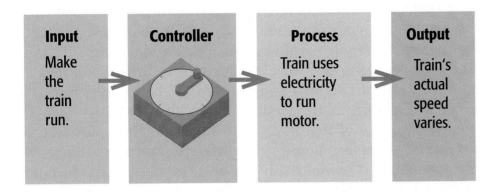

Input	Controller	Process	Output
Make the train run.		Train uses electricity to run motor.	Train's actual speed varies.

or the lights in your house can be set on timers. Although these systems are controlled by timers, they do not change when conditions change. The oven will turn off at the set time even if the food is not fully cooked.

Closed-Loop Systems

Some systems may require the use of feedback and control subsystems. A system that can respond to changes and adjust its inputs or outputs is called a **closed-loop system**. Closed-loop systems operate using feedback. **Feedback** is information about a system's output that is used to adjust the system. It "closes the loop" from output to input (Figure 6.8).

Figure 6.8 Open-loop systems can be changed to closed-loop systems by adding feedback.

Inferring *If the speedometer in a car is broken, is this an open-loop or a closed-loop system?*

Comparing Open-Loop and Closed-Loop Systems		
System	**Open-Loop**	**Closed-Loop**
Bicycle steering	Rider is blindfolded.	Rider watches road and adjusts pedaling or braking.
Car speed	Driver does not look at speedometer.	Driver looks at speedometer regularly and adjusts speed as needed.
Model train	Train is either on or off; there is no speed control.	Operator adjusts speed of train to maintain constant speed.

For example, a traffic light that uses feedback can adjust its output based on the amount of traffic.

Feedback Loops

Let's return to our car system example. How does the driver know when the car is traveling 30 miles an hour? He or she checks the speedometer. The speedometer, which shows how fast the vehicle is traveling, provides the driver with feedback—information about the car's speed. Based on this feedback, the driver monitors the speed and controls it manually by pressing the accelerator or the brake pedal. The combination of the speedometer, the driver's eye, and the driver's brain forms the feedback loop (Figure 6.9).

One purpose of feedback is to improve the operation of a system, therefore some type of feedback can be added to most systems. A model train can be equipped with a human-operated speed control, which allows the operator to control the speed when conditions change (as when the train goes up an incline). The human operator provides the feedback when he or she monitors and adjusts the speed of the train in order to keep it on track.

Similarly, a person or a thermostat can make an open-loop heating system closed. They provide data or information

Figure 6.9 A closed-loop system has a feedback loop that helps regulate the system's output.

Interpreting *How is the process different before and after feedback?*

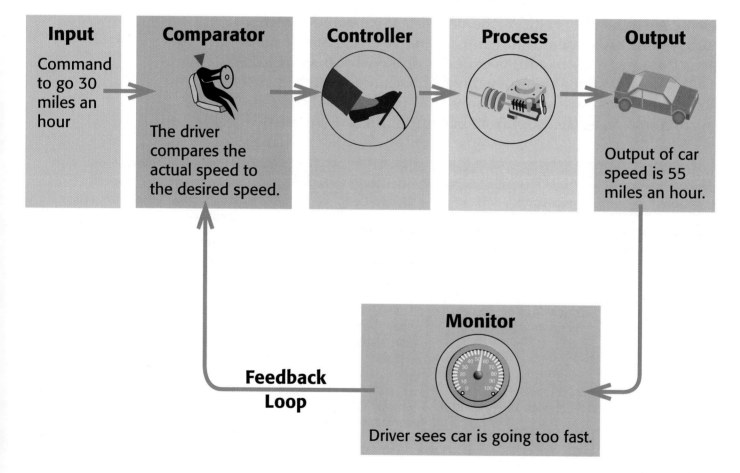

Input
Command to go 30 miles an hour

Comparator
The driver compares the actual speed to the desired speed.

Controller

Process

Output
Output of car speed is 55 miles an hour.

Monitor
Driver sees car is going too fast.

Feedback Loop

Figure 6.10 People who cannot see rely heavily on tactile (touching) and auditory (hearing) feedback.

Extending *What types of feedback have you received today?*

that adjusts the system. Feedback from the person or the thermostat turns the heater off or on, keeping the room at a constant temperature.

Types of Feedback

Feedback comes in many forms. It can come from other people, as when your parents ask you to turn down the volume on the television or radio. Feedback can come from touch or sight (Figure 6.10). It can also come from a mechanical device, such as a cash machine. Whenever you use your fingers to pick up or hold something, you are receiving tactile feedback. Feedback is also used in school. Your teacher uses homework and tests to monitor how well you are learning. The grades you receive on your work are your feedback, indicating how you are doing so that, if needed, you can change your study methods and work to improve your test-taking skills.

SECTION 2 Assessment

Recall and Comprehension

1. What is feedback? Explain feedback in the example of driving a car.
2. You turn on the water faucet in your bathroom and then leave for school. Is the running water an example of a closed-loop system or an open-loop system? Why?

Critical Thinking

1. **Comparing and Contrasting** How do open-loop and closed-loop systems differ? How are they similar?
2. **Hypothesizing** What systems can you think of that are open-loop systems? Would they be improved if feedback was added to the loop? Why or why not?

QUICK ACTIVITY

Perform a feedback experiment using a pencil. Place the pencil on a table. Close your eyes, turn around once, and try to pick up the pencil without opening your eyes. You may find that you need to feel around to find the pencil. You are receiving tactile feedback. Do some research to find out what other types of feedback exist.
For more related Design Activities, see pages 158–161.

Connecting to STEM

Human Feedback: Driving and Reaction Times

Imagine that you are driving and you suddenly see an animal in the road ahead. Your eyes act as a sensor. They send a signal to your brain (the controller), which sends a signal to your leg and foot muscles. You take your foot off the accelerator pedal and step on the brake. The car slows down.

Calculating Reaction Time

This process takes time before braking occurs, usually about 3/4 of a second. How far does the vehicle travel during this time? Calculate the distance for speeds of 20 miles per hour (mph), 40 mph, and 60 mph.

First, convert miles per hour (miles/hour) to feet per second (feet/second). Let's do this for 1 mph. There are 5,280 feet in 1 mile and 3,600 seconds in 1 hour. (Notice that the units for miles and hours cancel.)

$$1 \text{ mile/hour} = \frac{5,280 \text{ feet/mile}}{3,600 \text{ seconds/hour}}$$

$$= \frac{5,280 \text{ feet}}{3,600 \text{ seconds}}$$

$$= 1.467 \text{ feet/second}$$

At 20 mph, the car travels at twenty times this speed, or

$$(20)(1.467 \text{ ft/sec}) = 29.3 \text{ ft/sec}$$

In 3/4, or 0.75, of a second—the time it takes us to start braking—the distance traveled is

$$(0.75 \text{ sec})(29.3 \text{ ft/sec}) = 22 \text{ feet}$$

When we are driving at 20 mph, then, the car travels 22 feet before we begin braking.

Making a Graph

By performing similar calculations for 40 mph and 60 mph, we can create the graph shown. The graph describes how our driving speed affects the distance (number of feet) we travel before braking.

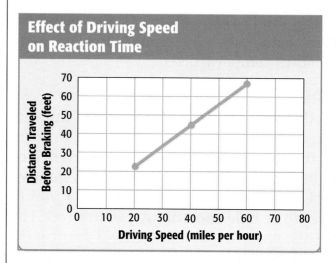

Effect of Driving Speed on Reaction Time

Critical Thinking

1. **Summarizing** What relationship does driving speed (along the bottom axis) have to the number of feet traveled before braking? Explain the information in the graph in your own words.

2. **Interpreting** Based on the graph, if a car is traveling at 60 mph, what distance will be traveled before braking?

3. **Calculating** You were given the number of seconds per hour: 3,600. How was this number calculated? Show your work and the units used to get this number.

Controlling a System

Reading Strategy

Listing As you read this section, make a list of the different parts of a feedback loop. For each part, describe in your own words what that part of the system does. Once you have finished the section, pick a new example of a system and add to your list how the parts of that system work.

Vocabulary

sensor   controller   limits of controllability
comparator

Benchmarks for Learning

- The stability of a system is influenced by components within the system.
- Complex systems have many layers of control and feedback.
- Nontechnological systems affect and are affected by social and environmental systems.

Three Elements of a Feedback Loop

You have learned that some systems use feedback to adjust their output based on changing conditions. Systems that use feedback are closed-loop systems. In these systems, feedback loops have three essential parts: sensors, comparators, and controllers. The three parts are sometimes combined into one mechanism, but all three must be present for the loop to work. Let's look at each one in more detail.

Sensor

A **sensor** is a system component that monitors a system's output. (A sensor can also be called a monitor.) Sensors with which you may be familiar include your eyes, a thermometer, a compass, and a pressure gauge (Figure 6.11). Many systems use several kinds of sensors at the same time. Our bodies, for example, sense the temperature, humidity, light, and noise around us. Computer systems can also monitor these conditions using electronic sensors.

Some sensors send back information directly to the system. Others change it into another form first. Information is most often changed into electrical signals, which can be sent easily and quickly.

A device that changes information from one form to another is called a transducer. Pressure sensors, for example, can change mechanical pressure readings into electrical voltage. In a stereo system, the laser of a CD player is a transducer that translates the information on the surface of a CD into music. Once translated, the information is used as part of a system's feedback loop.

Figure 6.11 Divers use a pressure gauge—a type of sensor—to measure the water pressure around them.

Extending *What types of sensors have you used?*

Comparator

A **comparator** compares the actual output of a system with the desired output or result. It makes this comparison using information from the sensor.

If a system's actual and desired outputs are not the same, the comparator indicates that there is a difference. For example, if a car has an automatic speed control system, such as cruise control, the driver does not need to monitor the speed. The cruise control device constantly compares the actual speed with the desired speed set by the driver.

Some comparators measure the difference between the actual and the desired results. For example, a thermostat in a home heating system decides whether the room temperature matches the setting. Another example is a computer in a car engine that monitors whether enough gas is going into the engine. It can decide how much to add in order to reach the amount needed. Comparators may be simple, or they may be complex devices that contain computers.

Figure 6.12 With manual control of a heating system, a person acts as sensor, comparator, and controller.

Hypothesizing *What will happen to the system as the room becomes warmer?*

Controller

A controller changes a system's process to make the actual output closer to the desired output. A **controller** is a device that turns a process on or off or that changes it in some way.

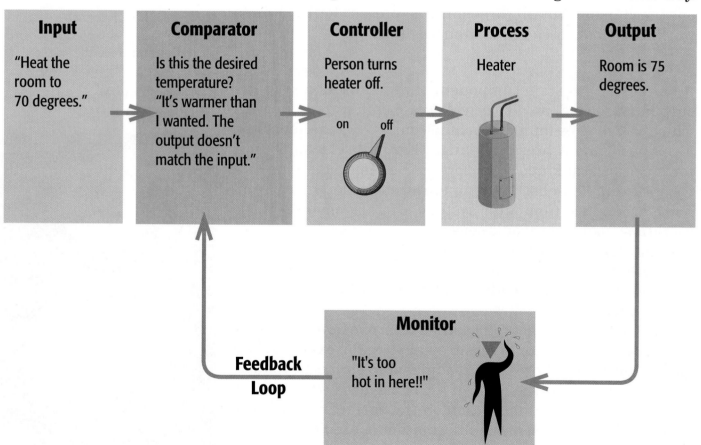

Input	Comparator	Controller	Process	Output
"Heat the room to 70 degrees."	Is this the desired temperature? "It's warmer than I wanted. The output doesn't match the input."	Person turns heater off. on off	Heater	Room is 75 degrees.

Feedback Loop

Monitor
"It's too hot in here!!"

Closed-loop systems can be controlled either manually or automatically. Consider a home heating system. If the system is controlled manually, a person acts as the sensor, comparator, and controller (Figure 6.12). Once the heater is turned on, the person may determine that the room is warmer than the desired temperature (the input command). The person must constantly monitor the temperature of the room to adjust the process (turn the heater on or off).

In an automatic home heating system, a thermostat determines whether the room temperature matches the temperature setting (Figure 6.13). The metal sensor, a bimetallic strip in the thermostat, changes when the temperature changes.

A bimetallic strip is made of two different types of metals. These metals expand or contract when heated or cooled, but no two metals expand or contract at the same rate. As a result, the bimetallic strip bends as one metal expands more than the other. The degree of the bend indicates the temperature. Using this simple mechanism, a thermostat continuously monitors a room's temperature and connects or disconnects with the furnace to turn it on or off.

Figure 6.13 With automatic control of a heating system, a thermostat regulates the system's process and output.

Interpreting *When the bimetallic strip loses contact with the heater, what happens?*

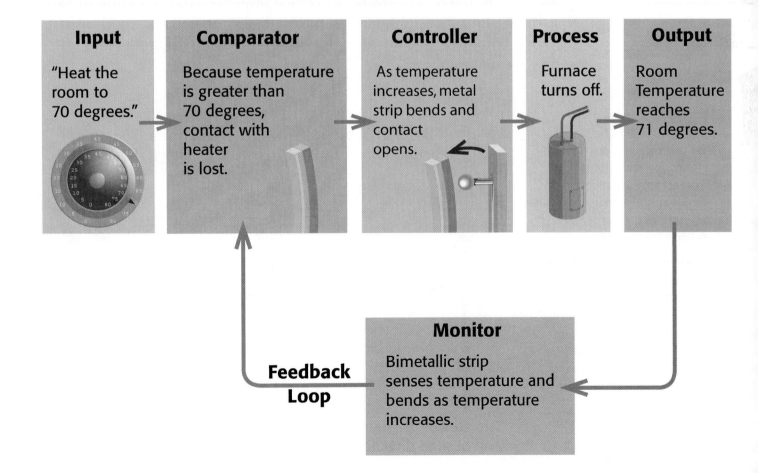

Input	Comparator	Controller	Process	Output
"Heat the room to 70 degrees."	Because temperature is greater than 70 degrees, contact with heater is lost.	As temperature increases, metal strip bends and contact opens.	Furnace turns off.	Room Temperature reaches 71 degrees.

Feedback Loop

Monitor
Bimetallic strip senses temperature and bends as temperature increases.

Types of System Controls

Because there are many types of processes, there are many types of controllers. Like the thermostatically controlled heating system you just read about, most systems are controlled automatically. Because of automatic controls, a jet can fly thousands of miles without any input from pilots. A greenhouse can continuously monitor and control the light, temperature, and humidity to ensure that conditions are optimal for its plants. A radio can automatically search for, select, and play one particular station. Robots can sort parts in a factory, which can be run and maintained by only a few people.

The controller chosen depends on the job to be done. Many modern systems are controlled by an electrical signal. The controllers in these systems change the electrical signals into mechanical, hydraulic, pneumatic, and other forms of action. Some system controllers, such as an electrical switch, a thermostat, or a water faucet, can only turn a system on or off. This single control is used to keep the output constant.

Other types of controllers respond proportionally to changing conditions. When steering a bicycle, for example, the rider—the controller—can make small or large adjustments based on how much the direction of the bicycle needs to be changed (Figure 6.14). The computer that controls an airplane in flight makes many different proportional adjustments based on wind speed, temperature, altitude, and other factors.

Figure 6.14 The steering on this bicycle is an example of proportional control. The greater the difference between the actual and desired directions, the harder the person has to steer.

Inferring *What types of feedback does the cyclist use to stay on track?*

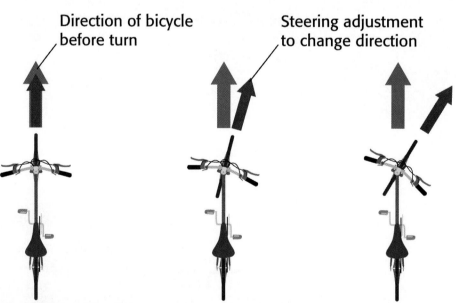

Direction of bicycle before turn

Steering adjustment to change direction

The greater the difference between the actual and desired directions, the greater the steering adjustment.

For: Systems Activity
Visit: www.mytechedkit.com

Systems and Subsystems Online

A bicycle is a system designed to enable people to travel quickly and easily. Go online to learn how a bicycle's components work together to get winning results.

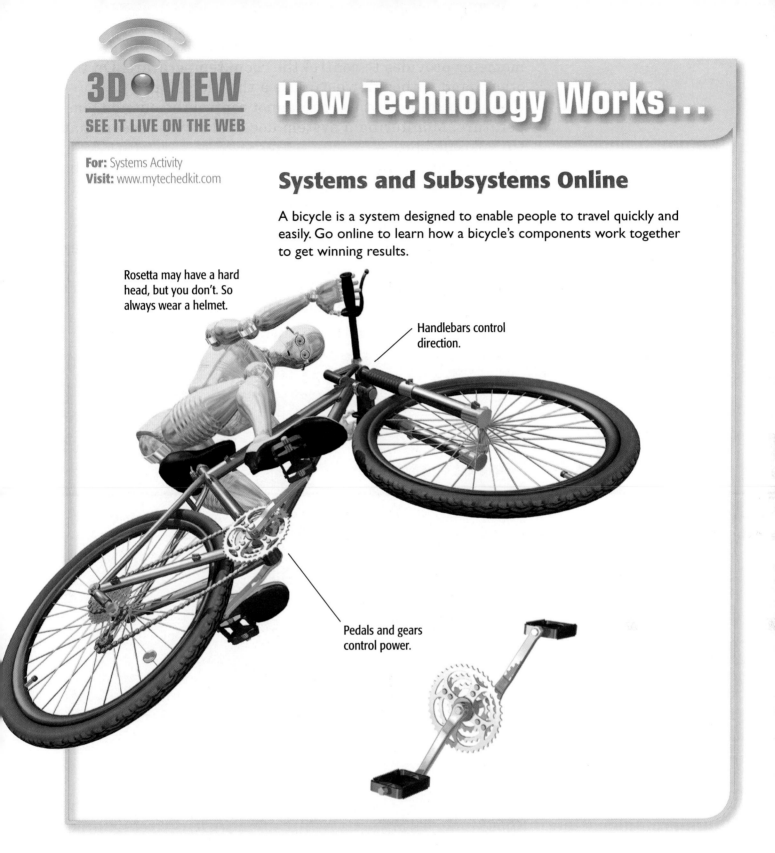

Rosetta may have a hard head, but you don't. So always wear a helmet.

Handlebars control direction.

Pedals and gears control power.

Controllability of a System

A system is controllable only if we can make the results, or outputs, what we want them to be. For example, we can use a thermometer to monitor the temperature in a house.

We can also use the information (or feedback) that the thermometer provides to control the room temperature. On the other hand, although we can use a thermometer to measure the outside temperature, we cannot control the outside temperature. Monitoring a system does not always mean that we can control it.

Most systems are controllable over a small range of outputs. By setting the thermostat of an air conditioner, we can keep the temperature inside a home anywhere from 65 to 85 degrees. We cannot, however, set the thermostat to lower our home temperature to 25 degrees. The lowest and highest temperatures that the air conditioner can maintain are called its limits of controllability. The **limits of controllability** for any system are the minimum and maximum outputs it can produce.

LIVING GREEN
Apply Your Limits

Controlling the settings on your home's thermostat is more than just knowing what the temperature is. The cooler your thermostat is set in the summer, the more the air conditioner works to meet that setting. On the other hand, the warmer you set the temperature in the winter, the harder your furnace has to work to keep the house warm. All this work requires the use of energy, and that work is reflected in your family's monthly energy bill. The more energy your home uses, the higher the energy bill!

MAKE A CHANGE
Remember that most electricity used to operate air conditioners is generated by burning coal or another fossil fuel, and heat is produced by electric heaters or furnaces that burn natural gas or oil. These are nonrenewable resources and have negative effects on the environment. Setting your thermostat a bit lower in the winter, and a bit higher in the summer not only saves your family's money, but it reduces the burning of fossil fuels and the toxins they release as well!

TRY THIS
Ask your family for a copy of the last utility bill for your home. Locate on the bill the portion for electric charges. Electricity is typically measured in kilowatt hours (kWh). To find out how much each kWh costs divide the total electric bill by the total kWhs consumed. If you wanted to reduce your monthly electric usage by 10%, how many less kWhs will you need to consume? What can you suggest to lower the usage?

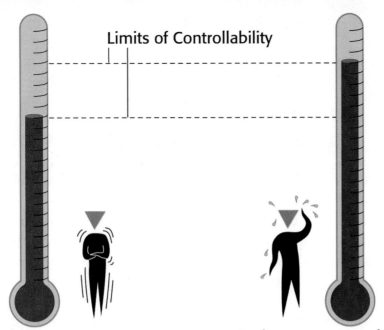

Limits of Controllability

Body shivers to warm up

Body sweats to cool down

Figure 6.15 The systems that regulate our body temperature must operate within their limits of controllability.

Comparing *How are shivering and sweating similar?*

Our body has many types of controlled systems. These systems regulate sugar levels in the blood, heartbeat, oxygen collection, and other important activities. They keep our body conditions just about the same, even though outside conditions may change. However, these systems must operate within the limits of controllability. For body temperature, these limits are the minimum and maximum temperatures our bodies can maintain (Figure 6.15).

Control of Nontechnological Systems

You have been learning how technological systems are controlled. Nontechnological systems must be controlled as well. Very often, information used to make decisions is processed using computers. However, some systems are so complex that the information is incomplete. Experience or even educated guesses can help people make decisions.

An economic system is an example of a system that relies on people's experience to control it. In a free-market system, people are free to decide what products to produce, whom to sell them to, and at what price to sell them. A nation's economic system is also controlled in part by the amount of money in circulation. A government may decide to print and distribute more money to help the economy grow. In that case, the sensor is a set of economic indicators that tell how the country's economy is doing (Figure 6.16). Economic indicators are statistics such as unemployment rates that are used to monitor and measure how well the economy is doing. The monitor is the group of people who study the indicators and

Figure 6.16

Figure 6.17

Figure 6.16 Prices rise and fall based on the strength of an economic system.

Inferring *What factors provide the monitor with feedback in a country's economy?*

Figure 6.17 Some states enact laws stating which gas drivers must use.

Extending *Who monitors compliance with the laws of a town or community?*

decide whether to print more money. The controller makes the decision to print more money. All of these people—and the information they use—make up a complex system.

In another type of system, such as a large company, managers are the decision makers, or controllers. They study company data, such as profits, debt, market forecasts, and many other factors. They make long- and short-term decisions. Should the company hire new people? Should it spend more money on research? The number of people hired, the amount spent on research, and the types of new products developed are some of the controlling factors in a company.

Laws and regulations also provide controls. Laws are made as a result of feedback. For example, when tests showed that our air was becoming polluted by the exhaust from trucks, cars, and buses, new laws were passed to require changes in fuel and design to reduce pollutants (Figure 6.17). The laws were used as a controller. Other laws govern toxic waste disposal, nuclear power plants, and genetic research.

SECTION 3 Assessment

Recall and Comprehension

1. What is a transducer? List two transducers, and describe what they do.
2. Explain what a nontechnological system is. Provide an example of your own.
3. Name the three parts of a feedback loop. What does each part do?
4. How are laws and regulations used in nontechnological systems?

Critical Thinking

1. **Evaluating** Does monitoring a system always allow control over it? Why or why not?
2. **Making Judgments** Which type of system would be more reliable for keeping a plane traveling at constant speed—an automatically controlled feedback loop, or a manually controlled feedback loop? Why?

QUICK ACTIVITY

The inside of a toilet tank contains an adjustable float that senses the proper water level. A fire sprinkler system has a heat sensor that activates it at a preset temperature. Using a clothespin and a sugar cube, design a sensor device that rings a buzzer when the sensor is submerged in water.
For more related Design Activities, see pages 158–161.

Technology in the Real World

The Hubble Space Telescope: Controlling a System from Miles Away

One amazing example of how a technological system uses controls is the Hubble Space Telescope. The Hubble orbits 600 kilometers above Earth, using a precise control system and state-of-the-art instruments to provide stunning views of the universe. These images cannot be made using ground-based telescopes or other satellites.

How Does It Work?

The telescope uses sophisticated instruments that serve as electronic eyes. These instruments include cameras and spectrographs. The Hubble's cameras do not use film. They collect light using electronic detectors similar to those in home video cameras. The spectrographs separate starlight into a rainbow of colors, just as a prism splits sunlight. By carefully studying the colors, astronomers can decode a star's temperature, motion, composition, and age.

An Incredibly Accurate System

The Hubble must be pointed at the object to be studied. It must hold very steady in position, or the image will blur. The system is so accurate that if the telescope were located in Washington, D.C., it could steadily focus a laser beam on a dime atop the Empire State Building in New York City. Once the Hubble locks onto an object, its sensors check for movement 40 times a second. Commands sent from Earth keep the telescope pointed at its target using a feedback control system. If movement occurs, constantly spinning wheels change speeds to smoothly bring the telescope back into position.

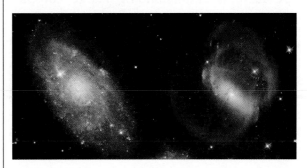

Hubble's photo of a galaxy that lies 150 million light years from Earth.

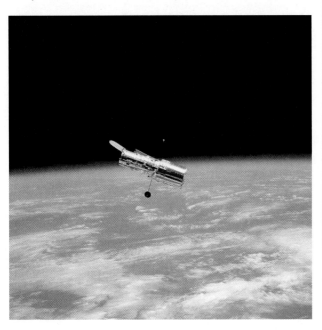

The Hubble Space Telescope in orbit above Earth.

Critical Thinking

1. **Calculating** How many miles above Earth is the Hubble's orbit?

2. **Summarizing** Draw a diagram showing the Hubble systems and its controls here on Earth. Show how these controls are used to obtain images and service the telescope.

3. **Extending** Visit **www.hubblesite.org** and download some of the images that the Hubble has taken of outer space.

CHAPTER 6 Review and Assessment

Chapter Summary

▪ All systems have inputs, processes, and outputs. The **input** is the command given to a system and also the desired result. The **process** is the action part of a system. The **output** is the actual result delivered by the system.

▪ Systems often have multiple outputs. A system design may have to be modified to reduce or eliminate undesirable outputs.

▪ Large systems are often made up of smaller subsystems. Examining each subsystem by itself can be useful in understanding a large, complex system.

▪ **Feedback** is used to ensure that the actual result of a system matches the desired result. Systems that have feedback are called closed-loop systems. Systems that do not have feedback are called open-loop systems. Open-loop systems cannot be controlled as well as closed-loop systems.

▪ Feedback consists of a monitor that observes the actual result, a comparator that compares the actual result with the desired result, and a controller that changes the process to make the output closer to the desired result.

▪ A system is controllable if we can make the outputs what we want them to be. Systems are controllable, but only within limits. People must know a system's limits and operate the system only within those limits.

▪ Nontechnical systems also have feedback loops made up of **sensors, comparators,** and **controllers.** These systems are most often dependent on information. People usually make the decisions, often with the help of computers.

Building Vocabulary

Your teacher may give you a crossword puzzle. Complete the puzzle using the following words from this chapter. Exchange puzzles with a classmate to check each other's answers.

1. closed-loop system
2. comparator
3. control
4. controllable
5. controller
6. desired results
7. feedback
8. feedback loop
9. input
10. monitoring
11. open-loop system
12. output
13. process
14. sensor
15. subsystem
16. system
17. thermostat
18. timer
19. transducer

See your teacher for the Crosstech puzzle.

Reviewing Content

1. What does the process part of a system do?

2. Name one subsystem of a large railway system.

3. Give an example of feedback you have received
   (a) in a class.
   (b) from a friend.
   (c) while using a computer.

4. How does a controller function in a feedback loop?

5. Give an example of a controllable system, and explain what makes it controllable.

Applying Your Knowledge

1. You want to play a movie at home on a VCR or a DVD player. Identify the input, process, and output in using the system you choose.

2. You are riding a bicycle and change gears to pedal uphill. Are your actions part of a closed-loop system? Why or why not?

3. Select a body system that is an example of a closed-loop system. Explain which body functions are controlled by the system, and how.

4. Research how an automatic lawn sprinkler system works as follows: Read the owner's manual and explain. Explain what the sensor, comparator, and control are and how they work together.

5. Explain how a rule at your school is an example of a control of a nontechnical system. What does the rule control, and why?

6. Explain why it is important for citizens to reduce the negative impacts and increase the positive impacts of their technologies on the lives of people in another area or on future generations.

7. Choose a system that you are familiar with (e.g., a toaster, a home heating system, an automobile cruise control) and diagram it using the universal systems model.

8. Investigate the legend of John Henry. Consider now the case of a hand saw and a power saw that can be used to cut wood. What are the advantages of each? Determine the time it takes for each to cut a two-by-six piece of lumber and comment on the production efficiency change using the power saw.

Critical Thinking

1. **Interpreting** When you use the computer to send an email message to a friend, what is the input command? What is the process for completing the output?

2. **Analyzing** Why is feedback important in a system? What is the feedback provided in this photograph?

3. **Contrasting** Explain how the feedback system for a thermostat makes it different from a thermometer, enabling it to control temperature.

4. **Solving** A printing system is set to print dark black letters on posters. The system has no feedback loop. After 30 minutes of printing, the ink becomes thinner and the letters are not as dark. How could the system be changed to a closed-loop system to automatically keep the letters uniformly dark?

 Connecting to STEM
science · technology · engineering · math

Advanced Airbag Systems

The airbag has proved to be a lifesaving device in an automobile collision. However, when airbags were first installed in automobiles, some people were injured by airbags, since the speed of an airbag's inflation sometimes reached 200 miles per hour.

Since then, airbag systems have been improved with sensors that can measure the weight and seat position of passengers as well as the severity of the crash. An airbag now opens in two stages with appropriate speed and intensity.

Conduct research to find out about advanced airbag systems. How do the sensors work to control the deployment of air bags? How does airbag deployment work as a system?

Design Activity 11

REVERSE ENGINEERING A TOASTER

Problem Situation

You can find a toaster in almost every home across the country. At breakfast, they are in great demand, turning bread into toast, browning frozen waffles, or heating prepared tarts and other tempting treats. If a toaster overcooks the food, how do you find out why? Reverse engineering is the process of examining existing products to see how they function, and to recommend design improvements that will make a product work better.

Your Challenge

You and your teammates are to perform a reverse engineering examination of a pop-up toaster.

> Go to your **Student Activity Guide, Design Activity 11.**
> Complete the activity in your Guide, and state the design challenge in your own words.

1 Clarify the Design Specifications and Constraints

To solve the problem, your design must meet the following specifications and constraints:
• Create a diagram showing how the bimetallic strip works as a thermostat.
• Provide two recommendations for improvement in the toaster's design.
• Construct one of the two recommendations you have made.

> In your Guide, state the specifications and constraints. Add any others that your team or your teacher included.

2 Research and Investigate

To better complete the design challenge, you need to first gather information to help you build a knowledge base.

> In your Guide, complete Knowledge and Skill Builder 1: Making a Bimetallic Strip.

> In your Guide, complete Knowledge and Skill Builder 2: Toaster Investigation.

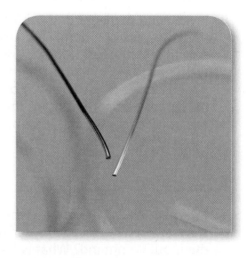

Materials

You will need:

- 0.010" brass and 0.005" steel shim stock
- 12" 22-gauge copper wire
- 12" 22-gauge Nichrome wire
- assorted screws
- digital temperature probe
- electrical switch
- multimeter wires with alligator clips
- plastic
- pliers
- screwdrivers
- sheet metal cutters (paper cutters also work well)
- small pieces of sheet metal
- small rivets
- toaster, with electrical plug removed

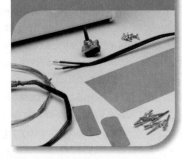

In your Guide, complete Knowledge and Skill Builder 3: Hot Wires.

In your Guide, complete Knowledge and Skill Builder 4: Reverse Engineering Your Toaster.

③ Generate Alternative Designs

In your Guide, describe two possible solutions that your team has created for the problem.

④ Choose and Justify the Optimal Solution

Refer to your Guide. Explain why you selected the solution you did, and why it was the better choice.

⑤ Develop a Prototype

Construct your improvement and incorporate it into the existing toaster. Provide a detailed drawing of your improvement, and show where it would be located in the toaster.

In any technological activity, you will use seven resources: people, capital, time, information, energy, materials, and tools and machines. In your Guide, indicate which resources were most important in this activity, and how you made trade-offs among them.

⑥ Test and Evaluate

How will you test and evaluate your design? In your Guide, describe the testing procedure you will use, recognizing that the toaster cannot be plugged in. Explain how the results show that the design solves the problem and meets the specifications and constraints.

⑦ Redesign the Solution

Respond to the questions in your Guide about how you would redesign your solution. Your redesign should be based on the knowledge and information that you gained during the activity.

⑧ Communicate Your Achievements

In your Guide describe the plan you will use to present your design to your class. Show what handouts and/or Powerpoint slides you will use.

Design Activity 12
TUG OF WAR

Problem Situation

The county fair is opening this weekend. One of the organized activities you have entered is the tractor pull. In this event, the tractor that pulls its opponent's tractor over the center line, or nearest to the center line, is the winner. Design and construct a model tractor that can pull an opponent's model tractor over the center line. The two opposing tractors will be positioned with their backs facing 4 inches from the line. A towline (string) fastened to the rear of the tractors will connect them.

Your Challenge

You and your teammates are to design and create a model tractor.

> Go to your **Student Activity Guide, Design Activity 12.** Complete the activity in your Guide, and state what the design challenge is in your own words.

① Clarify the Design Specifications and Constraints

To solve the problem, your design must meet the following specifications and constraints:
- The tractor pull will last 20 seconds.
- Your tractor can weigh no more than 18 ounces, including batteries.

> In your Guide, state the specifications and constraints. Add any others that your team or your teacher included.

② Research and Investigate

To better complete the design challenge, you need to first gather information to help you build a knowledge base.

> In your Guide, complete Knowledge and Skill Builder 1: Gear Interactions.
>
> In your Guide, complete Knowledge and Skill Builder 2: Constructing a Gear-Testing Machine.
>
> In your Guide, complete Knowledge and Skill Builder 3: Tractor Parameter Variation.

Materials

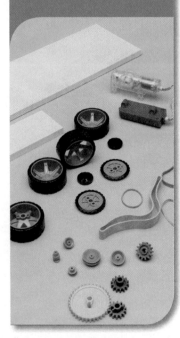

③ Generate Alternative Designs

In your Guide, describe two possible solutions that your team has created for the problem.

④ Choose and Justify the Optimal Solution

Refer to your Guide. Explain why you selected the solution you did, and why it was the better choice.

⑤ Develop a Prototype

Construct your model tractor. Include a drawing or a photograph of your final design in your Guide.

In any technological activity, you will use seven resources: people, capital, time, information, energy, materials, and tools and machines. In your Guide, indicate which resources were most important in this activity, and how you made trade-offs among them.

⑥ Test and Evaluate

How will you test and evaluate your design? In your Guide, describe the testing procedure you will use. Explain how the results show that the design solves the problem and meets the specifications and constraints.

⑦ Redesign the Solution

Respond to the questions in your Guide about how you would redesign your solution. The redesign should be based on the knowledge and information that you gained during the activity.

⑧ Communicate Your Achievements

In your Guide describe the plan you will use to present your design to your class. Show what handouts and/or Powerpoint slides you will use.

UNIT 3

Materials, Manufacturing, and Construction

Unit Outline

Processing Materials

In this chapter, you will learn about the following properties of materials and the techniques used to process them:

- casting
- pressing
- forging
- extruding
- shearing
- sawing
- drilling
- grinding
- shaping
- turning

Even something as "simple" as an in-line skate uses many different materials—plastic, metal, rubber, fabric. Each of these materials is processed from a simpler raw material.

- Rubber for the wheels is heated with sulfur. This process makes the rubber durable in hot or cold temperatures.

- Steel, which is used to make bearings, wheel parts, and screws, is made by heating iron ore. Liquid metal is poured into different molds to make the necessary parts.

- Plastic is processed from chemicals and then molded into shoe parts, buckles, and protective gear.

- Even the fabric inside the skate may be made from plastics. Synthetic fibers, such as polyester, are made from chemicals or recycled materials. They are comfortable and let your feet breathe.

SECTION 1
Types of Materials

Benchmarks for Learning

- Resources are either renewable or nonrenewable.
- Technological processes change materials from one form to another.

Reading Strategy

Mapping   Create a concept map to help you organize the different types of materials described in this section. As you read the section, list the materials under the appropriate bubbles, or headings.

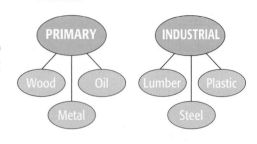

Vocabulary

primary material	softwood	ceramic
industrial material	manufactured board	glaze
vulcanize	ferrous metal	thermoplastic
hardwood	nonferrous metal	thermoset plastic

Figure 7.1   Primary materials, such as wood, can be processed into industrial materials, such as paper.

Identifying   *Can you name other primary materials that are turned into industrial materials?*

Primary and Industrial Materials

A **primary material** is a substance produced naturally on the Earth. Trees are cut for wood, ores are mined from the Earth, and stone is cut from quarries. Wood, metals, oil, clay, and stone are all examples of primary materials. Primary materials may be renewable, such as wood, or nonrenewable,

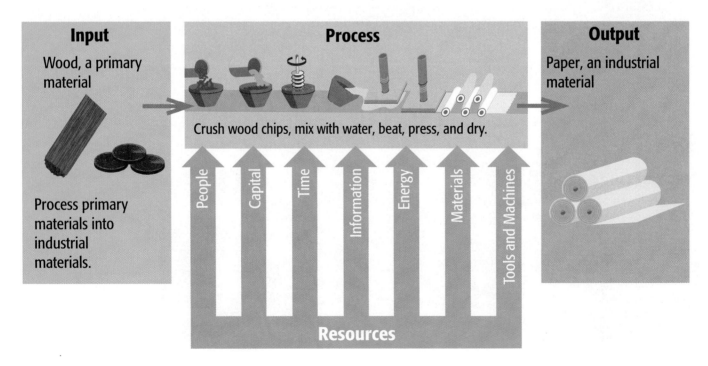

such as metal. Primary materials can be used as is, but usually they are the basis for other types of materials.

Through primary industries, such as sawmilling and metalworking, primary materials are processed into industrial materials (Figure 7.1). **Industrial material** is a substance that has been processed. The wood from trees—a primary material—is cut into wooden planks—an industrial material. Natural rubber, called latex, is **vulcanized**, or heated with sulfur, to make it more resistant to temperature changes. Iron is mixed with carbon to form steel. Crude oil is broken down into chemicals, which are then processed into plastics.

Types of Industrial Materials

Wooden planks, metal sheets and rods, ceramic powders, and plastic pellets are all types of industrial materials. Industrial materials are used to make final products such as furniture, buildings, and cars.

Wood

Wood is a renewable resource that is readily available. Most kinds of wood can be cut and shaped easily. Wood has beautiful colors and grain patterns, and can be quite strong.

Wood, or lumber, comes in different forms (Figure 7.2). **Hardwood** is the wood of broad-leaved deciduous trees, or trees that lose their leaves in the fall. Trees such as maples, oaks, and poplars are hardwoods. **Softwood** is the wood of coniferous, or cone-bearing, trees with needlelike leaves, such as pines and firs. Generally, as its name implies, hardwood is heavy compact timber. Softwood is generally light and easy to cut. However, these terms describe only the types of tree from which they come, and can be deceptive. For example, balsa and basswood are deciduous trees, but their woods are soft.

Manufactured board, such as plywood and particle board, is a construction material made from wood chips and sawdust. Manufactured board is often stronger than the original wood used to make the board. It does not warp or twist.

Metal

Iron and the steel that is made from it are probably the most important metals used today. They are turned into end products such as cars, skyscrapers, and machine tools. As we learned in Chapter 2, alloys are combinations of metals. A **ferrous metal** is an alloy, more than half of which is iron. Steel is a ferrous metal. Steel alloys can be made to have special properties. For example, stainless steel, which contains chromium, resists rusting.

Figure 7.2 Furniture makers often mix hardwoods and softwoods into a single design.

Summarizing *What is the difference between hardwoods and softwoods?*

Figure 7.3 Aluminum is a nonferrous common metal used for many different types of products.

Applying *Why do you suppose aluminum soda cans replaced glass?*

Any metal other than iron and any alloy without a large amount of iron is called a **nonferrous metal**. Nonferrous metals include aluminum (Figure 7.3), copper, magnesium, nickel, tin, and zinc. Brass is an alloy of copper and zinc. Bronze is an alloy of copper and tin. Pewter is an alloy of tin, antimony, and copper.

Ceramics

Ceramics are objects made from clay or similar inorganic (nonliving) material, such as plaster, cement, limestone, or glass. In the process of making clay products, the clay is first mined and then mixed with other substances. It is then fired (heated) in an oven called a kiln to about 2,000°F. Firing makes the clay very hard. The clay can then be coated with a **glaze**, which is a glasslike material that protects the surface of ceramics and gives them color. Ceramics are common household objects because clay is easy to obtain and is inexpensive.

Ceramics generally do not conduct electricity well, so they can be used as insulators. Wires carrying high-voltage electricity from power stations to your home use ceramic insulators. Ceramics are also used in light bulb sockets, switches, and other electric parts.

Ceramics can withstand high temperatures and remain strong. They keep their shape and strength and are unlikely to break. Ceramics are used in many types of cooking products (Figure 7.4). Combinations of metal and ceramics are also used inside some car and rocket engines.

Glass is a ceramic material. Most glass is made by melting sand, lime, and sodium oxide together at a temperature of about 2,500°F. This kind of glass is called soda-lime glass. It is used to make bottles, light bulbs, and windowpanes.

Fiberglass is a form of glass that is used for insulation. It is made by dropping molten glass onto a spinning steel dish with hundreds of tiny holes in it. Glass fibers are spun out of these holes as the dish turns.

Plastic

Plastic is made of many long, chainlike molecules called polymers. (In the Greek language, *poly* means "many.") Polymers can occur naturally. Plants, animals, and other living things produce the polymers they need from nutrients and other materials in the environment. A good example of this is a spider web. A spider web can stretch without breaking. The web is actually made from chemicals in the spider's body. These chemicals mix and react to form a silk-like polymer that is one of the strongest materials known.

Most plastics that you are familiar with are synthetic, or made by people rather than found in nature. Synthetic polymers are often used in place of natural materials that are too expensive or wear out too quickly. Any plastic item you use is almost certainly made from synthetic polymers.

Figure 7.4 Ceramics are used in many types of products.

Identifying *What features of ceramics make them ideal for heating up food in a microwave?*

Figure 7.5 PVC, which stands for polyvinyl chloride, is a type of plastic used to make pipes.

Inferring *What properties of PVC make it suitable to be used in plumbing?*

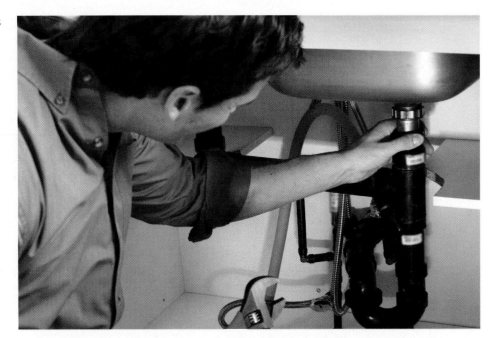

Plastics are divided into the following two categories:

Thermoplastics are plastics that soften when heated, so they can be melted and shaped. When they cool, they become hard again. Examples of thermoplastics are acrylic fibers (such as nylon and Orlon®), polyethylene (used for plastic bags), vinyl, and polyvinyl chloride (or PVC, used for plumbing pipes and electrical insulation) (Figure 7.5).

Thermoset plastic, such as Bakelite and Formica, is plastic that does not soften when heated. Thermoset plastics char and burn instead. These properties make them useful for handles for pots and pans. They are less common than thermoplastics. "Unbreakable" plastic cups and dishes are made from melamine, a thermoset plastic.

SECTION 1 Assessment

Recall and Comprehension

1. Make a list of five primary raw materials. For each one, indicate an industrial material that is created from it.
2. What are the different properties of ceramics?
3. What are hardwoods and softwoods? How are these woods different?

Critical Thinking

1. **Extending** Investigate the uses of fiberglass. What types of manufactured products contain fiberglass?
2. **Clarifying** Make a diagram showing the relationships between the terms primary material and industrial material.

QUICK ACTIVITY

Using a kiln to fire pottery is an ancient practice. Research how different cultures used kilns to make objects such as bowls, vessels, and hand tools. Note the seven resources used to make these items. How has modern technology affected both the way pottery is made today and how people use it? **For related Design Activities, see pages 200–203.**

Technology in the Real World

Synthetic Clothing

During most of human history, clothing was made from natural materials: animal skin, fur, and material derived from plants such as cotton and linen. Today, clothing is often manufactured from synthetic, or man-made material. Most synthetic clothing is derived from petroleum products. For outdoor activities, synthetic clothing has improved properties over natural material. It absorbs little moisture, dries quickly, and keeps you warm even when wet. Whereas cotton simply absorbs moisture and holds it, synthetics move the moisture away from the body to the outside of the fabric, where it quickly air-dries. A disadvantage of synthetic materials is that they do not decompose easily. Natural fibers are more environmentally friendly because they biodegrade, or break down naturally.

Fabrics made with synthetic fibers are ideal for outdoor activities.

History of Synthetic Fibers

The first synthetic fiber, rayon, was developed in 1891. It is highly absorbent, soft, comfortable, and easy to dye. Nylon, developed as a fabric in 1935, is exceptionally strong, elastic, abrasion resistant, easy to wash, and resistant to damage from oil and many chemicals. Polyester was invented in 1942. Today, polyester fleece is a fashionable fabric for outdoor wear. Acrylic fiber, first produced in 1950, has outstanding wicking ability.

Aqua Shell® membrane allows sweat and water vapor to pass through

Wind is completely blocked

UPF 30+ provides high sun protection

Tough outer skin is abrasion resistant

Membrane blocks cold-water exchange

Fabric provides neutral buoyancy

Comfortable inner surface traps body warmth; resists odor buildup

4-way stretch for freedom of movement

Synthetic fabrics work in several ways to block moisture and retain warmth.

Critical Thinking

1. **Inferring** Why is repelling moisture an important feature for outdoor clothing? Is it important for all clothing?

2. **Extending** List three specifications that you think are important for outdoor clothing.

Properties of Materials

Benchmarks for Learning

- Materials may be chosen on the basis of their mechanical, electrical, magnetic, thermal, or optical properties.
- Materials have characteristic properties that can be measured and observed.
- The management of waste is an important societal issue.

Reading Strategy

Table with Bullets Create a table with bulleted entries to describe the different properties that materials can have. For each entry, jot down a quick definition of that term.

MECHANICAL PROPERTIES		
• Ductility: can be twisted or bent • Strength		

Vocabulary

ductile	tension	toughness
brittle	compression	hardness
elasticity	torsion	conductor
plasticity	shear	insulator

Figure 7.6 Some materials need to be able to maintain their shape under strong forces.

Hypothesizing *What mechanical properties should a basketball rim have? Why?*

Mechanical Properties

The basic properties of a material include its strength, hardness, appearance, ability to conduct electricity, resistance to corrosion, and ability to transmit light.

We can use technology to change a material's properties—for example, heat can be applied during processing to make a metal softer or less brittle. Materials can be distinguished based on their specific properties. These include mechanical properties, electrical and magnetic properties, thermal properties, and optical properties.

Force applied to a material can make it bend, break, or change shape (Figure 7.6). A material's response to force is determined by its mechanical properties. A fishing rod made from graphite, for example, bends quite a bit without breaking. Many kinds of wood, however, break without bending much at all. Mechanical properties include ductility, strength, toughness, and hardness.

Ductility

A material that can be twisted, bent, or pressed without breaking is said to be **ductile**. Metals can be quite ductile. The manufacture of pots and pans begins with a flat sheet of metal. Adding pressure to part of the metal allows it to be formed into the desired shape. The metal may have to be deformed (changed in shape) a great deal. In making wire, for

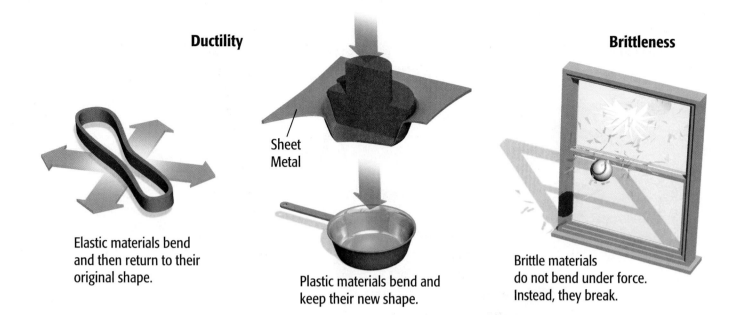

Ductility

Brittleness

Elastic materials bend and then return to their original shape.

Sheet Metal

Plastic materials bend and keep their new shape.

Brittle materials do not bend under force. Instead, they break.

example, a thick rod is pulled through a small hole to make it thinner. This process, called drawing wire, works because the metal rod used is made of a ductile material. It can be stretched very thin before it breaks.

A material that lacks ductility is said to be **brittle.** A brittle material is inflexible, easily broken or shattered, and will not deform without breaking. Window glass is a good example of a brittle material.

Ductile materials are either elastic or plastic (Figure 7.7). A material that can be bent or stretched and then returned to its original shape and size is said to be elastic. Rubber bands, springs, and fishing rods are made from materials with high elasticity. **Elasticity** is the quality of being flexible.

A property similar to elasticity is plasticity. Materials with **plasticity** stay deformed even after the force that shapes them is removed. Plastic materials can be bent and will then stay bent. The material that we call plastic was named for this property, but it is just one of a number of materials with that characteristic. Modeling clay is another plastic material, as are certain metals, such as steel.

Strength

A material's **strength** is its ability to keep its own shape even when a force is applied to it. The stronger the material, the less its shape can be changed under force. Strength varies from material to material. For instance, concrete is strong when compressed. It can withstand a great amount of force before it breaks up. Some stones, however, break easily under pressure.

Figure 7.7 Elastic materials change shape but bounce back to their original shape. Plastic materials keep their new shape. Brittle materials cannot change shape under force; they break instead.

Inferring *How would you describe modeling clay—elastic, plastic, or brittle?*

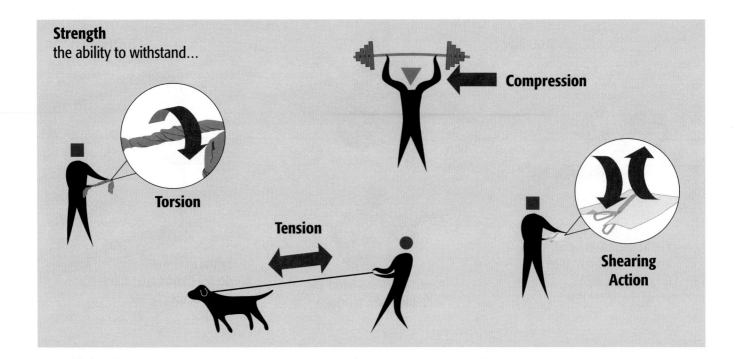

Torsion

Compression

Tension

Shearing
Action

Figure 7.8 Strength is the ability to withstand torsion, compression, tension, and shearing action.

Identifying *What type of force is used in torsion?*

Four kinds of force may be applied to a material (Figure 7.8). **Tension** is a force that pulls on a piece of material. When a spring is pulled, it is put under tension. **Compression,** the opposite of tension, is a force that pushes on or squeezes a material. Squeezing a sponge or walking on rubber-soled shoes causes compression.

Torsion is a force that twists a material. If we twist off a piece of licorice candy, the material is under torsion. The twisting force itself is called torque. When a wrench is used to turn a bolt, torque is used. A material comes under **shear** force when one part slides in one direction and the other part slides in the opposite direction. It acts on a material like a pair of scissors.

Toughness and Hardness

Toughness is the capacity of a material to absorb energy without breaking. For example, leather is tough. It can handle a great amount of wear and tear and still remain in good condition. Sometimes, meat is tough. It takes a lot of chewing to break down the fibers of the meat.

Hardness is a material's ability to resist being scratched or dented. It is an important quality for products that undergo daily wear and tear. A diamond is the hardest material known. Some metals are also very hard. Metal tools must be hardened to resist wear. Some synthetic materials, such as tungsten carbide, are very hard. The teeth of circular saw blades are often made of tungsten carbide (Figure 7.9).

Electrical and Magnetic Properties

All materials resist the flow of electricity to some degree. However, certain materials, called **conductors,** offer little electrical resistance and allow electricity to pass through. Most good conductors are metals. Wire is most often made from copper, which is an excellent conductor (Figure 7.10).

Materials that resist the flow of electricity most strongly and do not allow electricity to pass through are called **insulators.** Because they are good insulators, plastic and rubber are used to cover wire. This layer of insulation protects people from electric shock.

When electricity flows through a wire, a magnetic field is produced. An electromagnet can be made by wrapping wire around a piece of iron. When an electric current (for example, from a battery) flows through the wire, the iron piece becomes a magnet. Magnetic materials, which are attracted to magnets, include iron, steel, nickel, and cobalt. Copper, wood, glass, and leather are examples of nonmagnetic materials.

Figure 7.9 Because they are extremely hard, diamonds are used on the edge of some saw blades.

Predicting *What damage could this saw incur without a diamond edge blade?*

Thermal Properties

In the Greek language, *therm* means heat. Words such as *thermostat* and *thermos* name things related to heat. A **thermal property** refers to a material's ability to conduct heat. Metals, especially copper and aluminum, are good conductors of heat. Other materials, such as rubber and fiberglass, do not conduct heat well. In all types of buildings, insulation, a layer of material that does not conduct heat, is used to prevent the movement of heat through the walls.

Figure 7.10 Copper is an excellent electrical conductor.

Analyzing *Why is it important to insulate this wiring?*

Optical Properties

An **optical property** refers to a material's ability to transmit or reflect light. Some kinds of glass transmit light well. Window glass transmits light well enough to see through, but not well enough to be used for scientific tools such as telescopes.

Special optical glass is used for telescope lenses. Very pure glass is used to make fiber-optic cables, which carry information using light rather than electricity (Figure 7.11). Most of our telephone calls and emails are converted into pulses of light and carried by optical fiber cables. Optical properties are not limited to glass.

Plastic is lightweight and does not shatter, and clear plastic transmits light, making it useful for contact lenses and eyeglasses. Some metals reflect light very well. They are used in reflectors for headlights and flashlights.

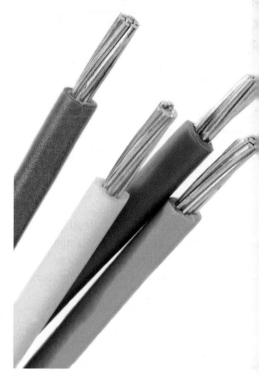

Figure 7.11 Fiber-optic cables carry information using light rather than electricity.

Inferring *Why is fiber-optic cable made of glass?*

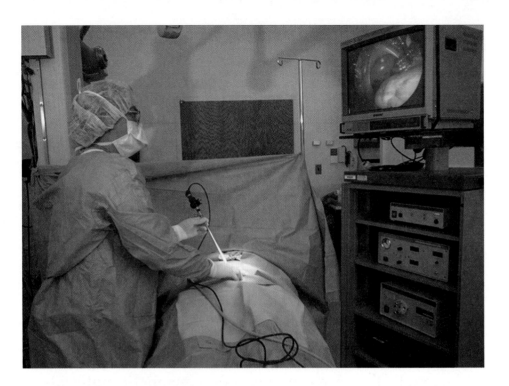

Testing of Materials

All products need to be tested to ensure that they meet certain standards for manufacturing and industry. The American Society for Testing and Materials (ASTM) was founded in 1898 to establish methods for testing materials

3D VIEW
SEE IT LIVE ON THE WEB

How Technology Works...

For: Steel Testing Activity
Visit: www.mytechedkit.com

Steel: A Chameleon Metal ... See How Online

Steel is an alloy of Iron and Carbon. The percentage of carbon is very small—between 0.02–3.0%, but the amount of carbon has a significant effect on how steel reacts to tensile (pulling) forces: it can stretch, or it can resist great forces. Go online to perform some tests to learn more about the effect of small changes in the percentage of carbon in steel.

Figure 7.12 Testing materials: Adding small amounts of carbon can affect the way steel materials behave.

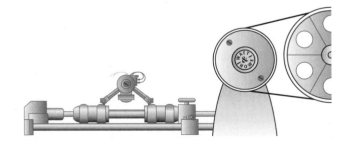

and products and for developing quality standards for systems on which industries could rely.

Now known as ASTM International, the Society's testing methods are used by companies all over the world to ensure that their products conform to recognized standards. ASTM defines technical terms and procedures so that materials testing is always carried out the same way, whether it is done in Japan, Germany, or the United States.

Materials can be tested for many different properties. For example, roads in cold, northern countries were being damaged by studded tires so engineers began testing wear-resistant concrete for resurfacing the roads (Figure 7.13).

Tests were created for manufacturers of bicycle helmets. By making the helmets to the ASTM's specifications, head injuries can be minimized.

Nanoscale Materials

Nanotechnology is science and engineering related to the design, manufacturing, and application of materials and devices on the nanometer length scale (1–100 nanometers). One nanometer is one billionth of a meter. A scientific and technical revolution has just begun based on the ability to systematically organize and manipulate matter at nanoscale. Nanostructure materials have attracted a great deal of interest in recent years because of their novel mechanical, electrical and optical properties.

The Nobel laureate chemist Richard E. Smalley said, "The impact of nanotechnology on health, wealth, and lives of people will be at least the equivalent of the combined influences of

For: Materials Testing Activity
Visit: www.mytechedkit.com

Impact Testing Materials Online

The properties of a specific material determine whether it is suitable for a given application. Tests are performed to find just the right materials to make products. Go online and perform some tests to find the mechanical properties of several types of materials.

Ball bearings are often used to test a material's response to impact.

microelectronics, medical imaging, computer-aided engineering, and man-made polymers [plastics] developed in this century." Nanoscale materials have the potential to affect many economic and societal endeavors including health-care, information technology, homeland security, and safeguard of the environment. Future innovative breakthroughs include significant advances in smart communication devices, human organ restoration, the development of novel materials, and the emergence of entirely new phenomena in chemistry and physics.

Nanoparticles have diameters less than 100 nanometers (nm). To give you an idea about that size, it's about the thickness of a human hair sliced into 10,000 pieces. The distance across the nail on your little finger (which grows at a rate of about one nanometer per second) is about 10 million nanometers.

Carbon nanotubes are cylindrical molecules with a diameter as small as 1 nm and a length up to several millimeters. They consist of only carbon atoms, and can be thought of as cylinders made of a single layer of a graphene sheet (Figure 7.14). They are very strong and conduct heat well. They have very interesting physical properties and, because of this, can be used for a wide variety of purposes including clothing, sports equipment, and even a theoretical "space elevator" that would consist of a cable made of carbon nanotubes anchored to the Earth's surface, reaching into space. The bicycle that Tour de France winner Floyd Landis used in 2006 was made of carbon nanotubes. Carbon nanotubes have a strength-to-weight ratio a hundred times better than aluminum, and are used because they are very strong and very lightweight. Landis' bicycle frame weighed only about two pounds.

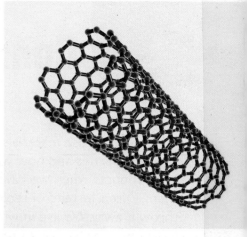

Figure 7.14 This nanotube has been created by rolling a single layer of carbon into a cylinder.

SECTION 2 Assessment

Recall and Comprehension

1. Explain the difference between elasticity and plasticity.
2. Identify the four different types of force. Draw a quick sketch to describe each one.
3. What are optical properties? What types of materials have these properties?

Critical Thinking

1. **Extending** Investigate fiber-optic cables. How does light travel through the cable? Does it leak out?
2. **Clarifying** What are the optimal properties for a thermos?
3. **Describing** Why is it important for companies to follow standards when making products?

QUICK ACTIVITY

Almost all materials, even the best conductors, have some resistance to the flow of electricity. They heat up when a high electric current is present. Superconductive materials, however, have no resistance to electricity when cooled to low temperatures. Research superconductivity on the Internet and report your findings to the class. **For more related Design Activities, see pages 200–203.**

Connecting to STEM

Disposal of Resources

When we choose materials, we must consider their properties and how we will dispose of end products when we have finished using them. Often, instead of repairing something, we throw it away. We live in what has been called a "throw-away" society. Our carbonated drinks come in cans or bottles that we throw away. Our food comes in disposable plastic or paper containers.

When we dispose of products, we bury them in landfills or burn them. Depending on the properties of the materials from which these products were made, however, disposal may cause pollution. Burning them may pollute the air. Storing them in landfills may pollute the groundwater and our own water supplies. Toxic materials such as lead, mercury, and radioactive waste can pollute the soil, water, and atmosphere.

Companies must work to make products out of materials that will decompose, or rot, to become part of the Earth again. Those materials that will not decompose, such as plastic, glass, or aluminum, should be recycled (reused).

Time for Material to Decompose

Material	Time to Decompose	Material	Time to Decompose
Orange peels	1 week to 6 months	Plastic bottles	50 to 80 years
Paper containers	2 weeks to 4 months	Aluminum cans	80 to 100 years
Paper containers with plastic coating	5 years	Plutonium	24,390 years (half-life)
Plastic bags	10 to 20 years	Glass bottles	Indefinite

Critical Thinking

1. **Evaluating** When choosing materials, why must people think about the disposal of end products?

2. **Calculating** At the end of a day, categorize the contents of the garbage bin in your home. How long will it take for all of it to decompose if it is put in a landfill?

3. **Making Judgments** Go on the Internet and find the value of recycled newspaper and recycled aluminum. Does recycling make "cents"?

Forming and Separating Processes

Reading Strategy

Outlining   Make an outline of the different processes described in this section. For every entry in your outline, write down an example of a material that is formed using that process.

> I. Forming Processes
>   A. Casting
>   1. ice cubes
>   B. Pressing
>   1. waffle iron

Vocabulary

pressing	blow molding	grinding
sintering	vacuum forming	shaping
forging	separating	turning
extruding	shearing	filtering

Benchmarks for Learning

- Different technologies involve different sets of processes.
- Forming includes bending, shaping, and stamping.
- Separating includes cutting, sawing, and shearing.

Processing

Material resources are made more useful through processing. **Processing** is the method by which materials are changed in form. Raw materials such as cotton fiber can be made into thread, and thread can be spun and woven into fabric, which is then sewn into clothing. Wood chips can be processed and used to make paper; animal hides can be made into shoes, handbags, and coats.

Generally, each time a material is processed, value is added. Steel is more valuable than iron, from which it is made, because it can be used to make more products that can be sold. Two ways of processing materials are forming and separating.

Forming

Forming is the process of changing the shape of a material without cutting it. For example, metal can be bent to change its shape, or it can be poured or pressed into different shapes. Different forming processes include casting, pressing, forging, extruding, blow molding, and vacuum forming.

Casting

One way of forming material is called casting. **Casting** is a process in which a liquid is poured into a mold, allowed to harden, and then removed from the mold. It takes on the

Figure 7.15 As you walk on the beach and press your foot into the sand, you make a mold of your footprint.

Extending *If you made a mold of your entire foot, how many pieces would the mold have to be made of? Why?*

Figure 7.16 In casting, a mold is filled with a liquid, which is allowed to harden. The outer mold is removed, revealing a solid casting.

Extending *What properties must the mold have in the casting process?*

reverse shape of the mold. As you walk on the beach and press your foot into the sand, you make a mold of your footprint (Figure 7.15). If you poured a liquid material, such as melted wax, into this sand mold and let it cool, you would have a casting of your footprint. Ice cubes and many different ceramic and metal objects are made using the casting process.

Molds can be composed of one piece or several pieces. A footprint is a one-piece mold. Another one-piece mold is a cake pan into which liquid batter is poured and then baked. A two-piece mold can be used to cast ceramic objects (Figure 7.16). First, a mold is made from plaster. Then, slip (liquid clay) is poured into the mold. It is allowed to set, or begin hardening, for a few minutes. The plaster absorbs water from the slip. This leaves a thin wall of clay inside the mold. The remaining slip is then poured out. When the clay dries, the mold is opened and the finished casting removed.

Pressing

Similar to casting, **pressing** is a process in which force is applied to change the shape of the material. It is used in many industrial processes. In pressing, a measured amount of material is poured into a mold. A plunger is lowered onto the material to force it to spread out and fill the mold. The material is forced into the shape of the mold at the bottom and into the shape of the plunger at the top. The plunger is then pulled out and the finished object removed. When we make hamburger patties, we press the meat into shape with our hands. Waffles are pressed by waffle irons.

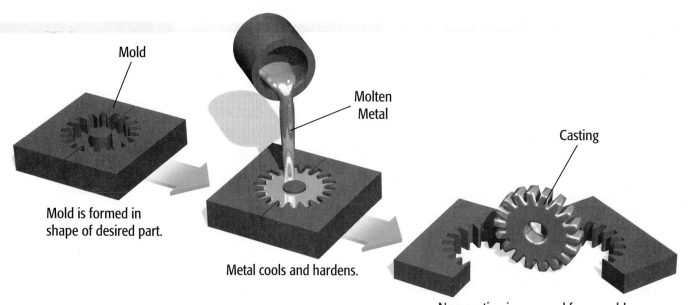

Mold

Mold is formed in shape of desired part.

Molten Metal

Metal cools and hardens.

Casting

New casting is removed from mold.

Sometimes, powdered metal is pressed to make objects. The powder is put into the bottom of a mold. The top of the mold is lowered. Great pressure is applied to change the powder into a solid. It is then heated to make the particles fuse together. This process of applying pressure and heat to metal particles is called **sintering.**

Forging

Forging is a process in which a metal is heated until it becomes ductile, but not melted, and then hammered into shape. Many metal parts for cars and planes are made by forging.

Long ago, forging was done by hand. Blacksmiths forged horseshoes by heating pieces of metal and hammering them into shape. Today, most metal forging is done using large machines. A piece of heated metal is placed in the lower half of a mold, but it does not assume the shape of the mold on its own. A powerful ram presses the metal downward into the mold with as much as 2,500 tons of force (Figure 7.17).

Extruding

Another type of forming is **extruding,** a process in which softened material is squeezed through an opening, much like squeezing toothpaste from a tube. The material takes the shape of the opening. The opening can be fitted with a device known as a die, which changes the shape of the material. If the die is square-shaped, then the extruded material will be square-shaped as well. Forming objects by extruding can save work because extruded objects do not require much more shaping or machining.

Figure 7.17 Most metal forging is done using large machines.

Identifying *What safety precautions is this man following?*

Figure 7.18 Vacuum forming is used to package products that hang on cards in supermarkets and toy stores.

Identifying *What are some advantages of this kind of packaging?*

Blow Molding and Vacuum Forming

Blow molding and vacuum forming are processes used to change plastic into desired forms. In both of these processes, plastic is heated until it is soft. In **blow molding,** a bit of heated plastic is placed in the center of a mold. Air is then injected so the plastic expands in a uniform thickness to form the desired shape. Plastic containers and bottles are made using blow molding.

In **vacuum forming,** a vacuum pulls a sheet of warm, soft plastic downward, and the plastic clings to whatever it is drawn against. Vacuum forming is used to package products that are displayed on hanging cards in supermarkets and toy stores. A plastic sheet is pulled down against the product on the cardboard. Consumers can see the product through the clear plastic without having to open the package. This is known as blister packaging (Figure 7.18).

Separating Processes

Separating is the process by which part of a material is removed, usually through cutting. Knives, saws, and scissors are used to separate one piece of material from another. We use a separating process when we cut food with a knife.

Many different kinds of tools and machines are used for cutting. Some are designed to cut specific materials, such as wood, metal, plastic, clay, leather, or paper. Tools and machines cut materials by shearing, sawing, drilling, grinding, shaping, and turning.

In other types of separating processes, some materials may be separated chemically or magnetically. Solids may be separated from liquids by filtering.

Shearing

Shearing is the process in which a knifelike blade is used for separating (Figure 7.19). One blade, as in a knife, or two blades, as in a pair of scissors, can be used. In fact, another name for *scissors* is *shears*. In shearing, the sharp edge of the blade compresses the material being cut. When the compression force reaches a certain level, the material separates along the line of the cut. The sharper the knife-edge, the greater the compression force.

Sawing

Sawing is the process in which material is separated using a blade that has teeth. Each tooth chips away tiny bits of material as the saw cuts.

Two kinds of sawing processes are used to cut wood. In ripping, wood is cut in the direction of the grain. In crosscutting, wood is cut across the grain. Handsaws that are used to rip or crosscut wood generally have 6 to 10 teeth per inch. Metal is cut by hand using a hacksaw, which has a very hard steel blade with about 18 teeth per inch.

Saw blades come in many forms. Machine saws, such as radial arm saws and table saws, use circular blades. These blades are made to exact specifications. Each tooth is very sharp. The blade spins rapidly, so that many cuts per second are made in the material.

Figure 7.19 Shearing, sawing, and drilling are all types of separating techniques that use cutting.

Extending *How would you decide whether to use shears or a saw to cut a piece of metal?*

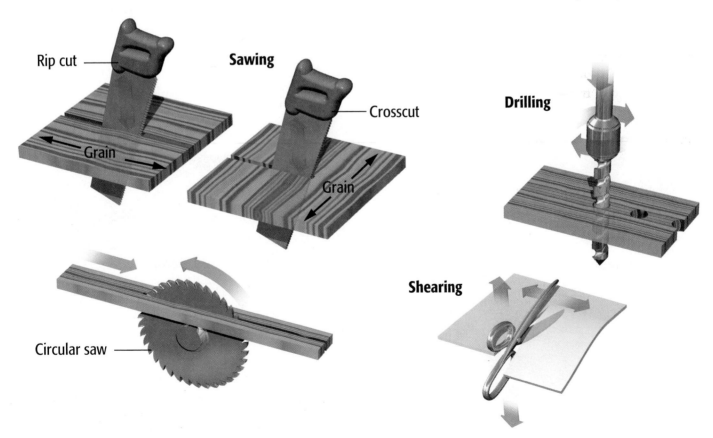

Rip cut — Sawing — Crosscut — Drilling

Grain — Grain — Shearing

Circular saw

Drilling

Drilling is a separating process in which round holes are cut in materials. A pointed tool with a sharpened end, called a twist drill, is turned very rapidly and pushed through the material to cut a hole. A hand drill or an electric drill can be used to spin the twist drill.

Grinding

Grinding is a process in which small particles of a material are used to sharpen or sand a surface. Grinders and sanding machines make use of pieces of very hard material called abrasives. Abrasives are crushed into very small particles. These particles are then glued onto a flat sheet of material to make sandpaper or onto a grinding wheel. As the grinding wheel turns, the abrasive particles rub against and cut away tiny pieces of the material being shaped (Figure 7.20).

Polishing is another form of grinding. Polishes generally contain some very fine abrasive powder. When you rub an object with polish, you remove tiny bits of it. Toothpaste is a fine abrasive material. When you brush your teeth, abrasives in the toothpaste help remove plaque.

Figure 7.20 Grinding, shaping, and turning are different types of separating techniques.

Applying *Which method would you use to sand the surface of a wood table?*

Shaping and Turning

Shaping is the process by which a material is chipped away to change its shape. Chisels and planes are hand tools used for shaping. Shaping tools have cutters with chisel-like edges that chip away material.

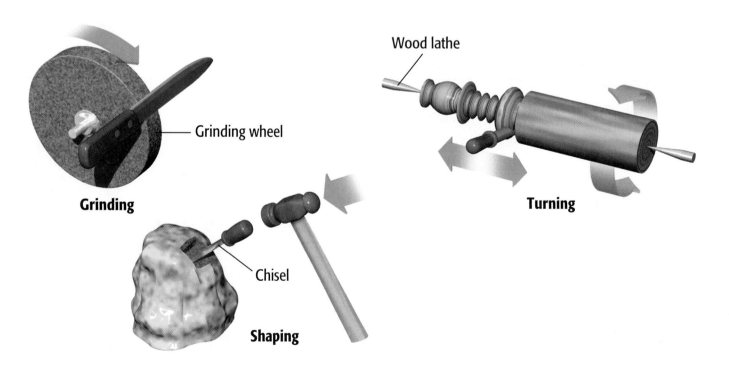

Grinding wheel

Grinding

Wood lathe

Turning

Chisel

Shaping

Figure 7.21 Filtering is a type of separating process. Here, it is being used in sewage treatment.

Clarifying *Why is filtering considered a separating process?*

Turning is another process used to shape materials. A turning tool is different from other shaping tools in that the tool itself does not move. In **turning**, the material to be shaped, or workpiece, is spun by a machine called a lathe. The cutting tool is held against the spinning workpiece.

Other Separating Processes

Filtering is a method of separating solids from liquids in a mixture (Figure 7.21). For example, you could separate the vegetables in a can of soup by pouring it through a strainer.

Some materials can be separated chemically, as when water is separated into oxygen and hydrogen. Salt can be removed from water by letting the water evaporate.

SECTION 3 Assessment

Recall and Comprehension

1. Define the process of forming. Give three examples in which forming is used.
2. Explain the different ways by which wood can be cut using a handsaw.
3. Describe the process of turning.

Critical Thinking

1. **Applying Concepts** If you wanted to make a baseball bat, which type of cutting machine would you use? Why?
2. **Contrasting** Pressing and casting are similar because they both use molds. How are these processes different?

QUICK ACTIVITY

Take some dry plaster of Paris and mix it with water, following the directions on the package. Pour it into a disposable aluminum pie pan. Notice how the temperature changes during the hardening process. Before it becomes solid, use the plaster of Paris as a mold for your hand. **For more related Design Activities, see pages 200–203.**

Combining and Conditioning Processes

Benchmarks for Learning

- Conditioning involves processing materials to improve their structures.
- Computer-controlled machines are capable of producing higher-quality goods than a craftsperson could produce.

Reading Strategy

Listing Make a list showing the different types of combining processes.

MECHANICAL FASTENERS	HEAT FASTENERS	GLUE FASTENERS
• Nails • Screws		

Vocabulary

rivet	soldering	hardening
nail	welding	tempering
screw	electroplating	annealing

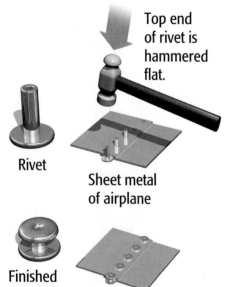

Figure 7.22 Rivets are metal fasteners used in airplane construction. They are also used to make jeans and other clothing.

Comparing *Why are rivets used in place of nails for metal fastening?*

Top end of rivet is hammered flat.

Rivet

Sheet metal of airplane

Finished rivet

Combining Materials Using Fasteners

Mechanical fasteners can be used to attach different materials together. Fasteners are made specifically for use with certain materials. Common fasteners include rivets, nails, and screws.

Rivets

Riveting is often used in building aircraft (Figure 7.22). A **rivet** holds pieces of sheet metal or other materials together. One end of a rivet is already formed. The other end is hammered and formed after it is placed through the two pieces of metal to be fastened.

Nails

A **nail** is a smooth metal piece used to fasten together two pieces of wood. Years ago, nails were forged one at a time, by hand. The end of an iron rod was heated until it became red-hot. It was hammered to a point, then snapped off at the desired length. Nails were costly because it took a long time to make them by hand. As a result, most houses were put together without nails. Today, automatic machines make nails by the thousands, and nails are much less expensive than years ago.

For the best holding power, nails should be driven into wood at a right angle to the grain. When you nail two boards

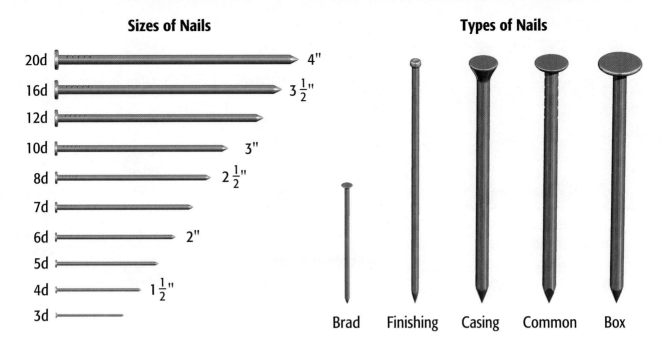

Sizes of Nails

20d — 4"
16d — 3½"
12d —
10d — 3"
8d — 2½"
7d —
6d — 2"
5d —
4d — 1½"
3d —

Types of Nails

Brad   Finishing   Casing   Common   Box

together, the nail should be long enough to travel two thirds of the way through the bottom piece of wood. If the wood is very hard, the nail might split it. To avoid this, a hole that is slightly smaller than the nail's diameter is drilled in the wood. This hole, called a pilot hole, makes it easier for the nail to enter the wood.

The five most common types of nails are brads, finishing nails, casing nails, common nails, and box nails (Figure 7.23). Nails are sold according to size. Nail size is measured in pennyweight, a term that once referred to the cost of 100 nails. An eight-penny nail (abbreviated 8d) once cost eight cents per 100. Now, *pennyweight* refers only to the length of nails.

Screws

A **screw** is a ridged metal piece that is used to pull one piece of material tightly against another. Screws provide more holding power than nails because of the ridges or threads that go down the end of the screw. They can be removed more easily so they are ideal for joining materials temporarily. Different types of screws are used for wood and metal (Figure 7.24).

Wood screws are tapered to a point and have threads that start at the point and wind about two thirds of the way to the head. Sheet metal screws, which are used to fasten pieces of thin sheet metal, are also tapered to a point, but they are threaded all the way to the head for more holding power. Pilot holes must be drilled for sheet metal screws because a sheet metal screw can't start itself in a piece of metal. A pilot hole should be the size of the body of the screw without its threads.

Figure 7.23 Different lengths and thicknesses of nails are used to join two pieces of wood.

Predicting *What type of nail would you use to build a birdhouse?*

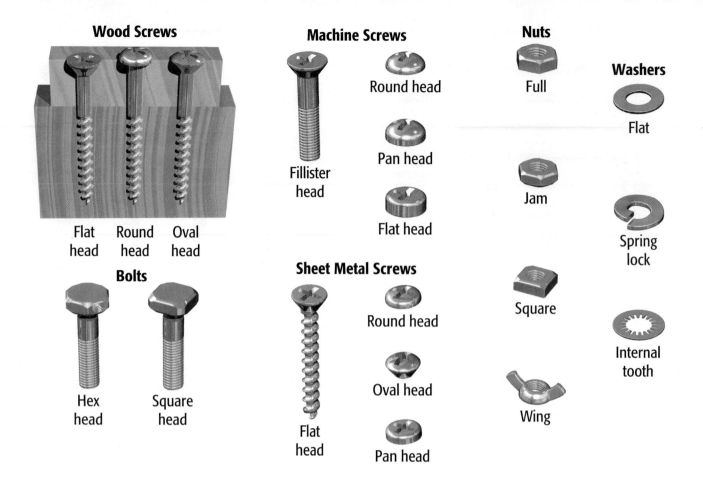

Wood Screws

Flat head   Round head   Oval head

Bolts

Hex head   Square head

Machine Screws

Fillister head   Round head   Pan head   Flat head

Sheet Metal Screws

Flat head   Round head   Oval head   Pan head

Nuts

Full   Jam   Square   Wing

Washers

Flat   Spring lock   Internal tooth

Figure 7.24 Screws are used to join wood, metal, glass, and other materials.

Contrasting *How do wood screws and machine screws differ?*

This is called the root diameter of the screw. Sheet metal screws cut a thread into the sheet metal as they are screwed in.

Machine screws and bolts do not taper to a point, and the thread is the same diameter from tip to head. Machine screws and bolts are held in place by a nut or a threaded hole. The nut has the same size thread as the machine screw or bolt. Washers are used between the nut and the material being fastened. Flat washers protect the material from being damaged by the nut. Lock washers keep the nut from loosening under vibration.

In driving screws, it is important to choose the right screwdriver. A screwdriver with too wide a blade will slip out of the screw slot and may damage the material being fastened. A screwdriver with too narrow a blade could damage the screw slot and not provide enough power to turn the screw. Never put your hand in line with the screwdriver blade because the screwdriver may slip and injure you.

Combining Materials Using Heat

Soft **soldering** is the process of joining metals with heat and soft solder. Soft solder is an alloy made from lead and tin. It melts at about 450°F. Soldering irons or guns are used to

Figure 7.25 In welding, metals are heated until hot enough to fuse together.

Comparing *How are soldering and welding similar?*

melt the solder. Soldering is the most common way to attach wires in electronic circuits.

Hard soldering involves the use of an alloy made from brass or silver. This material melts at about 1,400°F. An acetylene torch provides the heat source. Hard soldering is sometimes referred to as brazing or silver soldering.

Welding is another process in which heat is used to join metals. In **welding**, the metals to be joined are heated until they are hot enough to fuse, or melt, together (Figure 7.25). A material called welding rod is used like solder to help join the metals. Welding creates very strong bonds and requires temperatures of 6,000 to 7,000°F. In gas welding, the heat comes from a welding torch, which burns a mixture of gas and air. In arc welding, the heat comes from a machine that uses high electrical current.

Combining Materials Using Glue

Glue forms chemical bonds between itself and the materials being glued. Today, glues are manufactured for fastening almost every type of material. For wood, white or yellow glues are used. For plastic, metal, and ceramic, epoxy or superglue are best. For PVC (polyvinyl chloride) pipes, special glues that bond the atoms of the two pieces of pipe are used.

A good glue joint is stronger than the material it joins. Hot glues are applied with glue guns. These glues are very strong and set rapidly, allowing quick fastening. Such glues are now being used to hold airplane parts together. There is even a medical glue that doctors and surgeons use in place of stitches to close many types of wounds.

Combining Materials Using Coating

To beautify or protect its surface, a material can be coated with a specific finish. Paint, stain, wax, and glaze are different types of finishes that can be applied. Ceramic dishes are coated with glasslike glazes, which make the dishes easy to clean.

Metals can be coated with other metals by electroplating. **Electroplating** is a process in which electricity is used to form a thin metal coating on an object (Figure 7.26). Gold-plated jewelry and silver-plated tableware are made by this process. Aluminum is coated in a process called anodizing. Anodizing causes a thin oxide coating to form on the surface of the aluminum. Galvanizing is a process in which steel is coated with zinc to keep the steel from rusting.

Combining Materials into Composites

A composite is a material, formed by combining two or more other materials, that has distinctive properties possessed by none of the original materials. Wood, for example, is a natural composite. It is made of fibers of cellulose that are held together by a glue-like substance called lignin.

Thousands of years ago, the Egyptians added straw to the clay they used to make bricks. The straw made the bricks stronger. This was one of the first human-made composite materials. Today, people place steel rods or a steel reinforcing

Figure 7.26 Electroplating uses electricity to form a thin metal coating on an object.

Identifying *What characteristic makes copper ideal for electroplating?*

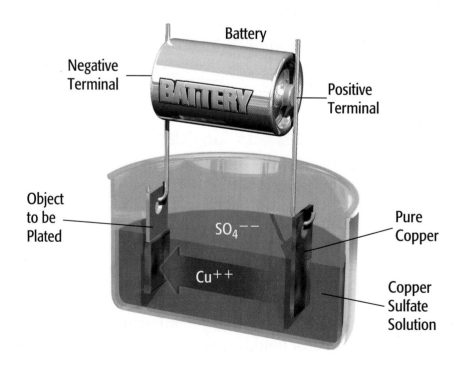

Battery

Negative Terminal

BATTERY

Positive Terminal

Object to be Plated

SO_4^{--}

Cu^{++}

Pure Copper

Copper Sulfate Solution

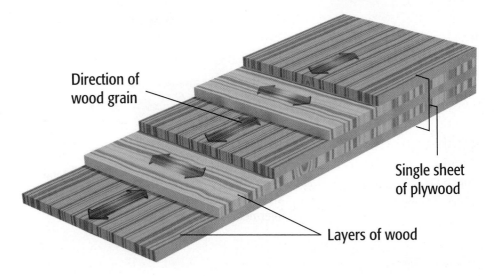

Direction of wood grain

Single sheet of plywood

Layers of wood

mat in concrete, which makes it much stronger when it is under tension. This is known as reinforced concrete. Reinforced concrete is a very popular material for foundations.

Technology has produced many other composites. Plywood is a commonly used composite (Figure 7.27). Because wood breaks more easily in the direction of the grain, plywood is made of layers of wood that have been glued together with the grain of each layer running in alternating directions. The composite material is much stronger than the wood from which it is made.

Fiberglass mats and cloth are composite materials made from glass and an epoxy or polyester resin. These composites may be stronger than steel but weigh much less and are easy to shape. Fiberglass-based composites are used to make boat hulls and the bodies of some automobiles.

Conditioning Processes

Conditioning materials changes their internal properties. For example, if a piece of steel is magnetized, the magnetizing force makes the molecules of the steel line up in one direction.

Heat-treating also causes changes within a material. When steel is heated red-hot and quickly cooled in water, it becomes harder. This process is called **hardening** (Figure 7.28). If the steel is heated again, not quite as hot, and again cooled quickly, the steel becomes less brittle. This process is called **tempering.** If the steel is heated red-hot and allowed to cool very slowly, the steel becomes softer. This process is called **annealing.**

When a piece of metal is hammered, it becomes harder in a process known as mechanical conditioning. The metal's crystal structure changes, getting longer and thinner.

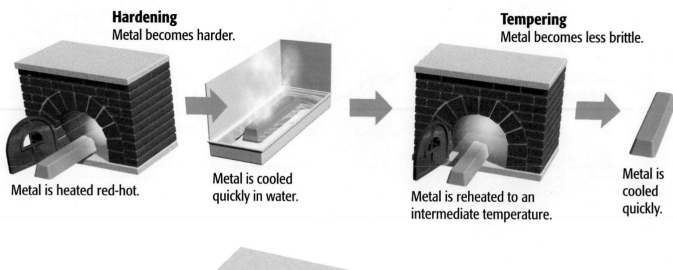

Hardening
Metal becomes harder.

Metal is heated red-hot.

Metal is cooled quickly in water.

Tempering
Metal becomes less brittle.

Metal is reheated to an intermediate temperature.

Metal is cooled quickly.

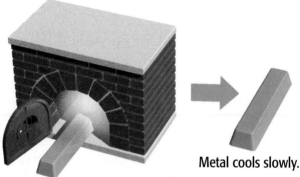

Annealing
Metal becomes softer.

Metal is heated red-hot.

Metal cools slowly.

Figure 7.28 Hardening, tempering, and annealing all use heat to change the properties of metal.

Interpreting *How many times has tempered metal been heated?*

When plaster and water are mixed, heat is given off because a chemical reaction takes place. The plaster hardens because of this chemical conditioning. When clay is fired in a kiln, it becomes harder and stronger.

In each example, the change takes place within the material itself. Here are some of the many other examples of conditioning:

- exposing photographic film with light
- developing an image on photographic paper
- baking a cake
- freezing or boiling water
- melting ice
- putting metal in liquid nitrogen to turn the metal into a superconductor
- using radiation on a tumor
- making butter from cream
- sending electricity through the filament wire of a light bulb

Each of these processes involves some change to the structure of the material.

Using Computers in Materials Processing

Computers can be used to control machines that process materials. When food is heated in a microwave oven, for example, the oven temperature and cooking time are controlled by a tiny computer.

In many factories, machines that process materials are also controlled by computers. Often, the computer is built right into the machine. Computers can be programmed to make a machine tool cut material along a specified path. They can also be reprogrammed. For example, if a company

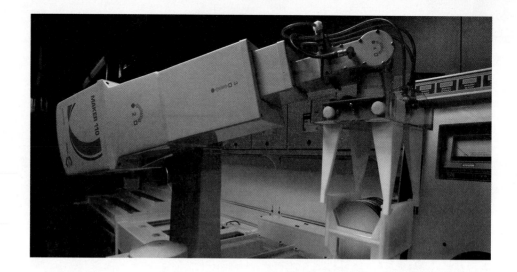

Figure 7.29 Robots can be programmed to carry out many of the tasks involved in materials processing.

Extending *What types of tasks involved in materials processing might be best suited for robots?*

decides to change a tractor's design, the machine that makes parts for the tractor can be reprogrammed to make the part differently. In the days before computer control, changing the way a machine worked could take a long time. A worker would have to spend hours or even days setting up and adjusting machines by hand.

Computer-controlled machines and robots are two examples of how computers are used for processing materials (Figure 7.29). Computers that direct the order of operations of a machine can often be programmed using simple computer languages. These languages use terms that are familiar to the people who use the machines.

Robots are sometimes programmed by having a person "show" the robot what it must do. With the robot in the "learn" mode of operation, the person uses a teach pendant, or remote controller, to take the robot through the movements needed to do a job (Figure 7.30). The robot stores the motions in its memory.

When commanded, the robot will go through the motions in the same order, again and again. One advantage of this programming method is that robots can be retrained to do a task by simply reprogramming the movements needed to do the job. Another advantage is that robots allow people to avoid handling dangerous materials. However, people are still needed to monitor and maintain the robots.

Stereolithography

Stereolithography, or printing/fabricating in three dimensions, allows custom manufacturing of products by taking a CAD rendering of the part. These can be new parts or those created by reverse engineering an old part that is no longer

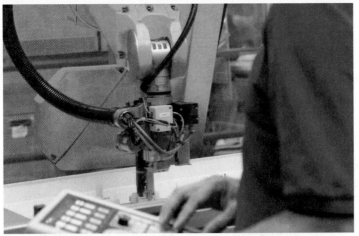

Figure 7.30

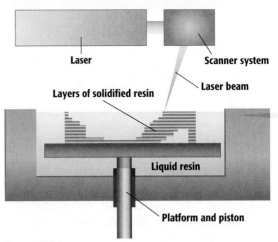

Laser
Scanner system
Laser beam
Layers of solidified resin
Liquid resin
Platform and piston

Figure 7.31

available. The images are sent to a digital fabricator, or printer. The fabricating software takes the CAD rendering and slices it into thin layers, perhaps one-tenth of a millimeter, and then each layer is printed, one upon the other. There are different techniques for the layering: one might use very fine powder that a binder is squirted onto, another might be layering with the material from a syringe-like object. Think of an inkjet printer, but instead of ink it uses a binder or gel material (Figure 7.31). Many, many passes are made as the layers are built up, creating the final object. The object is cured, depending on the materials, and ready for use. The objects can be made from metal, plastic, or food.

Figure 7.30 This engineer is using a remote control to program a robotic arm.

Identifying *How might a remote control increase safety when working with large equipment?*

Figure 7.31 An Illustration of a digital fabricator creating a new object by using many layers of material that are fused together.

SECTION 4 Assessment

Recall and Comprehension

1. What are nails and screws? Which are used to attach two pieces of wood?
2. Name different combining processes that use heat. Briefly describe each one.
3. Describe the end result of electroplating.

Critical Thinking

1. **Generalizing** When you want to combine two materials, how do you choose the best process to use?
2. **Describing** Why is plywood stronger than the original woods used to make it? Draw a diagram showing how a piece of plywood is constructed.

QUICK ACTIVITY

High-carbon steel can be hardened and softened using heat. Under your teacher's guidance, use pliers to hold a jigsaw blade over a torch until it becomes red-hot. Let it cool slowly. What happens when you bend the blade? Repeat the experiment, but this time quickly cool the blade in water. What happens? Why? **For more related Design Activities, see pages 200–203.**

Chapter Summary

- We process materials to make them more useful and more valuable. Technological resources such as people, capital, time, information, energy, and tools and machines are used to process materials.

- Primary materials are processed into industrial materials and then into end products.

- Materials are chosen on the basis of their properties. Mechanical properties include ductility, elasticity, plasticity, strength, and hardness. Other properties are electrical, magnetic, thermal, and optical.

- Materials are processed by forming, separating, combining, and conditioning.

- Forming a material means changing its shape without cutting it. Forming processes include casting, pressing, forging, extruding, blow molding, and vacuum forming.

- Separating processes separate one piece of material from another. Cutting is a separating process. We can cut materials by shearing, sawing, drilling, grinding, shaping, and turning.

- Some materials may be separated chemically or magnetically. Solids may be separated from liquids by filtering.

- Combining materials means putting one material together with others. We can combine materials by fastening, coating, or making composite materials.

- Conditioning materials means changing their internal properties. Conditioning processes include magnetizing, heat-treating, and mechanical and chemical conditioning.

Building Vocabulary

Your teacher may give you a crossword puzzle. Complete the puzzle using the following words from this chapter. Exchange puzzles with a classmate to check each other's answers.

1. annealing
2. blow molding
3. brittle
4. casting
5. ceramic
6. compression
7. conductor
8. drilling
9. ductile
10. elasticity
11. electroplating
12. extruding
13. ferrous metal
14. filtering
15. forging
16. forming
17. glaze
18. grinding
19. hardening
20. hardness
21. hardwood
22. industrial material
23. insulator
24. manufactured board
25. nail
26. nonferrous metal
27. optical property
28. plastic
29. plasticity
30. pressing
31. primary material
32. processing
33. rivet
34. sawing
35. screw
36. separating
37. shaping
38. shear
39. shearing
40. sintering
41. softwood
42. soldering
43. strength
44. tempering
45. tension
46. thermal property
47. thermoplastic
48. thermoset plastic
49. torsion
50. toughness
51. turning
52. vacuum forming
53. vulcanize
54. welding

See your teacher for the Crosstech puzzle.

Reviewing Content

1. Why must materials be processed?

2. Describe four ways in which materials are processed.

3. Give an example of a food that is processed from animal or vegetable material. Give two reasons why foods are processed.

4. List five ways of conditioning materials.

5. What happens to the internal structure of a piece of steel when it is magnetized?

Applying Your Knowledge

1. We plan to manufacture a jigsaw puzzle for 4-year-olds. We want to sell it for under two dollars. From what material might we make it?

2. Research the difference between stainless steel and regular steel. What are the properties of each? When would you use one type rather than the other?

3. What processes would you use to make hamburgers from raw meat?

4. What kind of fasteners would you use to attach a metal bracket to a wooden shelf?

5. Go to a hardware store and examine differently sized nails. Notice the difference between common nails and finishing nails. How many different sizes were you able to find?

Critical Thinking

1. **Generalizing** Explain why materials are chosen for products on the basis of their properties.

2. **Making Judgments** You need a screwdriver for electrical work. What kind of material should be used to make the handle? Why?

3. **Comparing and Contrasting** What is the difference between torsion and shear? Between tension and compression? How are these actions similar?

4. **Contrasting** How do sawing, drilling, and grinding differ?

5. **Analyzing** This person is using a welding torch. Why would you choose to weld two pieces of metal instead of gluing them or using screws?

 Connecting to STEM
science • technology • engineering • math

Abrasives, Up Close

Take a piece of metal and examine its surface under a microscope. Use emery cloth to rub the surface of the metal. Use different grades of emery cloth on different parts of the metal. Observe the surface again. What did you find? Why do you think different grades of emery cloth exist?

Design Activity 13

GAME BOARD TABLETOP

Problem Situation

You have received a new contract with a local restaurant. They want some unique and practical tables. The owner thinks more people would come to the restaurant if they could eat and socialize. How might you build a tabletop that could serve as a game board before and after food is served?

Materials

You will need:

- adhesives
- clear polyester resin
- plate glass, waxed and polished with a release agent
- resin wax release
- wood, plastics, or ceramics

Your Challenge

You and your teammates are to design and construct a scale model of a tabletop that can be used as a game board.

> Go to your **Student Activity Guide, Design Activity 13.** Complete the activity in the Guide, and state what the design challenge is in your own words.

① Clarify the Design Specifications and Constraints

To solve the problem, your design must meet the following specifications and constraints:

- The size of the scale model of the game board tabletop can be no larger than 12" by 12" and no smaller than 6" by 6".
- Use materials (wood, plastics, or ceramics) that are appropriate for the type of game board you select.
- Use polyester resin to cover the game board.

> In your Guide, state what the specifications and constraints are. Include any others that your team or your teacher included.

② Research and Investigate

To better complete the design challenge, you need to first gather information to help you build a knowledge base.

> In your Guide, complete Knowledge and Skill Builder 1: Tabletop and Game Board Investigations.

> In your Guide, complete Knowledge and Skill Builder 2: Materials Investigation.

In your Guide, complete Knowledge and Skill Builder 3: Scaling.

In your Guide, complete Knowledge and Skill Builder 4: Resin Requirements.

③ Generate Alternative Designs

In your Guide, describe two of your possible solutions to the problem.

④ Choose and Justify the Optimal Solution

Refer to your Guide. Explain why you selected the solution you did, and why it was the better choice.

⑤ Develop a Prototype

Make a scale drawing of the tabletop with the game board, including details of the game board. Then, construct your tabletop. Include a photograph of your final design in your Guide.

In any technological activity, you will use seven resources: people, capital, time, information, energy, materials, and tools and machines. In your Guide, indicate which resources were most important in this activity, and how you made trade-offs among them.

⑥ Test and Evaluate

How will you test and evaluate your design? In your Guide describe the testing procedure and explain how the results show that the design solves the problem and meets the specifications and constraints.

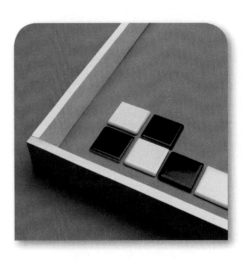

⑦ Redesign the Solution

Respond to the questions in your Guide about how you would redesign your solution based on the knowledge and information that you gained during the activity.

⑧ Communicate Your Achievements

In your Guide, describe the plan you will use to present your solution to your class and include any handouts and/or PowerPoint slides you will use.

Design Activity 14
CASTING

Problem Situation

Many articles that have to be reproduced over and over are made from molds. A liquid is poured into a mold, the liquid hardens, the mold is removed, and a solid replica of the object remains. This manufacturing process is used not only for intake manifolds on cars but also for ceramic objects and even jewelry.

Your Challenge

You and your team members are to design an artifact such as a ring, pendant, or bracelet from wood, plastic, or metal and create an exact replica of it using a casting.

Go to your **Student Activity Guide, Design Activity 14.** Complete the activity in the Guide, and state what the design challenge is in your own words.

① Clarify the Design Specifications and Constraints

To solve the problem, your design must meet the following specifications and constraints.
- The object should be no longer than 2 inches, nor wider than 1 inch, and no thicker than 1/2 inch.

② Research and Investigate

To better complete the design challenge, you need to first gather information to help you build a knowledge base.

In your Guide, complete Knowledge and Skill Builder 1: Investigate Molds We Use Every Day.

In your Guide, complete Knowledge and Skill Builder 2: Preparing the Cuttlebone Casting.

Materials

You will need:
- cuttlefish bone (for creating the mold)
- charcoal
- graphite
- lead-free pewter
- modeling clay
- #220 abrasive paper
- steel wire
- 3/16" dowel rod

③ Generate Alternative Designs

In your Guide, describe two of your possible solutions to the problem.

④ Choose and Justify the Optimal Solution

Refer to your Guide. Explain why you selected the solution you did, and why it was the better choice.

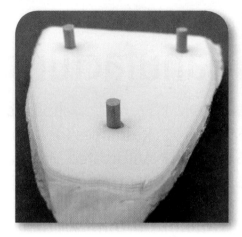

⑤ Develop a Prototype

Construct your design. Include a sketch or photograph of it in your Guide.

In any technological activity, you will use seven resources: people, capital, time, information, energy, materials, and tools and machines. In your Guide, indicate which resources were most important in this activity, and how you made trade-offs among them.

⑥ Test and Evaluate

Did your design meet the initial specifications and constraints? Indicate the tests you did and/or the experiments you performed to verify this.

⑦ Redesign the Solution

Respond to the questions in your Guide about how you would redesign your solution based on the knowledge and information that you gained during the activity.

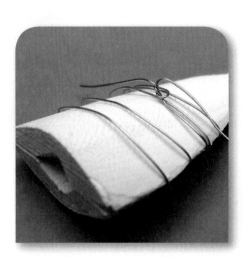

⑧ Communicate Your Achievements

In your Guide, describe the plan you will use to present your solution to your class and include any handouts and/or PowerPoint slides that you will use.

Manufacturing

In this chapter you will learn about the process of manufacturing as follows:
- the craft approach
- the factory system
- interchangeable parts
- mass production
- automated manufacturing

In the early twentieth century, Henry Ford used new technology to manufacture cars in a very short time. He set up one of the first assembly lines. People were trained to work in his factory. They learned how to put together the different parts that made up every car.

Even though technology has changed dramatically since then, the basic concept behind the Ford assembly line is still used.

• Parts are moved along a belt.

• Parts are put together one piece at a time.

• Many identical products are made.

What is most different today is who—or what—is assembling the parts. A century ago, people worked on assembly lines. Today, more and more of the manufacturing process has been taken over by robots. Have people been replaced? No, but the work they do has changed. Now, people program the computers and make sure all the machinery is running properly. They manage the entire manufacturing system.

From Workshops to Factories

Benchmarks for Learning

- Manufacturing is the process of making goods in a workshop or factory.
- Goods may be classified as durable or nondurable.
- Mass production and the factory system lowered the prices of goods, making them affordable to more people.

Reading Strategy

Listing   As you read this section, make two lists to compare craft manufacturing and mass production. Write down as many features as you can.

1. Craft Manufacturing
- Products made one at a time
- People work at home

2. Mass Production
- Products made in a factory

Vocabulary

manufacturing	craft approach	mass production
durable	interchangeable part	assembly line
nondurable		

Manufacturing

People have been using manufacturing technology for thousands of years. Evidence shows that people were making pottery from clay more than 30,000 years ago. The firing process was probably discovered by accident, when people found hardened clay under the ashes of their cooking fires. Once

Figure 8.1   These clocks are examples of long-lasting consumer products.

Distinguishing   *What properties are often associated with durable products?*

they discovered this process, people began to fire clay items on purpose. This is one of the first examples of **manufacturing**, which is the process of making goods by hand or machine.

Today, all types of products are manufactured in workshops and in factories (Figure 8.1). Some products, such as household appliances and vehicles, are **durable**—they last a long time. Others, such as foods and paper products, are **nondurable**—they are designed to be consumed in a short period of time. Both durable and nondurable products are manufactured in factories or in workshops, using the craft approach.

The Craft Approach

For centuries, people made objects only for themselves or their families. They made their own clothing and tools, and they built the houses in which they lived. Gradually, however, people began to specialize. Shoemakers made shoes not only for family and personal use but also for people in the community. In return, they received goods from farmers and other craftspeople in the area. Different people in the community made candles, cloth, wool, silverware, ceramics, and other items. The craft approach had begun.

The **craft approach** is the method of making products one at a time, from start to finish. Although most manufacturing is automated today, craft production continues. Most often, craftspeople work alone at home or in small workshops (Figure 8.2). Often, a master craftsperson hires a young

Figure 8.2 Many people are willing to spend extra money for handmade goods.

Inferring *Why are handmade products usually more expensive than machine-made items?*

person who wants to learn the trade. These young workers, called apprentices, are trained on the job.

Many people are still willing to pay for one-of-a-kind products. The craft approach allows products to be custom-made to meet the particular needs or desires of the purchaser. People with disabilities, for example, often buy custom-made furniture and other items that are designed to help them carry out daily activities.

The Factory System

Until the mid-1700s, all types of goods were made by craftspeople. People worked at home or in small workshops. As America grew, railroads, canals, and highways opened up new markets. More goods were needed, and transportation was needed to move goods to different areas of the country.

The Industrial Revolution, which started in the late 1700s, was a period during which many new machines and devices were invented. Eventually, machines such as the steam engine, the power loom, and the sewing machine reduced the time and effort needed to carry out many tasks. Businesspeople took advantage of machinery to improve production. Soon, goods were being made in factories that hired many workers.

Interchangeable Parts

In 1789, Eli Whitney signed a contract with the U.S. Army to make 10,000 rifles in 2 years. Ten thousand was a huge number to produce in those days! Up to that time, rifles had been made one at a time, with each part customized for each rifle. Whitney succeeded in his task by making large batches of each of the different rifle parts. These parts could be put

Figure 8.3 With interchangeable parts, entire machines can be assembled in minutes or even seconds.

Analyzing *What resources are saved by using interchangeable parts in an assembly line?*

Comparing Craft Manufacturing and Mass Production	
Craft Production	**Mass Production**
Workers are highly skilled.	Workers need only limited skills.
Workers make a product themselves, from start to finish.	Workers perform only one part of assembly. They do not see the finished product.
The work is varied and involves different tasks.	The work is routine and very repetitive.
Only one item is produced at a time.	Many items are produced during the production run.
Production is fairly slow.	Items are made relatively quickly.
Each part is handcrafted, so no two are exactly alike.	Parts are machine made and are interchangeable.
The products are usually expensive.	Mass-produced items are relatively inexpensive.
Quality depends mainly on the skill of the craftsperson.	Quality depends mainly on the accuracy of the machines and how well they have been set up by people.

together quickly and easily, thereby increasing production and saving time.

In Whitney's system, each batch of each part was kept in a separate bin. The parts were standardized—that is, all the parts in a batch were exactly alike. They were also interchangeable. **Interchangeable parts** are manufactured to be exactly alike so that any one can be substituted for any other (Figure 8.3).

Mass Production and the Assembly Line

Interchangeable parts became one of the most important characteristics of a mass-production system. **Mass production** is the production of goods in large quantities by groups of workers in factories (Figure 8.4). The production process is divided into steps. Each worker does one step and then passes the item on for the next step. Work is carried out on the assembly line itself and as one task is finished, another begins immediately so time is not wasted in setting things up. An **assembly line** is a system in which each worker performs a specific operation in assembling an item as it is passed along, often on a slowly moving belt, from one workstation to the next.

One of the first moving assembly lines was constructed at the Ford Motor Company in 1913. By applying the same principle that Eli Whitney had used for making rifles to the assembly of a complete car, Henry Ford speeded up automobile production. Auto parts were pushed from one worker to the next, and production time was reduced by about one half. A finished Model-T Ford came off the assembly line every 10 seconds. Today, modern automobiles are manufactured from interchangeable parts.

Figure 8.4 Both craft production and mass production are used today to make a variety of products.

Comparing *How are craft manufacturing and mass production similar?*

LIVING GREEN
Nuts about Packaging Shapes

Have you ever gone shopping and bought something you've been waiting forever to buy, just to get it home and struggle to get it out of the package? As manufacturers were able to increase their production of goods, they also wanted to protect their growing supply of stock. Also, as catalog and Internet shopping gained popularity, so did the need to safely package goods for shipment over long distances. This gave rise to an entirely new field of engineering—packaging design. This also meant an increase in the use of packaging resources such as cardboard, plastic bubble sheet, and Styrofoam packing peanuts.

MAKE A CHANGE
Though it may be difficult for any individual student to change the practices of large manufacturers, there are things you can do to reduce the effects of excessive packaging on the environment. First, recycle all the cardboard and plastics possible from the packages that your new goods come in, whether you buy them at the store or they are shipped to you in the mail. For the materials that can't be recycled, such as the packing peanuts, find a local business that is willing to take them and reuse them.

Also note the shapes of packages, such as cereal boxes. Quaker Oats oatmeal is packed in a cylindrical container. Cheerios are packed in containers that are rectangular in shape. The shape matters! If both boxes hold the same volume of cereal, the surface area of the cylinder is less than the surface area of the rectangle. Therefore, cylindrical containers use less material.

TRY THIS
Here are two things you can try: 1) The next time you're out shopping, look around for the item that appears to have the most complicated packaging design. Look carefully at how the package design protects the product from damage or even keeps it fresh, if it's a food product. Then, redesign the package using less material and/or more environmentally friendly packaging. Make sure to include in your design all your specification for size, shape, and materials. You can even build a prototype and test it out, or send your ideas to the manufacturer.

2) Note the shapes and measure the dimensions of several cereal boxes. Calculate the volume and the surface area of each using formulas learned in mathematics class. Of these boxes, which has the least surface area to volume ratio, and therefore, which would use the least materials?

Impacts of the Factory System

Over the years, the factory system replaced the craft approach in the manufacture of most goods. It also led to the development of other industries. The steel, car, and clothing industries provided millions of new jobs.

The factory system allowed more goods to be produced for less cost in a given amount of time. It also created wealth by adding value to the resources that were processed into goods. Luxury items became less expensive. More people could afford to buy the products of mass production. The standard of living improved for people who lived in industrialized nations.

During the early days of mass production, most businesspeople wanted only to improve the process to make a product cheaper and faster. They were not concerned with providing people a number of different choices when it came to the products. They wanted to offer a standard product at a low price. A suggestion was once made to Henry Ford that he paint his Model-Ts different colors. He replied, "Give it to them in any color, so long as it is black."

This attitude changed with competition. In the 1930s, General Motors started producing a new car model every year. The way cars were advertised and sold was suddenly as big a concern as how they were produced. Marketing and business management became important considerations in manufacturing. Today, companies that can deliver the right product, at the right time, at the right price have an advantage over their competitors.

SECTION 1 Assessment

Recall and Comprehension

1. List five durable and five nondurable manufactured products.
2. What are two disadvantages of mass production?
3. Why do manufacturers want their products to be uniform?
4. Who was Henry Ford and what did he contribute to manufacturing technology?

Critical Thinking

1. **Hypothesizing** Give two examples of manufactured goods that could be made in a home workshop. If thousands of each of these products were to be made, how would the production methods change?
2. **Summarizing** How did the factory system improve people's standard of living?

QUICK ACTIVITY

Because mass production has made many products less expensive, it is often cheaper to replace them than to have them repaired. With a little effort, these items can be recycled. Select a small product that has been discarded and, with your teacher's approval, disassemble it into its component parts. Identify as many of the parts as you can. **For more related Design Activities, see pages 234–237.**

People in Technology

A History of Craftsmanship
The C. F. Martin Story

"A good guitar cannot be built for the price of a poor one, but who regrets the extra cost for a good guitar?"

—F. H. Martin

If you have any doubt that craftsmanship is alive and well, you should visit Nazareth, Pennsylvania. This is the home of one of the most famous guitar manufacturers in the United States—the Martin Guitar Company.

Continuing a Family Tradition

The founder of the company, Christian Frederick (C. F.) Martin, Sr., was born in Germany in 1796 into a family of guitar makers.

Many musicians choose to have their guitars handmade because of the attention given to quality and detail.

He left his home country in 1833 and set sail for New York City, where he set up a small guitar-making shop on the Lower East Side.

C. F. Martin & Co

In 1838, Martin sold his shop in New York and purchased land for a new location in Nazareth. Six generations later, the Martin family is still in Nazareth, making guitars in a modern factory.

Handmade—With the Help of Computers

The early Martin guitars were totally handmade, one-of-a-kind products. In the late 1800s, however, the company evolved from a one-man craft manufacturing operation into a factory system that employed many craftspeople. Today, hand craftsmanship remains very important in building Martin guitars. New computer-aided manufacturing methods are also used to improve the processing and flow of materials so that more guitars can be made without sacrificing quality.

In the 1904 Martin catalog, C. F. Martin's grandson Frank Henry Martin wrote, "[Building] a guitar to give this tone takes care and patience. A good guitar cannot be built for the price of a poor one, but who regrets the extra cost for a good guitar?" One century later, the company's attitude toward guitar building remains the same.

Critical Thinking

1. **Clarifying** Why do you think musicians especially like handmade instruments?

2. **Contrasting** What does it mean when a product is one of a kind?

Resources for Manufacturing

Reading Strategy

Listing As you read the section, make a list of the seven types of resources used in manufacturing. For each resource, list as many examples as you can.

1. People
- factory workers
- managers
- entrepreneurs

Vocabulary

quality circle   venture capitalist   productivity
entrepreneur

Benchmarks for Learning

- Manufacturing systems use the seven technological resources.
- An entrepreneur comes up with a good idea and uses that idea to make money.
- Raw materials are made into basic industrial materials, which are made into finished products.

Resources

Manufacturing systems use the seven technological resources to make a product. As primary resources are processed and assembled into a final product, their value increases. The finished product is worth much more than the individual resources.

People

People are involved in every stage of manufacturing. They design the products. They decide how the products should be made. They choose the materials and the best tools and machines to manufacture the products. They organize the production lines and obtain the necessary capital. They advertise, distribute, and sell the products.

In the past, people provided more of the labor in manufacturing plants than they do today. Today, fewer people are needed to manufacture goods. Computers and more sophisticated machinery have reduced the number of factory workers needed.

Manufacturing has also become more competitive. Industries in the United States have had a hard time competing with foreign companies whose workers are paid much less. To remain in business, U.S. industries have been forced to either improve their production methods or lay off workers.

As a result, more than 300,000 American steelworkers have lost their jobs over the last 25 years. Workers who lose manufacturing jobs often find that without further training, they can get work only in lower-paying service industries.

Quality Circles and Teamwork

The roles of people in the manufacturing system have been changing. As workers have become more educated, they have been asked to take part in decisions that affect the company. **Quality circles** are groups of workers and managers who meet to discuss problems and ways of improving production. During these meetings, management can explain problems to workers about costs, profits, and competition. Quality circles allow workers and management to discuss issues openly. The result is a better work environment and improved production.

Quality circles use many attributes of effective teams, such as respect for one another, careful attention to others' opinions, and active participation by all members of the group. Whether a team is made up of employees, students, or athletes, a successful team shares leadership roles and responsibilities so everyone is empowered by the experience.

Thinking Up New Ideas

Entrepreneurs are a critical type of human resource for manufacturing. An **entrepreneur** is a person who comes up with a good idea and uses that idea to make money. He or she might start a business to fill a specific need, improve the way a product is made or sold, or even come up with an idea for a new product. A recent entrepreneur who has had great success is Michael Dell, who founded his own computer company (Figure 8.5).

Some entrepreneurs are inventors. An inventor comes up with a totally new idea—a safety razor, a laser beam, or a contact lens. Other types of entrepreneurs are innovators in that they improve a product or create a new way to accomplish something. An innovation is an improvement on an invention. Innovations can create new industries or change existing ones. The electric guitar, the diesel engine, and power steering are all types of innovations.

Capital

Capital is an important resource in the development of technology. The reason some technologies evolve and others do not can sometimes be related to the amount of available capital. Companies must have capital to finance their

Figure 8.5 Dell Computers was the first company to sell custom-built computers directly to consumers.

Analyzing *Why is Michael Dell considered an entrepreneur?*

operations. They must buy land, build factories, purchase equipment, pay workers, maintain machines, and advertise their products.

Capital is often obtained by selling shares of company stock to the public. Stockholders, or people who buy shares of stock, become partners in the corporation. If the company makes a profit, the value of its stock may go up.

Private companies may raise money from investors who contribute capital. A venture, like an adventure, is a trip into the unknown. A **venture capitalist** is a person who supplies money to finance the start-up of a new company. Venture capitalists sometimes take big risks because they expect to make large profits once a new company starts its operations (Figure 8.6).

Time

In manufacturing, time is money. For a given product and level of quality, the faster it is made, the more profit the company will receive. **Productivity** refers to how quickly and cost-effectively a product is made. A large part of a product's cost is the cost of paying workers. If workers do their jobs faster, productivity generally increases because more products are made for the same labor cost.

Information

Information is critical to all stages of manufacturing. To be successful, companies must find out what consumers will buy and how their tastes are changing. They must also

Figure 8.6 Venture capitalists are interested in the financial success of companies in which they invest.

Hypothesizing *Why might a venture capitalist choose to invest in a new company rather than the stock market?*

gather information about the costs of materials. This information helps managers choose which materials to buy. Companies use information about production to help them adjust the manufacturing process. They also use university research to learn about new materials and production methods.

Energy

About 40 percent of the energy consumed in the United States is used in manufacturing. Some of the biggest energy users are factories that make metals, chemicals, ceramics, paper, food, and equipment. Most manufacturers use electricity from fossil fuels (coal, oil, and natural gas). Others draw on alternate sources of electricity such as hydroelectric and nuclear energy plants. Factories are often built in places where energy resources are abundant and costs are low. In the late 1800s, natural gas became accepted as the best fuel for melting glass. As a result, the glass industry grew up in West Virginia because the state had plenty of natural gas. Small steel mills that use electric furnaces are built in places where electricity is cheap.

Some industries take advantage of the heat given off during manufacturing. The paper-making industry, for example, reuses the energy to heat water and make steam. The steam is used to turn steam turbines and produce electricity.

Figure 8.7 If goods have to be transported over great distances, their cost is greater.

Hypothesizing *Why is it sometimes necessary to transport materials over long distances?*

Materials

Through manufacturing, raw materials are made into basic industrial materials and then finished products. For example, the steel industry uses coal, limestone, and iron ore to make steel.

The cost of raw materials is important. The cost includes not only the material itself but any shipping or processing that is required (Figure 8.7). Suppose a steel company in Indiana needs iron ore. Should it buy the ore from a mine in Wisconsin, where it is only 20 percent pure, or should it pay the extra shipping charges to buy the ore from Brazil, where it is 65 percent pure? This is the kind of cost-benefit trade-off that manufacturers must consider in choosing materials.

Rapid Prototyping

A prototype is an original model that can be used as the basis for making more of the item, often using more efficient, mass production methods. Rapid prototyping is a method for creating these models that works from a design created by CAD software. There are a variety of rapid prototyping technologies, such as stereolithography, 3-D printing, and Fused Deposition Modeling. These methods use expensive machines to create an object by laying down multiple layers of material and building up thickness until the object is complete. The various methods use different materials. Three-D printing mixes ink and plastic and works like an ink-jet printer, spraying the ink and plastic mixture on a surface. Stereolithography uses light-sensitive plastic and lasers to melt thin layers of plastic together. Other methods use metal or ceramic powders to create the layers. Rapid prototyping is currently a very expensive technology. The machines and the plastics that they use are costly. The cutting edge of this technology, though, is machines that can make their own parts. Once this is accomplished, the technology should become much more affordable.

Tools and Machines

The tools used in modern factories are advanced and precise. Most of them are automatic. Computer programs and sensing devices provide feedback and guide their operation.

SECTION 2 Assessment

Recall and Comprehension

1. Pick an industry that manufactures a product of your choosing. Explain how each of the seven resources is used in the production of that item.
2. What are the main sources of energy for most manufacturers?
3. Explain how the saying "time is money" relates to manufacturing.

Critical Thinking

1. **Making Judgments** If you were part of a quality circle with your classmates and teacher, what improvements would you suggest for the technology laboratory?
2. **Extending** Identify an entrepreneur in your community. Describe how that person used a good idea to make money.

QUICK ACTIVITY

Entrepreneurs sometimes make money by changing existing products. Select a product that you think could be improved, and then write a brief explanation of what your modification would be. Use drawings and/or models to present your idea to the class.
For more related Design Activities, see pages 234–237.

Automated Manufacturing

Benchmarks for Learning

- Automation is a manufacturing system in which machines are controlled automatically.
- Robots are automated machines that are controlled by computers.
- Computer-aided manufacturing has improved product quality and lowered manufacturing costs.

Reading Strategy

Outlining As you read, make an outline of the section content. Write down each of the headings while organizing them into an outline. Be sure to pay attention to how different colored headings relate to each other.

Vocabulary

automation	computer-integrated	agile manufacturing
computer-aided	manufacturing	just-in-time
manufacturing	flexible manufacturing	manufacturing

Figure 8.8 Robots have sensors that allow them to "feel" and "see" things around them.

Extending *What other capabilities might be invented for robots in the future?*

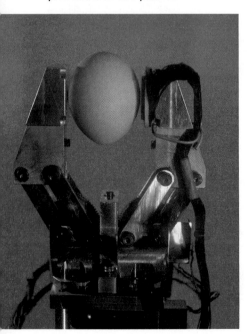

What Is Automation?

A robot can be programmed to pick up a part, move it a certain distance, and drop it into a bin. This simple process, done entirely by a machine, is an example of automation. **Automation** is a system in which machines are controlled automatically. In the example of the robot, a list of instructions is used to direct a machine's activity in a process called program control.

Using Feedback

Feedback control uses feedback to adjust the way a machine is working. Feedback control depends on a sensor, a device that gathers information about its environment. For example, a sensor can detect when a drill has drilled deep enough. At this point, it sends a signal to switch off the drill. A machine operator is no longer needed to turn the drill on and off. In automated factories, these types of sensors also provide quality control. The sensors make sure the machines perform their operations to exact specifications (Figure 8.8).

Lower Costs

Automation saves money on labor costs. To compete with lower-priced foreign products, more and more factories are being automated. Although this change requires extra capital at the start, the money is usually regained over

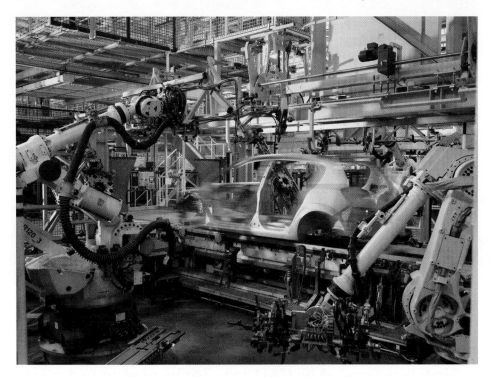

Figure 8.9 Robots and computers are at the center of automated manufacturing.

Contrasting *How do these robots save on the seven resources in comparison to people working the assembly line?*

time. Automated systems require fewer people to run and maintain them, which also helps conserve capital.

The Robot Revolution

You may have heard of the term "robot revolution." Why is the use of robots considered revolutionary? Because robots that exist today can actually take the place of many workers in a manufacturing plant, they are changing the way manufacturing is done. Using just the right amount of pressure, robots can hold and manipulate delicate parts. They have sensors that allow them to "feel" things around them. Other robots can "see" with television eyes to identify differently shaped parts.

Unlike people, robots can work 24 hours a day without tiring. Most of the work they perform relieves people from dangerous, boring, heavy, or unpleasant jobs (Figure 8.9). Robots are used for loading objects onto platforms and conveyors. Because robots are computer-controlled, they can also be reprogrammed. For example, a robot can be programmed to weld a fender on one kind of automobile and then reprogrammed to do the same job on a different model.

The use of robots has allowed manufacturers to make products of higher quality at lower cost. In many factories, robots are taking the place of people, although some workers are still needed to make sure that everything is operating the way it should. People are also needed to install, service, and program the robots.

Types of Automated Manufacturing

Some manufacturing methods lower costs and create a better product. The different types of manufacturing used today include different types of computerized manufacturing as well as flexible manufacturing, agile manufacturing, and just-in-time manufacturing.

CAD and Computer Modeling

Computer-aided design (CAD) is used in many stages of manufacturing. When a new product is being designed, CAD is used to create computer-generated models that can be viewed and moved around on screen.

CAD systems can also calculate manufacturing costs. Prices for individual components and materials are programmed into the computer. When the designer selects a component for the design, the computer automatically creates a manufacturing cost estimate.

Models generated by CAD can also be used to test a new design. CAD performs simulations, or tests, of the new design, using information on the operation of different moving parts. No physical model or prototype is needed. If the test shows that a part will not work as planned, it can be altered quickly on-screen and then retested. This reduction in design time helps companies get their products to consumers as fast as possible.

Computer-Aided Manufacturing

Computer-aided manufacturing (CAM) is a software application that uses computers to control factory machines. CAD/CAM is a technology that joins CAD and CAM applications. It lets a person create a design on a computer screen and then send it directly to a machine tool (Figure 8.10). The machine can make a prototype of the part. It can also make many copies of the final product.

Computer-Integrated Manufacturing

Similar to CAD/CAM technology, **computer-integrated manufacturing (CIM)** uses computers not only for design and manufacturing, but also for business needs. It is used to store information about raw materials and parts, to set times for the purchase and delivery of materials, and to report on finished goods. CIM is also used for billing and accounting. Managers involved in purchasing, shipping, accounting, and manufacturing can get a total picture of all the processes by looking at the computer screen.

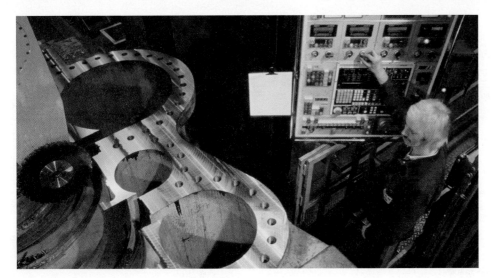

Figure 8.10 CAD/CAM technology allows designs to be sent directly to the machine that makes the part.

Inferring *How might using CAD/CAM help reduce manufacturing costs?*

Flexible Manufacturing

Today, many products are made in batches of a hundred or a thousand, rather than millions. Making products in smaller batches allows companies to cater to customers' special needs. For example, the General Electric Company in New Hampshire makes 2,000 different versions of its basic electric meter.

This type of customized manufacturing is possible with flexible manufacturing. **Flexible manufacturing** is the efficient production of small numbers of products. The same production line is used for each version of a product. To make the different versions, the machines are simply programmed to do different operations.

The John Deere factory in Iowa can make 5,000 different versions of its tractor for farmers with different needs. Each tractor may have different components, or it may be a slightly different size. Computers are used to program the manufacturing process so different parts can be assembled with each production run.

Agile Manufacturing

The word *agile* means quick, or possessing ease of movement. **Agile manufacturing** is an organizational strategy in which decisions are made quickly and a greater variety of products is produced quickly and inexpensively. Often, it includes a production system capable of making different products simultaneously.

One key influence on agile manufacturing is the role of the customer. From the beginning of the product-delivery cycle, the agile manufacturing strategy is to understand what customers want, how much they will spend on the product, and how the product can be improved and customized to suit customers better.

For: Assembly Line Activity
Visit: www.mytechedkit.com

Getting the Order Right Online

Making sure that products are uniformly made and of high quality is an important goal for manufacturers. Whether it is a complex product such as a computer or a simple product like this pen, consumers expect it to have no defects. Go online to see how sequencing an assembly line properly is the key to achieving this goal.

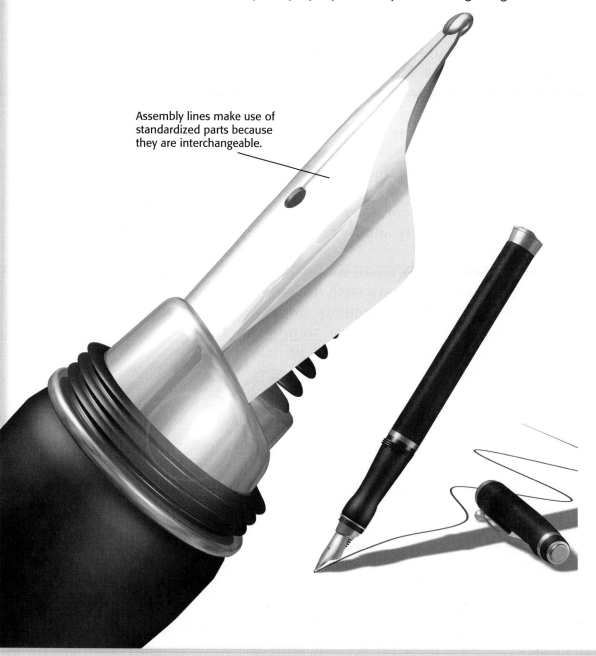

Assembly lines make use of standardized parts because they are interchangeable.

Figure 8.11 Just-in-time manufacturing keeps storage to a minimum. Materials arrive at the factory only when they are needed.

Predicting *What industries are best suited for just-in-time manufacturing?*

Just-in-Time Manufacturing

Just-in-time manufacturing uses careful scheduling to keep materials and products in storage for as little time as possible. In other types of manufacturing, raw materials are most often delivered in large amounts that take up a lot of costly storage space. Storage space can be costly because companies must pay to rent, heat, and light the space (Figure 8.11).

In just-in-time manufacturing, materials and parts are ordered so that they arrive directly at the factory only when they are needed. It also requires that equipment and people are made available when the materials arrive. Once the final product is made, it is shipped immediately to the customer. There is little need for storage space or for workers to stack and store parts in warehouses.

SECTION 3 Assessment

Recall and Comprehension

1. What is meant by automation?
2. Discuss two advantages and two disadvantages of using robots instead of people in manufacturing plants.
3. Explain how an agile manufacturing system works.

Critical Thinking

1. **Analyzing** How have robots improved product quality while bringing manufacturing costs down?
2. **Predicting** New production systems make increasing use of technologies such as robotics, CAD/CAM, and CIM. How will this affect you as a future worker and consumer?

QUICK ACTIVITY

For some applications, miniature robots may be a better choice than larger, more expensive devices. Using materials supplied by your teacher, build a miniature model robot that is specialized for a single task. Label the parts and write a brief explanation of how a working prototype would perform. **For more related Design Activities, see pages 234–237.**

Connecting to STEM

Lasers

Today, lasers are used in products as dissimilar as a CD player and surgical tools. Why have lasers become so important in technology?

A Different Kind of Light

All the light in a laser beam has the same wavelength and travels in the same direction. It is called coherent light. Because coherent light is so uniform, it is powerful and can travel long distances.

Just like a flashlight, a laser produces a beam of light. However, the types of light produced by a laser and a flashlight are vastly different. Ordinary light has many different wavelengths, all moving at various speeds and directions. This is called incoherent light.

Structure of a Laser

A typical laser consists of a gas-filled tube that is closed on both ends. Wrapped around the

and pass through the gases again. More atoms become energized and release light. At the other end of the tube is a mirror that is only partially reflective. Only light of a certain wavelength can pass through. This mirror acts to filter any incoherent light that is bouncing around in the tube. The coherent light particles that pass through make up the laser beam.

A Powerful Tool

Because the light in a laser beam is so uniform, the laser is a very powerful tool. In medicine, laser surgery is used to correct vision by reshaping the cornea of the eye. In communications, lasers are used inside optical fibers that transmit information. In supermarkets, lasers scan the bar code that contains a product's price. In manufacturing, lasers are used to find flaws or mistakes in parts. Lasers help manufacturers ensure that their products are of high quality and are made to specifications.

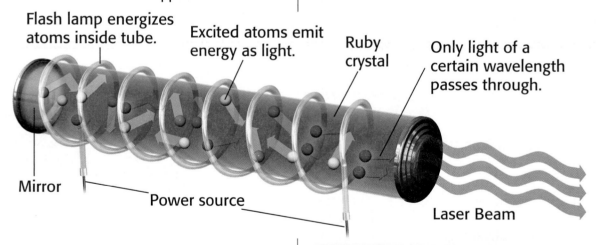

Flash lamp energizes atoms inside tube.

Excited atoms emit energy as light.

Ruby crystal

Only light of a certain wavelength passes through.

Mirror

Power source

Laser Beam

tube is a light source called a flash lamp. When the laser (and flash lamp) is turned on, the energy from the flash lamp causes tiny atoms to become energized. These energized atoms move around the tube, releasing energy as light.

On the back end of the glass tube is a reflective mirror. Photons bounce off the mirror

Critical Thinking

1. **Contrasting** How is coherent light different from incoherent light?

2. **Interpreting** What are the mirrors used for inside a laser? Why are the two mirrors different?

Managing Production Systems

Reading Strategy

Listing As you read this section, make a list of all the tasks completed during each phase of manufacturing. Use the headings to begin your list.

1. Design
- Clarify ideas
- Research design solutions

2. Development
- Add or change product features

Vocabulary

market research   total market   warranty
quality control   market share

Benchmarks for Learning

- The success of a company depends to a large degree on how well it is managed.
- Manufacturing processes include design, development, production, marketing, sales, and maintenance.
- Quality control ensures that quality stays high during all stages of manufacture.

Management

The success of a manufacturing company depends to a large degree on how well it is managed. A company must work within a schedule, within a budget, and within safety guidelines. It must deliver a product as well as be able to fix it.

Figure 8.12 Products are designed, developed, then produced for consumer use.

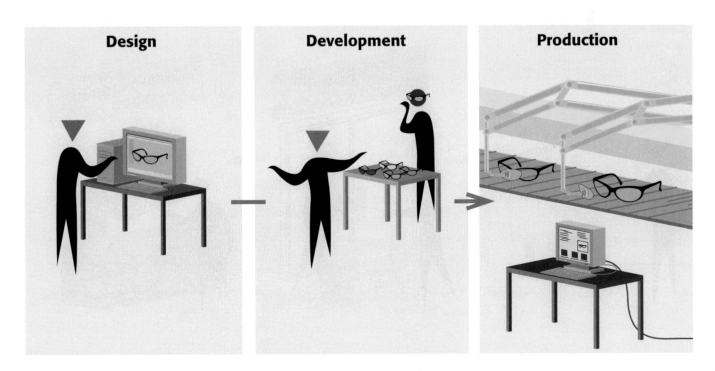

Design   **Development**   **Production**

Managing manufacturing systems is a complex job. Besides coming up with new products, it involves developing and manufacturing these products at the lowest cost, selling them, and servicing them. Although each of these tasks is carried out by specialists, the overall job must be handled carefully by a manager or a team of people.

Phases of Manufacturing

The business of manufacturing includes the management of all phases: design, development, production, marketing, sales, and maintenance (Figure 8.13). Because companies are in business to make a profit, they try to reduce their production costs during each phase of production. They also try to improve the results of each phase so that the quality of the manufactured project will be high.

Design

Designing a product involves using the informed design process. Through informed design, people clarify the design problem, research different ideas, and create new design solutions. During the informed design process, ideas are reviewed by more people. Research generates ideas for new products and ways of improving old ones.

One kind of research conducted before many copies of a product are made is market research. In **market research,**

Figure 8.13 Products are then made available to consumers by marketing and selling.

Summarizing *What are the phases of a manufacturing system?*

a company surveys a group of people to find out what consumers most desire in a product. The company tries to pick a sample group that represents the people who might buy their product. Market researchers ask the sample group about the product. They use feedback from the group to decide whether to make a new product, or how to improve an existing one. Using what has been learned from market research, product designers and engineers prepare drawings and develop finished design ideas.

Development

While a product is still in design, it may be changed so that it is easier to manufacture. If market research has found that a new feature is needed, it may be added at this time. During the development phase, management takes a close look at the proposed designs. There may be several good designs to choose from. The company must decide whether the product can be made at such a cost and sold at such a price that it will bring in a profit. If costs are too high, the product may be changed to use less expensive materials. New production ideas are sometimes used to lower costs.

When the engineers and management agree on a design, a prototype is usually made and test-marketed. A prototype is an original model of the product. Potential customers try out the product and give their reactions and comments. This additional feedback is used to refine the product's final design and path through production.

Production

Once a suitable prototype has been developed, many copies of the product are made. For production, managers set up the production and assembly lines needed to manufacture, assemble, and package the product. Forming, separating, combining, and conditioning tools must be chosen. A flowchart is used to diagram the sequence of operations in the production line. People working on the production line are able to see how their role fits in with the entire production process.

Companies want their products to be of the highest possible quality. Making sure that quality remains high during manufacturing is called **quality control.** As part of quality control, inspectors examine each of the materials that will be used in production. They also examine the product

Figure 8.14 Quality control inspections are made at several points in the manufacturing process.

Summarizing *Why is quality control necessary?*

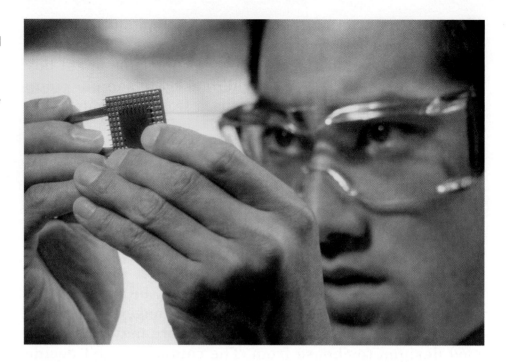

at each step of manufacture and before it leaves the plant (Figure 8.14).

Quality control also helps make sure that products are all alike, or uniform. Over the long term, higher-quality products increase customer satisfaction. Fewer items are returned or in need of repair, so maintenance costs are lower.

During production, management must also make sure that the manufacturing plant is safe and that laws and regulations are being followed. The Occupational Safety and Health Administration (OSHA) has set down rules for safety on the job. These rules must be followed carefully, or a company can be fined or even closed down.

The Environmental Protection Agency (EPA) is another agency that keeps an eye on manufacturers. EPA rules prevent manufacturers from polluting the water and air with waste materials or byproducts of manufacturing. Rules cover the kind and amount of substances that can be released from smokestacks or emptied into a sink drain. They describe proper storage of toxic wastes.

Marketing

The people in the marketing department are the ones who set up the market research studies to determine what features a product should have. The department identifies who might buy the product and how much they are likely to spend on it (Figure 8.15). It also estimates how much

the product can cost and still sell well. Based on these data, the marketing department makes an estimate of how many products will be bought in a year. This estimate is referred to as the **total market.** They also estimate what part of the public will buy their product rather than a similar one made by another company. This is called **market share.**

The marketing department also creates a plan for selling the product. This is known as a marketing plan. The marketing plan details how managers will position and sell their product to the public. People who work in advertising think of ways to sell the product. They create advertisements in newspapers and magazines and commercials for television and radio that will interest the public.

Figure 8.15 The packaging and pricing of CDs are the result of careful market research.

Summarizing *Why is market research necessary?*

Sales

Once a product has been made and advertised, a sales staff sells and distributes the finished product. Salespeople are an important part of the business side of manufacturing.

How a company sells a product depends on the kind of product it is trying to sell. In direct sales, a company sells its products directly to customers. In indirect sales, dealers and distributors sell products to people or companies by taking orders for them and then placing orders with the manufacturer. The dealer marks up the product—that is, he or she charges the customer more than what the manufacturer charged.

Retail sales are the kind of indirect sales with which you are probably most familiar. Retail stores buy products from manufacturers in large quantities. They display these products and sell them. A grocery store is one example. Like dealers and distributors, retail stores make money because they sell products to the public at a higher price than they paid for them.

Maintenance

A company's job does not end when the product is sold. Durable products normally need to be maintained. Parts may wear out or break down. Defective products may have to be replaced. A company's maintenance and repair departments service and repair products. Customer service departments also help with these important jobs.

A manufacturer must have a plan to take care of products that need maintenance or have broken after they were sold. Most products now come with a warranty. A **warranty** gives the terms under which a company will repair or replace a defective product at no cost to the customer. The company must have a repair department that can provide cost-effective, rapid repairs. Depending on the products that are

Figure 8.16 A PERT chart helps managers schedule and monitor complex projects. Each stage of the project is shown in a colored box.

Interpreting *What is the first stage shown in this PERT chart?*

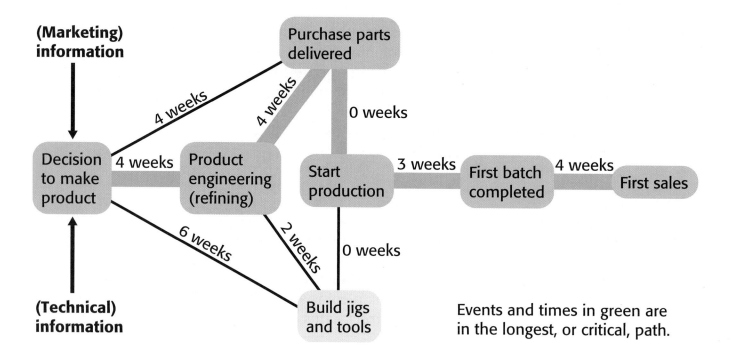

to be serviced, workers who perform repairs range in skill from entry level to highly skilled.

Scheduling with PERT Charts

To schedule manufacturing, managers use a PERT chart. PERT stands for Program Evaluation Review Technique. This type of scheduling was first used by the U.S. Navy during the Polaris missile and submarine project. This project was the most complex manufacturing project undertaken up to that time.

The PERT chart is a powerful scheduling tool (Figure 8.16). The events and times in green are called the critical path, which identifies the project's essential activities and shows the minimum length of time in which the project can be completed. It lists each step that is necessary to complete a project. It shows the order of the steps to be taken and the time needed to complete each step. PERT charts can also predict the effect of a late completion date on the rest of the manufacturing schedule. Using the PERT chart, a manager can keep careful track of what is happening. He or she can move resources to the critical jobs to stay on schedule.

SECTION 4 Assessment

Recall and Comprehension

1. Pick a company that manufactures a product you use. Research the company on the Internet and identify the people who manage it.
2. Describe what happens during each phase of car manufacturing. What is accomplished, and by whom?

Critical Thinking

1. **Hypothesizing** If a company decided to save money by eliminating quality control, what might happen to its products?
2. **Extending** Why might a company decide to extend the phase of development of a product?

QUICK ACTIVITY

Using a hexagonal piece of plastic or cloth, some string, and a large washer, design a small parachute that can be folded and tossed in the air. Once a design has been chosen, use a PERT chart as a time management technique to develop and maintain the work schedules for this design. Set up an assembly line with your classmates and mass-produce enough parachutes to provide one to each class member.
For more related Design Activities, see pages 234–237.

CHAPTER 8 Review and Assessment

Chapter Summary

- Manufacturing is the process of making goods in a workshop or factory. Craftspeople make items by hand, one at a time. Factories mass-produce goods, often using an assembly line.

- On an assembly line, production is divided into steps. Standardized interchangeable parts are made, and each worker does one job.

- Automation has made factories much more efficient. More goods can be made in a shorter time with less labor cost.

- Robots are used in modern manufacturing plants. They have taken over many dangerous or unpleasant jobs.

- CAD/CAM links engineering and design with the factory floor. In CAD/CAM, designs for parts are drawn using a computer and are then sent directly to a machine that makes the part.

- New production methods lower costs and often result in a better product. Methods include CIM, flexible manufacturing, and just-in-time manufacturing.

- The success of a manufacturing project depends on how well it is managed. A company's management team must ensure that resources are shipped at appropriate times and used carefully.

- Making sure products are uniform and of high quality is an important goal for manufacturers. Quality control lowers overall costs because it results in fewer product returns and repairs.

Building Vocabulary

Your teacher may give you a crossword puzzle. Complete the puzzle using the following words from this chapter. Exchange puzzles with a classmate to check each other's answers.

1. agile manufacturing
2. assembly line
3. automation
4. computer-aided manufacturing
5. computer-integrated manufacturing
6. craft approach
7. durable
8. entrepreneur
9. feedback control
10. flexible manufacturing
11. interchangeable part
12. just-in-time manufacturing
13. manufacturing
14. market research
15. market share
16. mass production
17. nondurable
18. productivity
19. program control
20. quality circle
21. quality control
22. total market
23. venture capitalist
24. warranty

See your teacher for the Crosstech puzzle.

Reviewing Content

1. Give five examples of how manufacturing technology has helped satisfy people's needs and wants.

2. Explain how craft production differs from mass production.

3. Name three things for which companies need capital.

4. How is feedback used to ensure good quality in manufactured products?

5. Identify several ways in which a manufacturer might sell products.

Applying Your Knowledge

1. Draw a flowchart of an assembly line for making a greeting card.

2. Suggest an invention that would help you with your homework.

3. How might you innovate a container for a soft drink?

4. Draw a system diagram for a system that manufactures chewing gum. Label the input command, resources, process, output, monitor, and comparison.

5. How might a designer change this robotic hand if its main purpose was to pull defective parts from a computer?

6. Select and research an innovative business idea; prepare a marketing plan for this idea.

7. Choose a manufactured product (e.g., a car, chocolate bar, bottle) and explain how the chosen product has changed from its historical beginning to its present state.

Critical Thinking

1. **Predicting** Many of the products you use every day are manufactured by machines and automated equipment. How would your life change if everything you use had to be made by hand?

2. **Applying Concepts** Suppose you wish to produce computer mice. Explain how the design and the development phases of that product would differ.

3. **Making Judgments** Do you think that robots should be used on assembly lines instead of people? Explain why or why not.

4. **Evaluating** How have computers affected the manufacturing industry?

5. **Contrasting** What are the differences between research and development?

 Connecting to STEM
science · technology · engineering · math

Constructing a PERT Chart

Draw a PERT chart for making hot cocoa, from the purchase of ingredients through cleaning up afterward. Show how each phase is related to the other phases. For each phase, estimate the amount of time that will be needed.

Design Activity 15
ROBOTIC ARM DESIGN CHALLENGE

Problem Situation

Robots are replacing people in many factories. They accurately perform repetitive tasks, such as welding car frames, assembling components, and packaging the completed product. Robots are often controlled by computers but can also be controlled by people operating them from a distance. Beanbag production is increasing, and you are directed to develop a robotic arm that can place them in containers. The first step in the development process is to design and construct a pneumatically controlled robotic arm that can grasp a beanbag and place it in a shipping container.

Your Challenge

You and your teammates are to construct a pneumatically controlled robotic arm.

Go to your **Student Activity Guide, Design Activity 15.** Complete the activity in your Guide, and state the design challenge in your own words.

1 Clarify the Design Specifications and Constraints

To solve the problem, your design must meet the following specifications and constraints:
- The robotic arm must be able to grasp the beanbag.
- The robotic arm must be able to place the beanbag in a container 6 inches high and located 6 inches to the right or left of where it was grasped.

In your Guide, state what the specifications and constraints are. Include any others that your team or your teacher included.

2 Research and Investigate

To better complete the design challenge, you need to first gather information to help you build a knowledge base.

In your Guide, complete Knowledge and Skill Builder 1: Pneumatics and Hydraulics.

In your Guide, complete Knowledge and Skill Builder 2: Levers and Linkages.

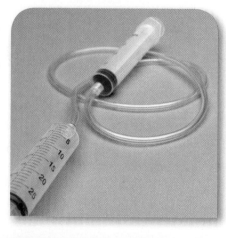

Materials

You will need:
- 1/8" dowel rod
- 2 plungers
- 6 pneumatic syringes and tubing
- 12 centimeter sticks, 60 cm long
- cardboard
- cardboard strips
- glue
- plywood base, 12" square, 3/4" thick
- rubber bands
- small beanbag

In your Guide, complete Knowledge and Skill Builder 3: Remote Rotation.

In your Guide, complete Knowledge and Skill Builder 4: Hinges.

In your Guide, complete Knowledge and Skill Builder 5: Getting a Grip.

③ Generate Alternative Designs

In your Guide, describe two possible solutions that your team has created for the problem. Your solutions should be based on the knowledge you have gathered so far.

④ Choose and Justify the Optimal Solution

Refer to your Guide. Explain why you selected the solution you did, and why it was the better choice.

⑤ Develop a Prototype

Construct your robotic arm. Include a drawing, either a side view or an isometric view, or a photograph of your final design in your Guide.

In any technological activity, you will use seven resources: people, capital, time, information, energy, materials, and tools and machines. In your Guide, indicate which resources were most important in this activity, and how you made trade-offs among them.

⑥ Test and Evaluate

How will you test and evaluate your design? In your Guide, describe the testing procedure. Explain how the results show that the design solves the problem and meets the specifications and constraints.

⑦ Redesign the Solution

Respond to the questions in your Guide about how you would redesign your solution. The redesign should be based on the knowledge and information that you gained during the activity.

⑧ Communicate Your Achievements

In your Guide, describe the plan you will use to present your solution to your class. Include any handouts and/or PowerPoint slides that you will use.

Design Activity 16

BLOWING IN THE WIND CHIME FACTORY

Problem Situation

Your technology class wants to have a party to celebrate finishing a big project. However, the class needs to raise money to pay for it. You look outside and the wind is blowing. Is it a gentle breeze, or a gale? Wind chimes! Wind chimes provide a pleasing way to capture the wind's energy and also tell how the wind is blowing. People like to buy them as gifts or for their own homes. Determine the best design for a wind chime, the cost for constructing it, and the selling price needed to make a profit.

Materials

You will need:

- 1/8" wood or plastic sheet for sail
- drill press
- electrical conduit, copper tubing, sink drain tubing, or other metal tubing
- hanging hooks
- heavy duty plastic string, such as fishing line or weed wacker line
- metal tubing cutter or hacksaw
- ring
- sander/grinder
- wood or plastic for clapper, 3/8" to 1/2" long

Your Challenge

You and your teammates are to design and construct wind chimes to sell for a profit.

> Go to your **Student Activity Guide, Design Activity 16.** Complete the activity in your Guide, and state what the design challenge is in your own words.

① Clarify the Design Specifications and Constraints

To solve the problem, your design must meet the following specifications and constraints:
- The wind chime must have 4 to 8 metal tubes.
- The wind chime's metal tubes must each have a diameter, length, and attachment point that will produce pleasant sounds.
- The wind chime must be constructed at a cost that will allow you to sell it at a reasonable price and make a good profit.

> In your Guide, state what the specifications and constraints are. Include any others that your team or your teacher included.

② Research and Investigate

To better complete the design challenge, you need to first gather information to help you build a knowledge base.

> In your Guide, complete Knowledge and Skill Builder 1: Wind Chime Investigation.

In your Guide, complete Knowledge and Skill Builder 2: Hitting the Right Note.

In your Guide, complete Knowledge and Skill Builder 3: Wind Chimes for Profit.

3 Generate Alternative Designs

In your Guide, describe two of your possible solutions to the problem.

4 Choose and Justify the Optimal Solution

Refer to your Guide. Explain why you selected the solution you did, and why it was the better choice.

5 Develop a Prototype

Make a drawing of your final design and construct the wind chime. Include the drawing or a photograph of your final design in your Guide.

In any technological activity, you will use seven resources: people, capital, time, information, energy, materials, and tools and machines. In your Guide, indicate which resources were most important in this activity, and how you made trade-offs among them.

6 Test and Evaluate

How will you test and evaluate your design? In your Guide, describe the testing procedure and explain how the results show that the design solves the problem and meets the specifications and constraints.

7 Redesign the Solution

Respond to the questions in your Guide about how you would re-design your solution based on the knowledge and information that you gained during the activity.

8 Communicate Your Achievements

In your Guide, describe the plan you will use to present your solution to your class. Include any handouts and/or PowerPoint slides that you will use.

Construction

In this chapter, you will learn how people design and build the following different types of structures:

- roads
- bridges
- tunnels
- buildings
- houses

As the saying goes, "There are two seasons: winter and construction." Construction is a big part of most people's lives. Cranes, scaffolding, and temporary walkways are familiar sights in every major city. People have become used to seeing construction around them.

This is especially true in the city of Boston, Mass., where a huge construction project called the Big Dig has been underway since 1991. City planners had estimated that, without this highway project, traffic in Boston would be jammed in stop-and-go misery for 15 to 16 hours a day by the year 2010. By the time construction is finished,

- 16 million cubic yards of dirt will have been excavated—enough to fill a football stadium to the rim 15 times.

- 3.8 million cubic yards of concrete will have been set—enough to build a 9-foot-wide sidewalk from Boston to San Francisco and back.

- The new Leonard P. Zakim Bunker Hill Bridge will be the widest cable-stayed bridge in the world.

Resources for Construction

Benchmarks for Learning

- People from many professions, such as architecture, engineering, and building, work in the construction industry.

- Structures are built using a variety of procedures, often depending on the types of material available.

Reading Strategy

Outlining   As you read, construct an outline of the material in this section. For each of the main headings of your outline, write down a short definition in your own words.

> **I. Resources for Construction**
> **A. People**
>   **1. Architects–**
>   **design buildings**
>   **2. Engineers–**
>   **3. Contractors–**

Vocabulary

construction	general contractor	tradesperson
architect	subcontractor	mortgage
engineer	project manager	thermal windows

People

Take a look around your neighborhood and think about what has been made by people. If you live in the city or the suburbs, that probably includes most of what you can see (Figure 9.1). The house in which you live, the school building in which you learn, the sidewalks and roads on which you walk and ride were all constructed by people.

Construction is the process of building a structure on the location where it will be used. Bridges, buildings, dams, harbors, roads, towers, and tunnels are all built structures. The U.S. construction industry produces over $100 billion in goods and services each year. Today, about 6 million people work in the construction industry in the United States. People are needed to design and engineer structures. They are needed to manage the business of construction and to do the actual building.

Figure 9.1   Cities and suburbs have many bridges, roads, and buildings that are the result of careful design and construction.

Extending   *Find out who the city planner is in the city closest to you. How does he or she determine the need for construction?*

The principal people involved in construction are architects, engineers, contractors, project managers, and tradespeople.

Architects

An **architect** is a person who designs buildings and other structures. Architects produce plans that show how a structure will be built and where it will be placed on a site. They choose the shape and form of the structure and the materials from which it will be made. They specify all the details of a structure's design (Figure 9.2).

Architects consider how the building will affect other buildings around it, and how the building will be affected if other buildings are constructed in the area. In most large cities, building codes do not permit a new building to block the view of the sky from nearby buildings. Architects deal with this kind of problem by making buildings smaller at the top than at the bottom. This design allows more light to enter surrounding buildings.

Architects are trained for their work in college. They take courses in mathematics, architectural design, technical and architectural drawing, and art.

Engineers

An **engineer** is a person who applies principles of science and mathematics in the design and construction of a project. There are many types of engineers. Civil engineers work with architects to make sure that a bridge, road, or building is structurally sound. They prepare exact drawings and plans for a structure's framework and foundation (the underground part of a building), choosing the size of each column and beam to make sure they will carry the load of the building and its contents. Engineers often use computer-aided design (CAD) to do this kind of work.

Other types of engineers design the major systems within a building. These systems include the electrical and plumbing systems and the heating, ventilating, and air conditioning (HVAC) system. Engineers also plan fire detection, telephone, and Internet access systems.

Engineers learn their profession in college. They study mathematics, science, engineering design, the technology of materials, and technical drawing.

Contractors and Project Managers

During the construction of any project, many different jobs must be coordinated. Small projects may be managed by the owner of the building or complex. If a job is more complex,

Figure 9.2 Architects and engineers work together to design structures that will be structurally sound.

Extending *Why do architects and engineers need to be involved in the later stages of construction?*

Figure 9.3 Contractors and project managers work with a team of subcontractors on a construction project.

Hypothesizing *What types of skills would a contractor need to have?*

however, the owner usually hires a contractor or project manager.

A **general contractor** is a person or company that takes overall responsibility for a construction project. General contractors work with an estimator, who determines a project's cost. They schedule the work so all parts of the job are completed in the right order. They purchase materials and arrange for them to be delivered near the time they are scheduled to be used. A general contractor may also hire subcontractors to complete the work on a project. A **subcontractor** is a person or company with the skills needed for a particular portion of a project. On a large building construction project, there may be dozens of subcontractors, each working on a different part of the building (Figure 9.3). Subcontractors often provide skill in areas such as plumbing, carpentry, and masonry.

Some large projects are so complex that special skills are needed to manage them. A **project manager** oversees the contracts, scheduling, material deliveries, and overall progress of a large construction job. Project managers make sure that construction meets building codes. They hire and supervise workers. Some colleges have programs to train construction project managers.

Tradespeople

A **tradesperson** is the person who does the actual work on a project (Figure 9.4). Tradespeople work on projects ranging from houses to skyscrapers to swimming pools. They are usually hired as subcontractors on a job. Tradespeople use special tools and materials. Some run heavy equipment such as bulldozers or are skilled at laying brick or tile. Tradespeople include carpenters, electricians, plumbers, and masons. Tradespeople study their trade in high school or at a technical or vocational school. Often, they learn on the job as apprentices to more experienced tradespeople.

Figure 9.4 Tradespeople, such as this mason, do the actual building on a construction job.

Identifying *What building materials do masons work with?*

Capital

Construction is expensive. Machines and tools, building materials, and labor are all costly. Money must be raised to finance construction projects. Large projects often use money from both private and government sources. A bank will lend money only if it is fairly certain that the money will be repaid. A loan for the purchase of any building is called a **mortgage.**

A loan is paid back over a period of years with interest added to each payment. Interest is the fee that the bank charges for the loan. Home mortgages usually must be paid back within 15 to 30 years.

Land is often the most costly part of a construction project. In the middle of a major city such as New York or San Francisco, land can cost as much as $1,000 per square foot—or $40 million per acre. Because land costs are so high, people who build in big cities must charge high rents for offices and apartments. Skyscrapers make the best use of expensive land.

Time

Construction takes a lot of time. Usually, houses are built in a matter of months. A bridge or tunnel, however, may take years to complete. Modern tools and equipment save time. For example, air-driven staple guns are replacing hammers and nails for fastening shingles to roofs. New building techniques also can reduce construction time.

Information

In construction, essential information is provided in the form of plans, bids, and specifications. It comes from the people who want the construction done and from those who will do it. The specifications on a project include information about the materials to be used, the way the foundation will be built, and even the kinds of trees and bushes to be planted around the structure.

Many of those involved in construction must be able to make and read mechanical and architectural drawings and blueprints. Tradespeople must have information about building techniques. Engineers must have information about the building materials and the loads that the structure will support.

Energy

Construction projects require energy to operate machines and tools at the site. Even more energy is used in the

Figure 9.5 I-shaped steel beams are used commonly in construction. They are as strong as solid beams but use less steel.

Identifying *What are the advantages to using less steel on a construction site?*

Figure 9.6 Concrete, used to build this skateboarding park, can be poured into many different shapes.

Assessing *Why is concrete such a popular construction material?*

industries that support the construction industry. The production of materials such as concrete, bricks, and steel requires vast amounts of energy. Transporting these materials to building sites also uses energy.

Materials

The materials used in construction include steel, concrete, lumber, brick, and glass. Materials are chosen for a particular project based on properties such as appearance, cost, and strength.

Steel

Steel is used for the framework of skyscrapers, bridges, and towers. It is a strong, elastic material, which means that it can be stretched and compressed with little deformation. These properties make it valuable for construction, especially of tall buildings (Figure 9.5).

Concrete

Concrete is a mixture of stone, sand, water, and cement (a mixture of limestone and clay). Concrete is inexpensive, strong, and can be poured into many different shapes (Figure 9.6). Concrete is used in dams, roadways, tunnels, and buildings—just about every kind of structure.

Wet concrete looks and feels like mud. It can be poured into lumber molds to make any shape needed. Once mixed and poured, concrete sets (gets hard). Because concrete is brittle, steel reinforcing rods are set in place before the concrete is poured to give it more strength in large structures. Sometimes wire screens, called reinforcing mesh, are used.

Lumber

Lumber (wood) is a building material used in the framework of homes. Wood is easy to work with and is fairly inexpensive. Some composite materials used in building, such as plywood and particle board, are made from wood. These are strong, and their cost is reasonable.

Brick

Brick is made from clay. The clay is fired (heated) in an oven called a kiln—a process that hardens the clay. Brick houses are expensive. Labor costs are high because each brick must be set in place by a mason. Although they are more expensive, brick houses last longer and are more fire-resistant than wood houses.

Glass

Glass not only lets in light but also adds great beauty to a structure. Windows can be installed as a single pane of glass or with several panes of glass sandwiched together. Windows that are made from two or three panes of glass are called **thermal windows.** The different panes of thermal glass are separated by a gas or a vacuum. Thermal glass reduces the flow of heat through windows and has proven to be an effective way to conserve energy.

Tools and Machines

Carpenters, plumbers, and other tradespeople use hand and electric tools for some of the smaller tasks on a construction job. Heavy equipment, including cranes and bulldozers, is used in large construction jobs. The largest pieces of equipment are usually used in earth-moving jobs. There is specialized equipment for many of the phases in a construction job, such as

- bulldozers and backhoes for excavating or digging.

- cranes for heavy lifting.

- powerful drills and welding tools used in building bridges, dams, roads, pipelines, and tunnels.

Robots are one machine that is being used more and more often in construction. They are useful in dangerous jobs such as working high in the air, below ground, and deep under water.

SECTION 1 Assessment

Recall and Comprehension

1. What is the role of a general contractor?
2. Name three forms of information used in construction.
3. What materials are used for the framework of a house? Of a skyscraper?
4. How might you finance the construction of a private home?

Critical Thinking

1. **Contrasting** In building design and construction, what are the different responsibilities of architects and civil engineers?
2. **Cause and Effect** What type of coordination occurs among subcontractors on a construction job? Why is this important for meeting the highest quality standards and the schedule?

QUICK ACTIVITY

Working in a group of 2 or 3 students, build the tallest structure you can using only 20 strands of uncooked spaghetti, one sheet of 8-1/2" × 11" copy paper, and 12" of 1"-wide masking tape. Make some plans and sketches of your ideas before you begin.
For more related Design Activities, see pages 272–275.

Connecting to STEM

How Safe Is Safe?

The next time you step into an elevator, consider how safe it is. Look around for a safety certificate that indicates the maximum number of passengers and the maximum weight the elevator can carry—for example, the posted number of passengers (also called the posted load) is 10 and the weight limit is 1,500 pounds. These loads assume that the average weight of each person is 150 pounds. What if, on that particular day, the 10 people are all football players who, together, weigh more than the posted limit? Or, what happens if 12 people crowd into the elevator? How safe would that be?

Putting a Number on Safety

The safety of any kind of structure is indicated by something called a safety factor. A **safety factor** is the ratio between the load that will cause the device to fail, and the posted load. For example, if the load that will cause the elevator to fail is 15,000 pounds and the posted load is 1,500 pounds, the safety factor is 10 (15,000 ÷ 1,500 = 10). The greater the safety factor, the less likely the device is to fail. Fast passenger elevators have safety factors of 11.9. The cables that support these elevators are designed to hold a weight

All elevators have weight limits.

at least 11.9 times greater than the posted load. Other safety factors are listed in the table.

Bridges and buildings are also assigned safety factors. They are essential to ensuring safety when the loads are extremely heavy. During rush hour, for example, many heavy trucks may be stuck on a bridge at the same time.

Is Safer Always Better?

If a greater safety factor is better, why don't engineers build structures with higher safety factors—say 100 or more? Because cost is a key factor as well. If a structure is made from stronger materials, it costs more to build and maintain. A product that is overly strong (and expensive) for its purpose will not be purchased by customers, so it will not survive in the marketplace.

Type of Structure	Safety Factor
Cable of a fast passenger elevator	11.9
Cable of a slow passenger elevator	7.6
Cable of a crane	6
Wooden building	6
Leg bone of a galloping horse	4.8
Steel building or bridge	2

Critical Thinking

1. **Calculating** A fast passenger elevator has a floor that measures 8 feet by 6 feet. A designer allocates 4 square feet for each person in the posted load. Assuming that each person weighs 150 pounds, what is the operating load of the elevator? What is the maximum load before failure?

2. **Hypothesizing** Why do you think the safety factor for wooden buildings is much greater than that for steel buildings?

Structures

Reading Strategy

Mapping As you read the section, make a map to show the different types of structures that exist. For each type of structure, either write a definition or draw a simple sketch to help you remember that term.

Vocabulary

bridge
beam bridge
arch bridge

cantilever bridge
suspension bridge
wind drift

renovation
macadam
tunnel

Benchmarks for Learning

- Form and function are complementary aspects of systems in both natural and designed worlds.
- The designs of structures and their methods of construction have evolved over time.
- Roads and bridges make up the framework of our cities.

Bridges

People have been building bridges since prehistoric times. A **bridge** is a structure that crosses over water, a valley, or a road. Bridges can also be used to carry water. This type of bridge is called an aqueduct. The type of construction used in building a bridge depends mainly on its size (Figure 9.7).

Beam Bridge

A bridge that uses a single beam to cross a distance is called a **beam bridge.** Beam bridges are the simplest kind of bridge and are used to cross the shortest distances. The strength of a beam bridge is determined by the strength of the beam used in its construction (Figure 9.8).

Arch Bridge

An **arch bridge** uses a curved structure above an opening to support the load of the bridge. This type of bridge is in common use because the material that makes up the arch remains very strong even when under compression. The pressure of the bridge acts to squeeze the pieces of the arch together so that the weight is transferred to the ground. The first arch bridges were built from stone by the Romans. Arch bridges are now made from concrete or steel because of the strength of these materials.

Cantilever Bridge

A **cantilever bridge** works like two diving boards facing each other. The sections are firmly attached at their ends. If they do not meet, another section may be added to link

Figure 9.7 A suspension bridge, such as the Golden Gate, can have a span thousands of feet long.

Analyzing *Why is this called a suspension bridge?*

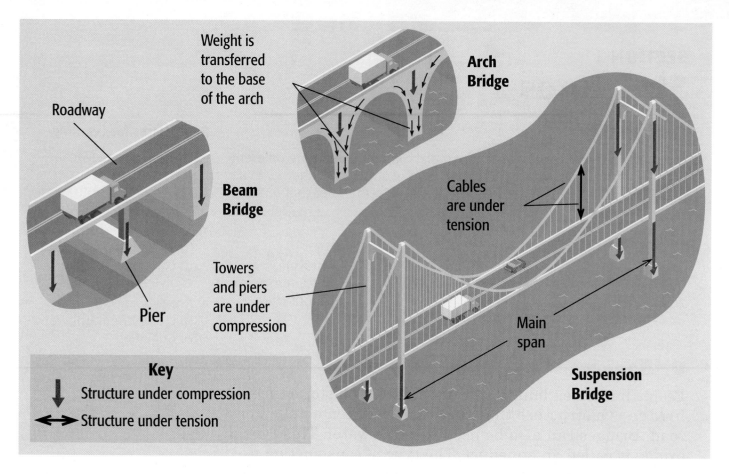

Weight is transferred to the base of the arch

Arch Bridge

Roadway

Beam Bridge

Cables are under tension

Towers and piers are under compression

Pier

Main span

Suspension Bridge

Key

↓ Structure under compression

↔ Structure under tension

Figure 9.8 The construction of a bridge determines how much weight it can support.

Interpreting *Is a beam bridge based on compression or tension?*

them. A cantilever bridge requires a huge structure to support each end. Double cantilever bridges have been designed to overcome this problem. The two sides of the cantilever section balance each other.

Suspension Bridge

Suspension bridges are the longest type of bridge. They are used to bridge wide spans. A **suspension bridge** uses steel cables to hang the deck, or roadbed, from towers. Modern suspension bridges have two tall towers to which the cables are attached. The supporting cables are under tension from the weight of the bridge and its traffic.

Some of the world's most famous bridges are suspension bridges. Here are some examples:

- The Golden Gate Bridge in San Francisco is 4,200 feet long. It connects southern San Francisco and Sausalito, California.

- The George Washington Bridge, which links New Jersey and New York, is 3,500 feet long.

- The Sunshine Skyway Bridge is 29,040 feet long and spans the mouth of Tampa Bay, connecting St. Petersburg and Bradenton, Florida.

Buildings

A **building** is any structure with walls and a roof that is made for permanent use. Buildings are designed to fulfill different needs. Therefore the size and shape of a building depends on its purpose.

There are four basic types of buildings.

1. Residential buildings are made for people to live in. They include apartment buildings and houses (Figure 9.9).
2. Commercial buildings are structures in which people carry out business. They include banks, stores, malls, and offices.
3. Institutional buildings are made for large organizations such as schools and hospitals.
4. Industrial buildings such as factories and warehouses house equipment and materials for making things.

The design requirements are different for each of these buildings. Comfort and aesthetics are important for people's homes, whereas functionality and space are more important in industrial buildings. A factory must house many kinds of equipment as cheaply and efficiently as possible.

Locations also vary for different types of buildings. Some structures do not require as much land as others. In addition, land costs can vary based on location. Schools, offices, and hospitals must be located near residential areas. Factories can be built in rural areas, where land is cheap and possible noise and pollution would not affect as many people.

Figure 9.9 These homes in San Francisco were built in the Victorian style.

Inferring *How are the criteria for building these homes different from those for the buildings in the background?*

Materials for Buildings

Some of the earliest buildings were made from materials such as wood, stone, and mud. Later, buildings were constructed of bricks made from straw and mud.

Today, buildings are constructed using a variety of materials. Most houses are made of lumber, plywood, plasterboard, insulation, and vinyl, wood, or aluminum siding. In commercial buildings, which are larger than houses, reinforced concrete and steel are used for framing. These materials are stronger, longer-lasting, and more fire-resistant than wood. Floors may be made of concrete or steel. The outside walls are usually made of brick or panels of concrete, metal, glass, or plastic.

Building for Windy Conditions

A tall building is like a long lever connected to a pivot in the ground. The farther away from the pivot point that force is exerted, the greater its effects will be. Wind, for example, has a much greater effect at the top of a skyscraper than it does at ground level.

When wind blows against the top of a tall building, the building sways. The **wind drift** is the distance a building moves from its vertical center because of wind (Figure 9.10). Tall buildings are actually made to be flexible to minimize the possibility of damage in strong winds. They are also designed to be stiff to keep wind drift as small as possible. Concrete panels might be used to cover the exterior, or steel trusses might be built into the framework of the building.

Figure 9.10 Some buildings sway several feet during strong winds. Imagine how that feels on the top floor!

Analyzing *Why is swaying with the wind better than completely resisting it?*

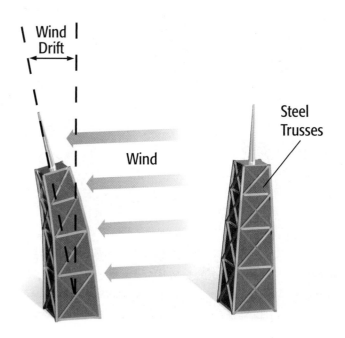

Figure 9.11 Renovating an old building can be cheaper and more rewarding than constructing an entirely new building.

Extending *What types of changes might need to be made when renovating a factory into apartments?*

Renovating Older Buildings

As years pass, dust, wind, and water can cause damage to a building. **Renovation** is the process of rebuilding an existing building (Figure 9.11). Buildings are renovated to change or update their style and to carry out major repairs. When a structure is renovated, old materials, such as windows and plaster, are replaced or renewed. Renovation is often less costly than demolishing a structure and rebuilding it.

Roads

Before the Roman era, roads were simply narrow paths used for walking or pushing two-wheeled carts. Roman engineers, however, became masters at road construction. They built more than 40,000 miles of roads—many of stone—for the empire's armies to travel. Some of these roads are still in existence today.

Improving Road Materials

Modern road building was strongly influenced by the ideas of a Scotsman named John Loudon McAdam. McAdam, who lived in the eighteenth century, made roads that lasted longer because of good drainage. He placed a layer of stone on a base of hard soil and topped it with a layer of tar. Like the Roman roads, McAdam's roads were higher in the center than at the edges, making water flow away from the road. McAdam's name was given to the material that is now used as a surface for many roads: **macadam.**

Figure 9.12 Roads are constructed of layers of different materials.

Interpreting *In this road, which material is the surface that cars drive on?*

Asphalt

Stone

Packed Soil

Road Construction

Since the early 1900s, most roads have been built with macadam, concrete, or asphalt. They are built to support heavy loads carried by cars and trucks. Safe, well-designed roads have linked our cities and opened up new markets for business.

Road construction begins with choosing the desired route. When possible, the route avoids existing structures. Sometimes, however, buildings must be removed to make way for a road because it would cost too much to go around them.

Next, the ground is smoothed by bulldozers. The soil is pressed down by heavy rollers. It is then covered with stone, which spreads out the load and provides drainage (Figure 9.12). The roadbed is usually made from blacktop materials such as macadam or asphalt. Center barriers, good lighting, and traffic control devices are also part of the road-building system.

Today, roads and highways are built to support the heavy loads carried by many high-speed vehicles. Bridges and tunnels allow roads to go in a straight line from one place to another. Safe, well-designed roads allow people to live miles from their workplaces. They have linked our cities and opened up new markets for business.

Tunnels

A **tunnel** is a covered passageway through or under an obstruction. Since ancient times, tunnels under water and through mountains have been used to shorten travel routes. The first known tunnels were dug by the Babylonians about 2000 B.C.

Early tunnels were dug by hand because little machinery was available. The removed earth was hauled away in carts. Workers always feared that the tunnel would cave in around them. In 1818, the tunneling shield was invented. This device held the earth up while the tunnel was being dug.

Figure 9.13 Large drilling machines are used to drill tunnels through rock or soft ground.

Inferring *When do you think drilling is preferable to blasting?*

Modern techniques for building tunnels depend on where the tunnel is to be built. Tunnels dug into rock are drilled and blasted by explosives. The opening is supported by steel arches. A new technique uses a concrete mixture called shotcrete to spray the walls. Shotcrete prevents water from seeping through the rock.

Tunnels drilled in soft ground can be dug with machines (Figure 9.13). These machines have rotating cutters as large as 15 feet across. The cutters rotate and push against the earth with the force of nearly 1 million pounds. As the earth is cut, conveyor belts carry the rocks and soil out the back of the machine. Precast concrete or steel rings are put into place to support the opening.

SECTION 2 Assessment

Recall and Comprehension

1. Describe the four basic types of buildings.
2. What materials are used in the construction of modern roads?
3. Identify the different methods of building tunnels.

Critical Thinking

1. **Summarizing** Explain how an arch can support a great amount of weight.
2. **Analyzing** Why does wind affect tall buildings more than it does short buildings?

QUICK ACTIVITY

Some materials are stronger than you might think. Cut a 1" × 3" piece of copy paper. Pinch the paper between your thumbs and index fingers and try to pull the paper apart. (NO twisting, bending, tearing, or using your fingernails allowed!) Did the paper rip? Why or why not? Repeat the experiment with a twisting movement. What happened? Why?
For more related Design Activities, see pages 272–275.

Construction of a House

Benchmarks for Learning

- Decisions related to construction have impacts on individuals, society, and the environment.
- Structures rest on foundations.
- Structures are built using a variety of processes and procedures.
- There are many types of interior and exterior building materials.
- Structures can include prefabricated materials.
- Buildings generally contain a variety of subsystems.

Reading Strategy

Listing Suppose that you are in charge of the construction of a house. As you read this section, make a list that shows all the different tasks that need to be completed.
1. Select Site
2. Prepare Site
3.

Vocabulary

construction site	mortar	sheathing
foundation	superstructure	roof truss
footing	stud	voltage
foundation wall	subfloor	National Electrical Code®

Selecting the Construction Site

An important first step in house construction is choosing a site. The location where a structure is to be built is called a **construction site.** Conditions on the site must be taken into consideration. If it is hilly, much time and money must be spent on leveling the soil. If the ground is rocky, it will be hard to dig a foundation. The cost of removing trees, hauling in fill dirt, and constructing an access road, if needed, will all add to the cost of building on the site.

The site itself must also be priced reasonably. Land costs and taxes can vary greatly from one place to another. If the site is near a famous ski resort or has a beautiful view, the land may be very costly. Often, only a few miles can make the difference between high and low land costs.

A construction site must also be near roads, railroads, or ports so materials and equipment can be delivered. Water and electricity must be available, as well as a way to dispose of wastes. At the same time, the environment and the community must be considered. For example, an airport ideally should not be located in the middle of a quiet residential community because of the noise.

Many neighborhoods will allow only one style or type of architecture. If the site is in a historical area, the style of the structure must fit in with the surrounding buildings and landscape or permission to build may be withheld.

Preparing the Construction Site

Next, the construction site is prepared for building (Figure 9.14). If unwanted buildings are on the site, they are removed using wrecking balls or explosives. Bulldozers are used to clear trees or brush from the area. Unwanted materials are hauled away by dump trucks.

The structure is also laid out. A person called a surveyor marks the site to show where the structure will be built. A surveyor uses an instrument called a transit to measure and lay out angles, as well as an engineer's level to set the elevation (height above ground) of different points.

Building the Foundation

Building a house begins with the foundation. A **foundation** is an underground structure that spreads the weight of a building over a larger area of ground (Figure 9.15). The greater the size of the foundation, the less likely the building is to sink. In cold climates, foundations also prevent frost damage to buildings.

A house foundation has two parts: the footing and the foundation wall. The **footing** is the base of a foundation. In most houses, it is made of concrete. It is dug to a depth below the frost line, the depth to which the ground freezes in winter. A more shallow footing could crack because when the ground freezes, it expands.

The **foundation wall** is built on top of the footing and supports the weight of the house. It is most often made of concrete blocks. The concrete blocks are attached to the

Figure 9.14 Before construction can begin, the building site needs to be cleared and prepared for laying the foundation.

Contrasting *How is this process different from a renovation process?*

Figure 9.15 The foundation is the first part of a house to be constructed.

Extending *Which parts of a house are included in its superstructure?*

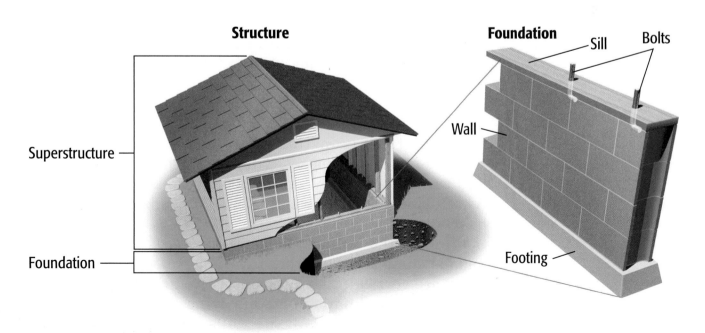

Structure

Superstructure

Foundation

Foundation

Sill   Bolts

Wall

Footing

footing and to each other with mortar. **Mortar** is a mixture of cement, lime, sand, and water. It acts like glue, bonding the concrete blocks to the footing and to each other.

The horizontal parts of a foundation must be level— precisely horizontal—and the vertical parts must be plumb— precisely vertical. If the foundation is not level and plumb, the house built on it will be crooked.

Building the Superstructure

The **superstructure** is the part of a structure that is aboveground. The superstructure of most houses is a framework made of lumber. The lumber that is used to frame houses comes in many thicknesses and lengths. Most walls, for example, are framed with 8-foot lengths of two-by-four lumber. These vertical pieces, called **studs,** are spaced 16 inches apart to allow room for insulation, electrical wiring, and plumbing.

Building the Floor

The floor is built on top of the foundation on a framework of long boards called floor joists. Floor joists are usually spaced so that the center of one joist is 16 inches from the center of the next joist. Because the foundation is concrete, the wooden joists cannot be nailed directly to it. Instead, bolts are used to secure a wooden sill plate to the concrete foundation, and the joists are nailed to the sill plate. The sill plate is a piece of lumber, approximately 2 inches thick.

After the joists have been nailed onto the sill, the subfloor is nailed to them. The **subfloor** is the bottom layer of flooring in a building. The subfloor provides the surface to which the finished floor will be attached. The subfloor is often made from plywood panels that are 4 feet wide and approximately 8 feet long.

Framing the Walls

After the floor is installed, the walls are constructed. Walls serve two purposes. They carry the load of the roof and ceiling, and they serve as partitions for the rooms. A wall that supports weight from above is called a load-bearing wall.

Framing the walls means cutting pieces of wood to size and fastening them together to form a framework. Although walls are vertical, they are constructed horizontally. They are laid out and nailed together lying flat on the subfloor. When the framing for all the walls is complete, with door

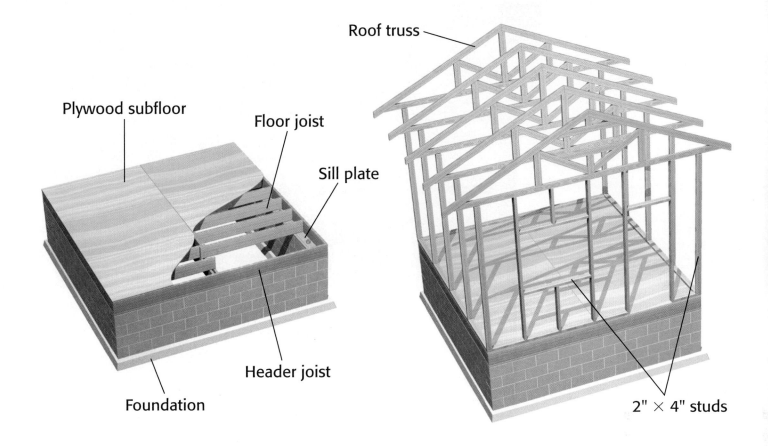

Plywood subfloor

Floor joist

Sill plate

Header joist

Foundation

Roof truss

2" × 4" studs

and window openings in place, the walls are raised to their final position and nailed into place (Figure 9.16).

Sheathing

After the wall has been framed and nailed into place, it is covered with wall sheathing. **Sheathing** is a waterproof layer of plywood, particleboard, or rigid foamboard that closes in the building and protects it from the weather. Most often, 4-foot by 8-foot sheets of sheathing are used. Sheathing makes the frame of a house stronger. It also provides a surface to which outside wall coverings are fastened.

Constructing the Roof

The roof protects the house against the weather and prevents heat from escaping when it is cold outside. The roof also affects the way a house looks. The most common kind of roof is a gable roof, which has an "A" shape.

In hot climates, roofs are often flat. But in most countries, they are sloped so rain and melted snow can drain off easily. The climate determines the degree of slope, or pitch, that is needed. In very cold climates with heavy snowfall, the pitch needs to be very steep. Snow is more likely to slide off a steep roof instead of piling up, melting, and leaking into the house.

Figure 9.16 Once a foundation is laid, the floor is constructed of floor joists and plywood.

Clarifying *Why is it important to build the foundation underground rather than above?*

LIVING GREEN
Put Some White in Your Green

Earth-friendly or "green" construction has advanced dramatically in the last few years. Construction companies are increasingly building new homes that use energy-efficient heating and cooling systems, installing high-efficiency appliances and installing flooring made from recycled fibers or easily renewable woods. Current homeowners are also making green improvements such as painting with products that release little or no hazardous fumes (called low-VOC paints because they contain few volatile organic compounds), and installing solar panels and individual windmills for generating their own energy.

MAKE A CHANGE

One of the simplest ways to reduce energy costs in a home is to use roof shingles that are light in color. When sunlight reaches a traditionally dark roof, the heat is absorbed, warming the air inside the home. When sunlight reaches a light-colored roof, it's reflected back into the environment. Studies show that whereas dark-roofed buildings can trap and retain up to 80% of heat from sunlight, a lighter roof can reflect up to 75% of this energy. This way, the home stays cooler and air conditioning cost remains lower. Of course, in a cold climate, the opposite would be true—you would want a darker roof, to retain heat.

TRY THIS

Using cardboard or an old shoebox, design and build two model homes. Pay attention to roof design (whether it's a sloped or flat roof). Be sure to keep the design the same for both models. Then paint the roof of one model black and the other white. Install an inexpensive plastic thermometer in each and set them up outside your home or school building in an area that gets an average amount of sunlight for your area. Create a data table and record the indoor temperatures at regular intervals. How does the temperature change compare in your two models?

Roof construction is easier now than ever before. Today, carpenters do not have to measure, saw, and nail every board into place. Instead, most of a roof may be made in a factory in the form of large wooden triangles called **roof trusses.** The trusses are trucked to the construction site and then hung upside down across the width of the house,

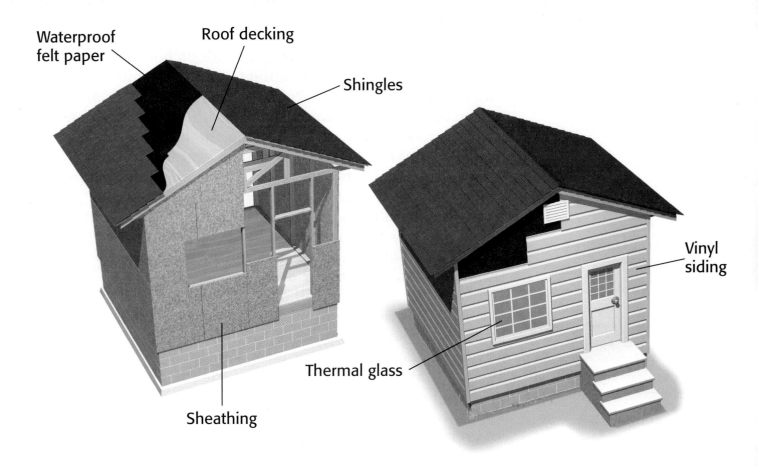

Waterproof felt paper

Roof decking

Shingles

Vinyl siding

Thermal glass

Sheathing

like a V. Then, a team of carpenters (possibly using a small crane) swings the trusses upright and nails them into place. Plywood sheathing called roof decking is then nailed over the trusses.

To prevent rain and snow from getting into the house, once the roof has been framed and decking nailed on, it is covered with a weatherproof roofing material (Figure 9.17). Usually, this material is a felt paper that has been soaked with asphalt. Asphalt is a waterproof, tarlike substance—the same material that is used to pave roads. On top of this waterproof layer, shingles are nailed or stapled into place. Shingles are made from asphalt, fiberglass, or wood. Each course, or row, of shingles is put on so that it overlaps the one below it. There are also metal and tile roofs.

Installing the Utilities

Before the inside walls are finished, parts of the plumbing and heating systems and the electrical wiring are installed. The major portion of these systems is hidden in ceilings, floors, and walls.

Figure 9.17 The roof of a house is constructed of layers of plywood, waterproof paper, and then shingles.

Identifying *How does the shape of a roof help repel the elements?*

Plumbing Systems

The plumbing system of a house includes a hot water supply, a cold water supply, and drainage. The freshwater coming into the house and the wastewater leaving the house are carried in different pipes and are never allowed to mix. Freshwater comes from a well drilled on the property or from a community water supply. Drainage systems, often called sewer systems, discharge into the house's septic system or into a community sewer system.

The water in the house moves, under pressure, through pipes. Cold water is carried directly to sinks, showers, tubs, toilets, washing machines, and outside faucets (Figure 9.18). Cold water also goes to a hot water heater. Hot water heaters burn oil or gas or use electricity to heat the water. Once the water is heated, it circulates through a separate system of pipes. Water supply pipes are generally made of copper. Copper does not rust and has a smooth surface that lets water flow easily. Plastic pipe made from polyvinyl chloride (PVC) is being used in many plumbing applications. It is easy to work with, relatively inexpensive, and resists corrosion.

Figure 9.18 A plumbing system circulates water through an entire house. Once water is heated in a water heater, it is kept separate from the cold water supply.

Interpreting *Where do the cold and hot water finally mix?*

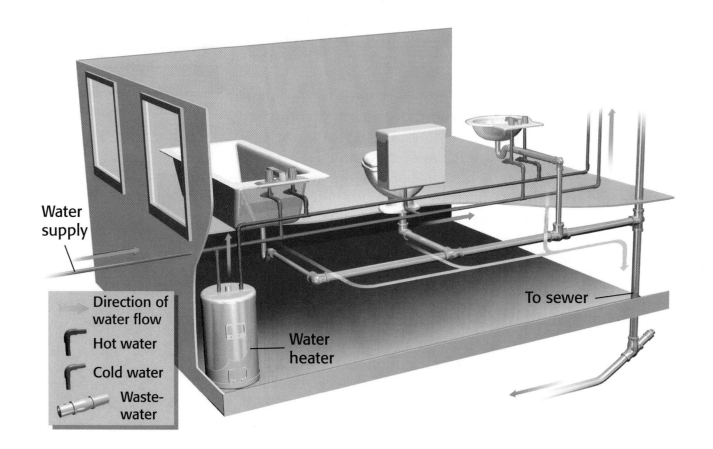

Water supply

Direction of water flow

Hot water

Cold water

Waste-water

Water heater

To sewer

Heating Systems

Most home heating systems are fueled by oil, natural gas, or electricity. In these systems, a furnace heats air or water and circulates it to the areas to be heated. In a forced-air system, oil or gas is burned, causing steel fins to get hot. Fans blow over the steel fins, sending heated air through ducts (sheet metal or plastic pathways). The ducts carry the warm air to vents in each room, heating the house.

Some furnaces heat water in a boiler. This water is pumped through pipes to radiators located in different rooms of a house. The radiators become warm and heat the air around them. The water then returns to the boiler for reheating.

Homes focusing on "green" systems may use heat pumps, which take heat from the outside air or a body of water near the home and blow the heated air into the living spaces. A conventional furnace backs up the system.

Electrical Systems

Electrical service is supplied to a house by a power company (Figure 9.19). Generally, these companies provide 120-volt and 240-volt service. **Voltage** is a measure of the force with which electric current is pushed through wires. The 120-volt service powers lighting and most appliances. The 240-volt service is used for stoves and appliances such as air conditioners,

Figure 9.19 Electricity enters through an electricity meter, goes to an electrical panel, and then travels to individual circuits.

Identifying *Where are the circuit breakers located in this illustration?*

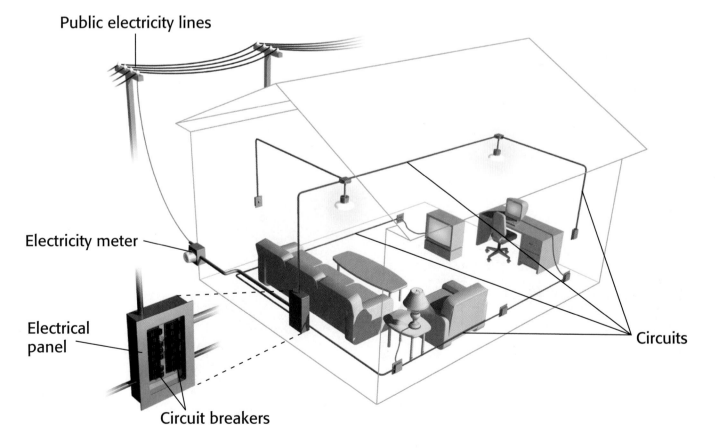

Public electricity lines

Electricity meter

Electrical panel

Circuit breakers

Circuits

For: Construction Activity
Visit: www.mytechedkit.com

Earthquake-Proofing a Building online

When earthquakes hit, the results can be disastrous for both property and people. The intense shaking produced by earthquakes can damage or destroy buildings. Go online to learn how new construction technology can help minimize damage caused by earthquakes.

Modern buildings in areas prone to quakes are made stronger with more flexible materials.

As the earthquake waves move underground, they can put enough stress on buildings and other structures to cause significant damage.

which use a lot of electric current. The **National Electrical Code**® (NEC)® sets standards for the safe installation of electrical systems. It determines electrical codes for houses as well as the correct voltages for electrical appliances.

Electricity enters a house through an electricity meter and then an electrical panel. An **electrical panel** distributes electricity to different branch circuits in a house. The branch circuits supply power to different areas of the house. One circuit might serve the kitchen, whereas another sends electricity to a bedroom and a nearby hallway. Each branch circuit has its own circuit breaker in the electrical panel. A circuit breaker controls the amount of current that can safely pass through household wiring. If the current exceeds this amount—even momentarily, from an overload—the breaker trips, shutting off the current. The breaker can be reset if current usage returns to a safe level.

Finishing the House

When the framing, sheathing, roofing, and utilities are installed, the inside and outside of the house can be finished. The finish work on a house includes all the details that need to be added, such as siding, trimwork, and paint.

Exterior finishing involves painting or adding the final layer of materials to the outside walls of a house. The outermost layer of a house is usually siding. **Siding** is a layer of wood, aluminum, or vinyl boards that waterproofs and protects a house. Wood siding needs to be painted or stained every few years, but aluminum and vinyl siding need little care.

Inside, insulation is added after the utilities have been installed. Insulation is material that does not conduct heat well. In a house, it is usually placed within the exterior walls, between the inside and outside surfaces of the house. It slows the movement of heat through the walls. On cold days, heat escapes more slowly to the outside. On hot days, heat enters the house more slowly. Insulation helps keep the house temperature constant and reduces fuel bills. Fiberglass and some kinds of plastic are used to insulate houses.

Interior finishing involves completing the inside walls and ceilings of a house. It is often called drywall construction because the inside walls and ceilings are finished with drywall (Figure 9.20). **Drywall** (also known as gypsum wallboard or sheetrock) is a sheet of plaster covered with heavy paper. It comes in 4-foot by 8-foot panels and is fastened to the studs using special drywall nails or screws.

Figure 9.20 The interior walls of a house are covered with drywall and then smoothed over with spackling compound.

Contrasting *How are the materials used for exterior finishing different from those used in interior finishing?*

Figure 9.21 Modules reduce the cost of home construction because much less labor is needed.

Comparing *How might the construction materials be different in a modular home, as compared to a standard home?*

Building Houses from Modules

Modular construction is a construction technique in which buildings or other structures are made by fitting together similar standardized units (Figure 9.21). A **module**, which is a section of uniform size, can include one or more rooms. Modular homes are turned out in factories that work much like car plants. Walls are framed, sheathed, and insulated on assembly lines. Windows, plumbing, electrical wiring, and even kitchen and bathroom cabinets are installed. Interior finishing can be completed in the factory.

SECTION 3 Assessment

Recall and Comprehension

1. What is modular construction? What are some of its advantages over standard construction?
2. What is a roof truss? When is it installed?
3. How does insulation decrease the cost of owning a house?
4. What are the different subsystems of a house? Name as many as you can.

Critical Thinking

1. **Evaluating** Briefly explain how the seven technological resources are used in the construction of a house.
2. **Comparing and Contrasting** How are the footing and the foundation wall alike? How are they different?

QUICK ACTIVITY

Earthquakes can cause tremendous damage to people and property. Investigate earthquakes by visiting U.S. Geological Survey sites, http://earthquake.usgs.gov/regional/neic/ and http://quake.usgs.gov/prepare/factsheets/SaferStructures/. Investigate techniques that are being developed to make more earthquake-resistant structures and how these might be employed in developing countries in a sustainable fashion.
For more related Design Activities, see pages 272–275.

People in Technology

Clarence Elder
Adding Intelligence to Buildings

Using electronic beams aimed across a building's entrance, the system monitors how many people enter and leave.

The use of electronics and computer controls in all sorts of ways has led to the development of "intelligent" buildings. These controls can monitor systems to increase a building's energy efficiency and safety. They can also be used to set up networks for television, phone, and Internet access. More and more, electronic intelligence is being built into new homes and office buildings.

One person who has spent a lot of time thinking about intelligent buildings is Clarence Elder. Elder patented a device called an Occustat. The name Occustat combines the terms *thermostat* and *occupants*.

Elder's Occustat adjusts a building's lighting and heating conditions based on the number of people who occupy the building. Using electronic beams aimed across a building's entrance, the system monitors how many people enter and leave.

Clarence Elder

Energy Saving Technology

When a building, room, or floor no longer has any occupants, the Occustat reduces the heating/cooling and lighting load, saving energy. This system is especially useful in schools and hotels, which may have rooms left empty for long periods.

The entrepreneurial spirit was alive and well in Clarence Elder, who was born in Georgia in 1935 and attended Morgan State College. He later established his own business, Elder Systems Incorporated. Elder's company, which is still in business today, develops energy monitoring and conservation systems. It has received 12 additional patents in the United States and abroad.

The Occustat is especially useful in a large building, such as this theatre.

Critical Thinking

1. **Inferring** Motion detectors are sometimes used to turn room lights on and off. Why is this not as accurate as a system such as the Occustat?

2. **Extending** Imagine that you are designing a device like the Occustat. What information (such as ease of use) would you need to make the device as useful as possible?

3. **Analyzing** Does an Occustat use a closed-loop system or an open-loop system? Explain your answer.

Managing Construction Projects

Benchmarks for Learning

- Project management involves planning a construction timetable and supervising members of a construction team.
- The selection of designs for structures is based on factors such as building laws and codes.
- Structures require maintenance.

Reading Strategy

Listing List all the different tasks involved in managing a construction project. Show which person is responsible for completing each task.

1. **Management**
   - architects
   - engineers
2. **Scheduling**

Vocabulary

| contract | Gantt chart | zoning board |
| union | building permit | Certificate of Occupancy |

Figure 9.22 An important job in managing construction projects is making sure that the right materials arrive.

Predicting *What are the consequences of materials arriving late to a site?*

Who Manages a Construction Project?

It is not enough to know how to construct a good building. The success of a project depends to a large degree on how well it is managed. A building must be built within a schedule, within a budget, and safely. Many people work together to make sure that this happens.

During construction, architects and engineers visit the site often to check the work of the builders and other tradespeople. They check to see that the construction matches the original drawings and plans for the building.

General contractors and project managers oversee the work of the builders and tradespeople. They make sure that a project stays on schedule. They order materials and schedule workers to complete specific tasks (Figure 9.22).

One of their jobs is to negotiate the terms of the project in a contract. A **contract** is a written agreement that states what the contractor will do for a given project and how much the owner will pay. It specifies the scheduling and responsibilities for each part of the project. It specifies who is responsible if there is an accident and who must obtain the building permits. It notes any penalties if the job is not finished on time. In some parts of the country, all construction workers are represented by unions. A **union** is a labor organization that works with companies to set pay and working conditions for its members. Union workers take care in choosing the correct trade for each job. Carpenters, for example, may not be permitted to do even minor electrical work. A manager's skill negotiating with union workers is often important for getting a job done quickly and well.

Scheduling and Project Tracking

Scheduling is a big part of the general contractor's or project manager's job. Different jobs must be completed in the right order. Careful scheduling ensures that the construction site is safe and that systems already in place are not damaged by later work. For example, a floor must be structurally sound and properly prepared before finishing work can be started. Electrical wiring cannot be done until the walls are framed.

With so many tasks and workers on a single job site, special management tools are needed to set a schedule and stay on it. A **Gantt chart,** shown in Figure 9.23, is a scheduling tool that shows when each part of a project begins and ends. It indicates which jobs must be completed before another is begun. It also shows the impact of any delays in the schedule. Gantt charts are used in managing both construction and manufacturing systems.

Project meetings are held regularly during construction. The people who manage the project—the owner, the architect, the contractor, and others—meet to talk about progress on the project. Workers present any problems they have encountered, and group members together discuss ways of resolving these problems. They also discuss any potential problems and ways of preventing them before they happen.

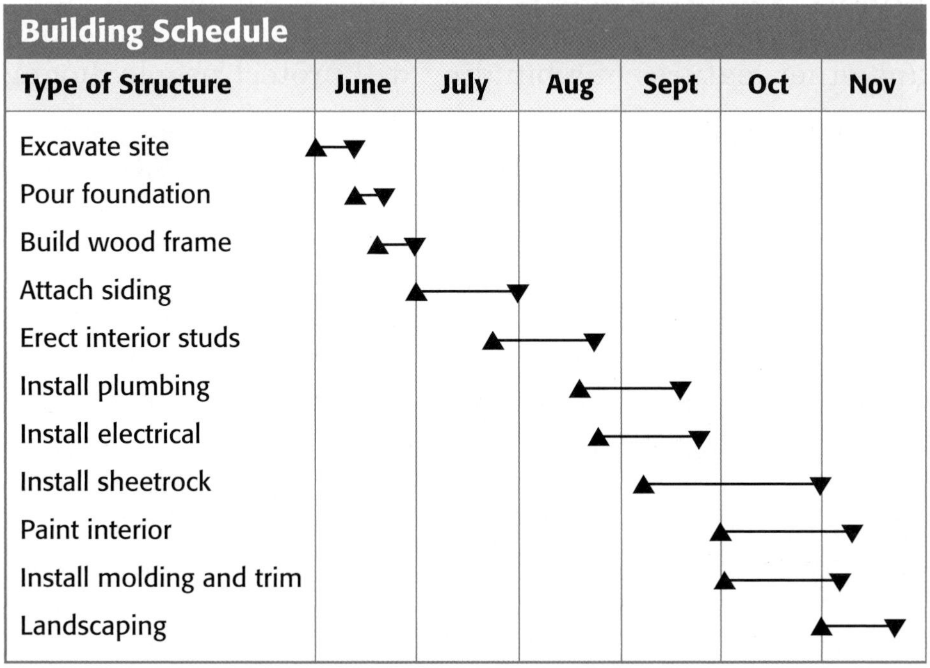

Building Schedule						
Type of Structure	June	July	Aug	Sept	Oct	Nov
Excavate site						
Pour foundation						
Build wood frame						
Attach siding						
Erect interior studs						
Install plumbing						
Install electrical						
Install sheetrock						
Paint interior						
Install molding and trim						
Landscaping						

Figure 9.23 A Gantt chart is a type of bar graph. Each bar indicates when a project begins and ends.

Interpreting *According to this Gantt chart, which system is installed first—plumbing or electrical?*

Written notes, called minutes, record the important issues discussed and any decisions made. Project meetings are usually held weekly or twice a month.

Building Permits and Inspections

Before, during, and after the construction of a building, permits must be obtained, and inspections must be carried out (Figure 9.24). Permits and approvals are necessary to protect the workers on a project as well as the people who will pass by or work near the construction site. They also protect the health and safety of people who will later occupy the building. People will often hire an independent inspector to ensure a building has met all the correct specifications.

Before construction begins, a building permit is issued. A **building permit** indicates that construction can take place on a specific piece of land. Building permits are issued by a community's planning, or zoning, board. A **zoning board** controls what types of construction can be undertaken and how the land is used.

In one area of town, for example, zoning laws may allow only homes to be built. Another area may be zoned for factories and yet another for stores. The zoning board controls how new construction will affect a community. It makes sure that there will be enough schools, roads, fire protection, and parking spaces for new residents or tenants. It tries to keep residential areas quiet and pleasant to live in.

Other permits are needed during later stages of construction. In a busy area, a builder might need a permit to build a covered walkway to protect passersby. For commercial buildings, a life safety plan must show how stairways, exit doors, and other features of a building will protect people during

Figure 9.24 Regular inspections are an important part of construction.

Describing *What kinds of things do you think building inspectors look for at large construction sites?*

emergencies. Permits must also be obtained to dig up the street, to hook up electrical, water, sewer, and gas services.

When the building is almost finished, the major systems of the building are tested to make sure that they have been installed properly. Elevators are tested and approved for safety. Fire detection and sprinkler systems are reviewed and tested. After all systems are checked and approved, the local building department issues a document known as a **Certificate of Occupancy,** which states that the building can be occupied.

Building Maintenance

After a building has been completed, the owner takes on new management jobs. If the building contains rental property, available spaces are advertised, and a rental agent is hired. To generate income, the building must be occupied with renters as soon as possible. This income is used to make payments on construction loans or on a mortgage.

All the systems of the building must be kept working properly. People can either maintain the existing systems or upgrade them to something new and modern. Plumbing systems may need to be cleaned out or have parts replaced. Gutters need to be cleaned out, and trees on the property may need to be trimmed. Inside, floors may need replacing, and appliances may need to be repaired. Weather and the daily wear and tear of living in a building can create a need for repair and maintenance. For instance, sometimes water can seep in through cracks in the wall or foundation, causing unseen damage. The maintenance of a house or other building is an ongoing job.

SECTION 4 Assessment

Recall and Comprehension

1. How are a general contractor and project manager different?
2. Why is maintenance so important for all structures?
3. Explain the need for building permits during a construction project.

Critical Thinking

1. **Describing** Describe four career opportunities provided by the construction industry.
2. **Evaluating** How does a Gantt chart help managers keep a construction project on schedule?

QUICK ACTIVITY

Almost all structures are susceptible to fires that can cause a loss of property and lives. What are the main causes of structural fires? What can be done to lessen the damage from a structural fire? Make a list of all the fire safety and fire prevention ideas you can find and share them with your class.
For more related Design Activities, see pages 272–275.

Chapter Summary

- Construction is the building of a structure on a site. Construction systems use the seven technological resources. Architects design a structure's shape and choose basic materials for its construction. Engineers design the structure's framework and its major systems. A general contractor directs the work of many different people on a construction job.

- Building techniques have improved over many centuries. Today, modern techniques are combined with a knowledge of materials to build roads, bridges, tunnels, buildings, and many other structures.

- Home construction involves selecting and preparing a site, then building the footing, foundation, floors, walls, and roof. It involves installing utilities and insulation and finishing the structure on the outside and the inside.

- As in manufacturing, the success of a construction project depends largely on how well it is managed.

- Scheduling large building jobs often requires the use of a Gantt chart. A Gantt chart shows the start and completion dates for different parts of a construction job.

- Permits and approvals are necessary before building begins, during construction, and upon completion. Permits include zoning approval, a building permit, permits to close sidewalks and dig up streets, and a Certificate of Occupancy.

- Buildings must be maintained. All the systems must be kept working properly and inspected regularly.

Building Vocabulary

Your teacher may give you a crossword puzzle. Complete the puzzle using the following words from this chapter. Exchange puzzles with a classmate to check each other's answers.

1. arch bridge
2. architect
3. beam bridge
4. bridge
5. building
6. building permit
7. cantilever bridge
8. Certificate of Occupancy
9. concrete
10. construction
11. construction site
12. contract
13. drywall
14. electrical panel
15. engineer
16. exterior finishing
17. footing
18. foundation
19. foundation wall
20. general contractor
21. Gantt chart
22. interior finishing
23. macadam
24. modular construction
25. modules
26. mortar
27. mortgage
28. National Electrical Code®
29. project manager
30. renovation
31. roof truss
32. sheathing
33. siding
34. stud
35. subcontractor
36. subfloor
37. superstructure
38. suspension bridge
39. thermal windows
40. tradesperson
41. tunnel
42. union
43. voltage
44. wind drift
45. zoning board

 See your teacher for the Crosstech puzzle.

Reviewing Content

1. Name five different types of structures.

2. What do architects do on a construction project? What do engineers do?

3. Explain the responsibilities of subcontractors on a construction project.

4. Describe some building techniques that reduce construction time.

5. List three things that need to be considered when choosing a construction site.

Applying Your Knowledge

1. Design a tower using rolls of newspaper as your building material.

2. Do research on the Golden Gate Bridge. Find out how the weight of this type of bridge is supported and in what situations it would be the best type of bridge to use.

3. Investigate how the Appian Way was constructed by the Romans.

4. If you were to choose a site for a movie theater, what are five things you would have to consider?

5. As a project manager who is planning the construction of a small office building, make a list of the permits and approvals you will need before, during, and after construction.

Critical Thinking

1. **Comparing** What are some similarities between manufacturing and construction systems?

2. **Contrasting** How is the job of a project manager different from that of a general contractor?

3. **Describing** Describe four career opportunities provided by the construction industry.

4. **Evaluating** How does a Gantt chart help in keeping a construction project on schedule?

5. **Drawing Conclusions** Why is maintenance so important for all structures?

Connecting to STEM
science · technology · engineering · math

Develop a Gantt Chart

Gantt charts can be drawn on graph paper or created with computer software. Each horizontal bar represents one project task. The length of each bar represents the time the task takes and runs from the beginning time (or date) to the ending time (or date).

In your technology class, you are involved in design activities that take a matter of days or weeks to complete. Choose one of these design activities and use the Gantt chart as a time management technique to develop a work schedule for your activity.

Assign people to each task, and include the amount of time each task will take. Compare your finished Gantt chart with those developed by two other individuals or groups.

Design Activity 17

BRIDGE DESIGN

Problem Situation

You have probably driven over or under many different kinds of bridges. Bridges are used to cross bodies of water and to span ravines, canyons, highways, and railroad tracks. Imagine that you have to design a bridge between two embankments. The sides are 2 feet apart, and the distance to the ground is approximately 8 inches. The bridge that holds the most weight for the least construction cost—holding a minimum weight of 20 pounds—will be the optimum design.

Your Challenge

You and your teammates are to design and construct a bridge.

> Go to your **Student Activity Guide, Design Activity 17.** Complete the activity in your Guide, and state the design challenge in your own words.

① Clarify the Design Specifications and Constraints

To solve the problem, your design must meet the following specifications and constraints:
- Your bridge must be constructed only with materials from this list: oaktag, string, wooden rods, brass fasteners, craft sticks and/or Popsicle sticks, and glue.
- The bridge must be wide enough to allow a 4" x 8" brick to be placed on the midspan section.

> In your Guide, state what the specifications and constraints are. Include any others that your team or your teacher included.

② Research and Investigate

To better complete the design challenge, you need to first gather information to help you build a knowledge base.

> In your Guide, complete Knowledge and Skill Builder 1: Beam Bridges.

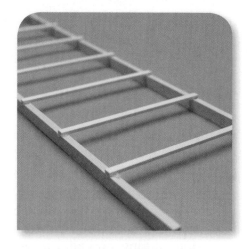

Materials

You will need:
- 24" wooden rods
- brass fasteners
- building bricks (to use as embankments)
- craft sticks (basswood strips, 1/8" x 2" x 24")
- large weights (1/2 lb)
- modeling clay
- oaktag sheet, 30" x 30"
- Popsicle sticks (optional)
- small weights (metal washers)
- string
- tape
- white glue

In your Guide, complete Knowledge and Skill Builder 2: Suspension Bridges.

In your Guide, complete Knowledge and Skill Builder 3: Arch Bridges.

In your Guide, complete Knowledge and Skill Builder 4: Truss Bridges.

③ Generate Alternative Designs

In your Guide, describe two possible solutions that your team has created for the problem. Your solutions should be based on the knowledge you have gathered so far.

④ Choose and Justify the Optimal Solution

Refer to your Guide. Explain why you selected the solution you did, and why it was the better choice.

⑤ Develop a Prototype

Construct your bridge. Include a drawing, either a side view or an isometric view, or a photograph of your final design in your Guide.

In any technological activity, you will use seven resources: people, capital, time, information, energy, materials, and tools and machines. In your Guide, indicate which resources were most important in this activity, and how you made trade-offs among them.

⑥ Test and Evaluate

How will you test and evaluate your design? In your Guide, describe the testing procedure. Explain how the results show that the design solves the problem and meets the specifications and constraints.

⑦ Redesign the Solution

Respond to the questions in your Guide about how you would redesign your solution. The redesign should be based on the knowledge and information that you gained during the activity.

⑧ Communicate Your Achievements

In your Guide, describe the plan you will use to present your solution to your class and include any handouts and/or PowerPoint slides you will use.

Design Activity 18

ONE-ROOM SCHOOLHOUSE

Problem Situation

In rural areas of the United States in the early 1900s, children were often educated in one-room schoolhouses. One teacher had to teach children from kindergarten to eighth grade in one room. Your class has been given the challenge of constructing a scale model of a one-room schoolhouse, showing appropriate structural elements. It will have electricity, so lighting must be provided.

Your Challenge

You and your teammates are to design and create a scale model of a one-room schoolhouse.

> Go to your **Student Activity Guide, Design Activity 18.** Complete the activity in your Guide, and state the design challenge in your own words.

Materials

You will need:
- basswood or wooden strips
- battery
- electrical bell wire
- flashlight bulbs and holders
- glue
- oaktag
- switches

1 Clarify the Design Specifications and Constraints

To solve the problem, your design must meet the following specifications and constraints:
- The scale for the model one-room schoolhouse will be 1 inch = 1 ft.
- The schoolhouse base should be approximately 20" x 30".
- The walls will have 2" x 4" studs on 16" centers built to approximate the scale.
- See more specifications and constraints in your Guide.

> In your Guide, state what the specifications and constraints are. Include any others that your team or your teacher included.

2 Research and Investigate

In order to better complete the design challenge, you need to first gather information to help you build a knowledge base.

> In your Guide, complete Knowledge and Skill Builder 1: Ratio and Proportion.

> In your Guide, complete Knowledge and Skill Builder 2: Roof Shapes.

In your Guide, complete Knowledge and Skill Builder 3: Framing Doors and Windows.

In your Guide, complete Knowledge and Skill Builder 4: Electrical Circuits.

3 Generate Alternative Designs

In your Guide, describe two possible solutions that your team has created for the problem. Your solutions should be based on the knowledge you have gathered so far.

4 Choose and Justify the Optimal Solution

Refer to your Guide. Explain why you selected the solution you did, and why it was the better choice.

5 Develop a Prototype

Construct your scale model of a one-room schoolhouse. Include an isometric drawing, or drawings showing front, top, and side views, or a photograph of your model in your Guide.

In any technological activity, you will use seven resources: people, capital, time, information, energy, materials, and tools and machines. In your Guide, indicate which resources were most important in this activity, and how you made trade-offs among them.

6 Test and Evaluate

How will you test and evaluate your design? In your Guide, describe the testing procedure. Explain how the results show that the design solves the problem and meets the specifications and constraints.

7 Redesign the Solution

Respond to the questions in your Guide about how you would redesign your solution. The redesign should be based on the knowledge and information that you gained during the activity.

8 Communicate Your Achievements

In your Guide, describe the plan you will use to present your solution to your class and include any handouts and/or PowerPoint slides you will use.

UNIT 4

Communication and Information Technology

"The Internet is here to stay. It really is one of those revolutionary technologies that has the ability to change so much of how we live, how we communicate, how we plan, how we get information, how we buy things."

—Meg Whitman, former CEO of eBay (March 1998 to March 2008)

Communication Systems

In this chapter, you will learn about the following different parts of the communication process:
- source
- encoder
- channel
- decoder
- receiver

Our world depends on communication systems.

- We keep in touch with our friends and families by telephone.

- Companies use the Internet to send information to branch offices throughout the world.

- News is distributed worldwide via live television.

- During emergencies, communication systems are used to send urgent messages to fire and hospital personnel.

Much of this information is sent via satellites. Right now, literally hundreds of satellites are circling the Earth. These satellites can transmit information from one part of the globe to another in a matter of seconds.

Communications satellites are designed, built, launched into orbit, and maintained solely to allow for more efficient communication. In this chapter, you will learn how different parts of a communication system work together.

What Is Communication?

Benchmarks for Learning

- In communication systems, information is transferred from a source to a destination.
- For communication to occur, a message must be sent, received, and understood.
- Humans, animals, and machines can all communicate.

Reading Strategy

Listing   Make a list of the different types of communication that exist. Next to each item on the list, write down an example of your own.

Types of Communication
1. person-to-person—talking with friends
2. written communication—

Vocabulary

communication

person-to-person communication

machine-to-machine communication

The Ways We Communicate

Humans have always needed to communicate. **Communication** is the process of sending any type of message from one place to another. We communicate to have our needs met, to stay safe and well, and for social interaction with others. Most often, we communicate using speech or writing, but there are many other ways we get our message across.

The communication systems we use today—live television, satellite radio, the Internet—evolved from our ancestors' primitive speech, gestures, and drawings. Although methods of communicating have changed dramatically, the purpose of communication systems has always been the same: to transfer information from a source to a destination.

Effective communication requires three things: sending a message, having it received, and having it understood. The sending and receiving of messages between two people is called **person-to-person communication.** Most often, we communicate with other people by speaking or writing a language.

Speaking and Writing

The earliest written communication known is cuneiform. It was first used by people who lived in the Middle East about 6,000 years ago (4000 B.C.). Cuneiform symbols were made by pressing a wedge-shaped tool into wet clay. Each

symbol, which was made up of several wedge-shaped marks, stood for a word or an idea. Later, the Egyptians used a writing method known as hieroglyphics (Figure 10.1). Hieroglyphic symbols are pictures of things. Hieroglyphics were carved into stone or painted on stone or paper.

The first alphabet came from people living in the Middle East. The Hebrew alphabet started with the letters "aleph" and "bet." Later, the first two letters of the Greek alphabet were "alpha" and "beta." These letters together give us the word *alphabet*.

When people speak, they use the words and expressions of their own language. Hundreds of different languages are

Figure 10.2 Former Japanese Prime Minister Keizo Obuchi and Former French Prime Minister Lionel Jospin are using an interpreter to translate their ideas from one language to another.

Interpreting *What messages are also being sent with these politicians' body language and expressions?*

spoken and written throughout the world (Figure 10.2). In the United States, Canada, Great Britain, and South Africa, the common language is English. Yet, there are differences even in the English spoken in these countries. A "truck" in the United States is a "lorry" in Great Britain. A "flat" in South Africa is an "apartment" in the United States.

Communicating Without Words

Sometimes, we communicate without speaking. We use facial expressions, such as smiling or winking, to show how we feel. We also use body language to communicate. When we are angry, we might cross our arms in front of our body. If we are bored, we might shift about in our seat and look around. Often, we are not even aware that our body language is communicating how we really feel.

Of course, sight and hearing are important in communication, but so are other senses, such as touch and smell. A child's mother communicates love and security with a hug. Pleasant smells and touch are used in massage to relieve stress and to help people relax.

Communicating with Animals

People can also communicate with animals. A dog, for example, knows when its owner is ready to take it for a walk. Perhaps it knows the word "walk," or what to expect because of its owner's tone of voice. The dog's owner knows when the dog is hungry. The dog may stand next to its food bowl and whine or wag its tail.

Figure 10.3 This chimp learned sign language and could "talk" with her trainers.

Extending *How are visual signals used to communicate with other animals, such as dogs or horses?*

Animals can also communicate with other animals. When bees find a source of food, they move in a figure-eight pattern called a waggle dance. This "dance" communicates to other bees that there is food nearby. Chimpanzees and other primates have more sophisticated ways of communicating. Some have been taught a human language called sign language, in which they have learned a vocabulary of hundreds of words (Figure 10.3).

Communicating with Machines

With technology all around us, people communicate with machines on a daily basis (Figure 10.4). When you want to

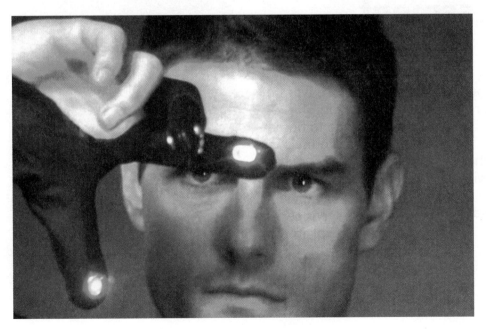

Figure 10.4 In the futuristic movie *Minority Report,* Tom Cruise uses hand movements to communicate with a computer.

Making Judgments *Why is using hand movements considered more futuristic than using a typical input device, such as a mouse?*

call a friend, you communicate with the phone by keying in your friend's number. The movement of your fingers is changed into electrical signals that are received by the phone.

Machines can communicate with people, too. A computer relays a message when it displays text or graphics on its monitor. Its output can also be a print, sounds from a voice synthesizer, or the display of lights on a screen.

Machine-to-Machine Communication

In **machine-to-machine communication,** machines send messages to and from each other. For example, a robot arm that spray-paints automobiles is controlled by a computer in the factory. The computer "tells" the robot arm what to paint, in what direction, and for how long.

Another example of machine-to-machine communication is the movement of information between a thermostat and a furnace. Once a person sets a thermostat to a desired temperature, the thermostat and the furnace send information back and forth to keep the building at that temperature.

You have seen that people communicate with each other, with animals, and with machines. What other forms of communication might be made possible by technology?

SECTION 1 Assessment

Recall and Comprehension

1. What three things must happen to a message for communication to occur?
2. Name three ways that humans communicate.
3. Identify the source and the destination for the following systems:
   - **(a)** televising a baseball game
   - **(b)** writing a letter
   - **(c)** retrieving information from a Web site

Critical Thinking

1. **Applying** Give an example of each of these types of communication:
   - **(a)** machine to human
   - **(b)** human to machine
   - **(c)** machine to machine
2. **Extending** Is communication accomplished every time a message is sent or received? Explain your answer.

QUICK ACTIVITY

A simple code uses the number 1 to represent an A, the number 2 to represent a B, 3 a C, and so on. As long as the recipient knows the code, the message can be deciphered. Try developing a code of your own. Translate a message into that code, and then challenge your classmates to "break" your code.
For more related Design Activities, see pages 298–301.

SECTION 2
Communication as a System

Reading Strategy

Mapping Draw a concept map to show the different parts of a communication system. Make sure that you can explain the map to your classmates.

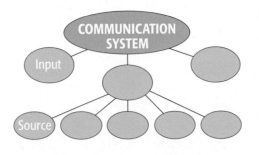

COMMUNICATION SYSTEM

Input

Source

Vocabulary

transmit
source
encoder
channel

decoder
receiver
graphic
   communication

electronic
   communication
communication medium
mass media

Benchmarks for Learning

- Effective communication systems have an input, a process, an output, and a feedback loop.
- The communication process includes a message, a source, an encoder, a channel, a decoder, and a receiver.
- Communication systems are used to inform, persuade, educate, and entertain.
- The design of a message is influenced by its intended audience, medium, and purpose.

Parts of a Communication System

Like all systems, a communication system has an input, a process, and an output. Effective communication systems also include a feedback loop.

Input, Output, and Feedback

A communication system can be as complex and large as a message sent via satellite. Or it can be as simple as two people having a conversation. In the example below, Max and Parker are talking about the baseball game they are watching (Figures 10.5 and 10.6). Max says something to

Figures 10.5 and 10.6
A spoken conversation is an example of a communication process.

Interpreting *How might nonverbal communication provide feedback in this system?*

Figure 10.5

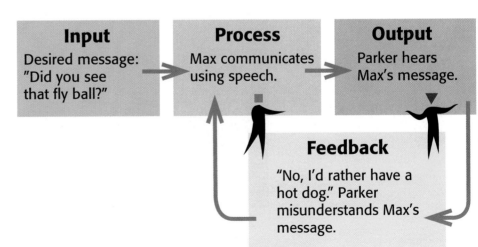

Figure 10.6

Input
Desired message: "Did you see that fly ball?"

Process
Max communicates using speech.

Output
Parker hears Max's message.

Feedback
"No, I'd rather have a hot dog." Parker misunderstands Max's message.

LIVING GREEN
Sensing a Change?

The development of high-tech satellites provides up-to-the-minute news, weather, and nonstop communication of information. Many of these tools are placed directly into your hands by visiting a number of available portals on the Internet. Satellite information collected by government programs such as the National Oceanic and Atmospheric Administration (NOAA) is used to track ocean temperatures and weather events and to monitor the destruction of coral reefs. The U.S. Environmental Protection Agency uses satellites to evaluate air pollution. NASA, the National Aeronautics and Space Administration, uses satellites to observe changes in Earth's forests and land masses. Lastly, Google Earth® gives anyone with an Internet connection a view all the way from space right down to the house next door.

MAKE A CHANGE
Satellite information provides information through the use of remote-sensing technology to inform our understanding about the surface of the Earth, the atmosphere, and celestial bodies. Remote sensing means gathering data using techniques such as radar, sonar, photographic imagery, lasers, and instruments to collect radiation, and sending that data from a great distance through a wireless connection to computers that process the data into useful information. Many science and technical fields including archaeology, agriculture, astronomy and cosmology, geology, oceanography, and medical technology use some form of remote sensing. The best part is that every student has the ability to use remote sensing to investigate any number of environmental concerns. Visit some of the sites and see what's available.

TRY THIS
Incorporating remote-sensing images and data into your next engineering and technology presentation can really communicate a powerful message to your audience. Choose an issue that is a concern of your community, such as water or air pollution, mining of natural resources, or flooding in local rivers and lakes, and download satellite images to help communicate these issues. Then, create a public information message in a brochure, video, or blog to get the word out.

Parker: "Did you see that fly ball?" The input command is the desired message, or result. In a communication system, the message is the information that we wish to send. The process, which is how the input is communicated, is Max's speech. The output is the message that Parker actually receives.

The Communication Process

For whatever type of message we wish to send, or **transmit,** the communication process consists of five elements: a source, an encoder, a channel, a decoder, and a receiver.

The **source** is who or what produces the information or message to be sent. The **encoder** converts the information into a form that can be sent easily. The **channel** is the medium through which the message moves. Just as a source needs an encoder to convert the message for transmission, a **decoder** is needed to convert the message back to a form that can be understood by the **receiver.**

In the example below, a written message is typed into a computer, where it is encoded digitally by the computer modem (Figure 10.7). The encoded message travels through

Figure 10.7 During the communication process, messages are encoded and then decoded.

Interpreting *What machine in this system encodes and decodes the message?*

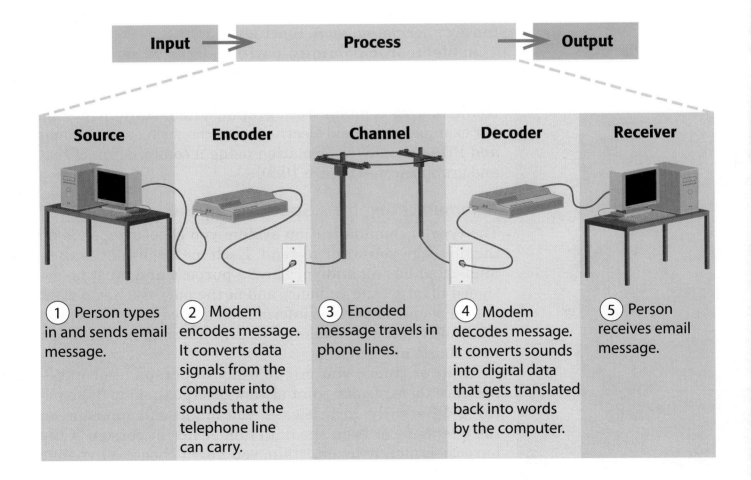

Input → Process → Output

| Source | Encoder | Channel | Decoder | Receiver |

① Person types in and sends email message.

② Modem encodes message. It converts data signals from the computer into sounds that the telephone line can carry.

③ Encoded message travels in phone lines.

④ Modem decodes message. It converts sounds into digital data that gets translated back into words by the computer.

⑤ Person receives email message.

phone lines to another person's computer. When it is downloaded to that computer, that computer's modem decodes the message. The person receives a written message displayed on the computer screen.

Sending email through a computer over a phone line is just one example of a communication process. Messages can also be sent using ink and paper (writing), cellular telephones (speech), digital cameras (photography), and in many other ways. The process for each of these systems is different, but they are all forms of communication.

Types of Communication Systems

Communication systems serve many different purposes. They can be used to inform, persuade, educate, and entertain. Communication systems are developed with these specific purposes in mind.

Graphic and Electronic Communication

Today, all forms of communication fall into one of two categories: graphic communication and electronic communication. In **graphic communication,** printed pictures or words carry the message. Small pictures or icons are often used to convey information more concisely than text.

In **electronic communication,** electrical signals carry the message. Computers use electronic communication by sending and receiving binary code. Systems that use electronic images combine both graphic and electronic media. For example, handheld electronic devices, such as cell phones and PDAs, display information using a combination of text and graphic icons (Figure 10.8).

Communicating for a Specific Purpose

The type of communication system you choose depends on the message you want to send. Each message that is communicated has an audience and a purpose and must be designed to target the audience and fit the purpose. If you want to give people important information and educate them, you might distribute a newsletter or give a formal presentation with charts and graphs.

At other times, you might want to persuade people—convince them of your point of view or urge them to do something differently. In this case, you might call someone on the telephone or even speak to him or her in person. If the purpose is simply to entertain, you might choose a dynamic format such as a video game or a television show.

Figure 10.8 Many forms of communication today use a combination of electronic and graphic media.

Interpreting *What type of communication are you using as you read this book?*

For: Emotions Activity
Visit: www.mytechedkit.com

How to Express Your Emotions Online

Scott E. Fahlman is the recognized inventor of emoticons, the symbols you create with keyboard characters to illustrate your emotional state in a message. His first emoticon—a simple sideways smile made up of punctuation marks :-)—gave us a new way to communicate in our Information Age. Go online to see how some cleverly placed emoticons can express a great deal.

Figure 10.9 Mass media are forms of communication that reach many people.

Extending *When would mass media be useful for sending a message? When would another form of communication be more useful?*

Communication Media

The method that we use to communicate our message is called the **communication medium.** Common communication media are letters, newspapers, magazines, books, television, radio, email, chatrooms, and Web pages. Some media can be used effectively for more than one purpose. For example, television can be used to inform, persuade, and entertain. A cartoon in a magazine might entertain and persuade at the same time.

An especially useful way of communicating is through mass media. **Mass media** are forms of communication that reach large numbers of people (Figure 10.9). Mass media include newspapers, magazines, television programs, and radio.

SECTION 2 Assessment

Recall and Comprehension

1. Describe at least three ways that communication systems are used.
2. What are the different parts of an effective communication system?
3. How is feedback received in a communication system using spoken words?
4. Name three different forms of mass media. Provide a specific example of each.

Critical Thinking

1. **Clarifying** The design of a message is influenced by its intended audience, medium, and purpose. Explain this principle using a magazine ad as an example.
2. **Applying** Choose a communication system. Draw a diagram of that system and identify the source, encoder, channel, decoder, and receiver.

QUICK ACTIVITY

Your ability to communicate a particular message can even save someone's life. Design a poster that encourages your classmates not to smoke or use tobacco products. As a class, select the poster that is the most effective. Why was this poster selected?
For more related Design Activities, see pages 298–301.

People in Technology

Tim Berners-Lee
Inventor of the World Wide Web

"There have always been things which people are good at, and things computers have been good at, and little overlap between the two."

—Tim Berners-Lee

"There have always been things which people are good at, and things computers have been good at, and little overlap between the two."

This statement by Tim Berners-Lee explains much of his mission. Credited as the person who invented the World Wide Web, Berners-Lee has been continually focused on increasing the usefulness of computers for people.

An Early Fascination with Math
Berners-Lee inherited his fascination with computers from his parents, who both worked on one of England's first computers. As a child, he made a model of a computer out of cardboard boxes.

Tim Berners-Lee

Later, in college, he built a computer out of spare parts and an old TV.

In 1980, Tim Berners-Lee was working as a software engineer for a physics laboratory in Europe. Surrounded by masses of information, Berners-Lee developed a software program that could, as he put it, keep "track of all the random associations one comes across in real life."

The system could create links among all the files on Berners-Lee's computer. This simple network allowed people to share data quickly and easily.

Developing a Web Language
To set up a larger, more useful network for sharing information, Berners-Lee developed a coding system that made linking easy. He called it Hypertext Markup Language (HTML). In HTML, a specific code is attached to every piece of information in a system. The address of each person's file, or Web page, is called a *universal resource locator,* or URL. To connect all this information, Berners-Lee developed a simple browser that enabled people to access his web of files.

Nearly 2 billion people, or 26% of the world's population, now have access to the Internet.

Critical Thinking

1. **Summarizing** What does the term World Wide Web refer to? How did it grow from Tim Berners-Lee's early work on computers?

2. **Extending** Do research on the Internet to find out which countries have the most Internet users. Make a list of the top ten.

SECTION 3
Resources for Communication Systems

Benchmarks for Learning

- Communication systems use the seven technological resources.

- Information technologies make it possible to analyze and interpret data, including text, images, and sound in ways that are not possible with human senses alone. These uses may result in positive or negative impacts.

Reading Strategy

Listing   Make a list of the different resources that are specific to communication systems. Use the headings throughout the section to organize your list.

Resources Used in Communication Systems
1. People—run and repair systems
2. Capital—

Vocabulary

gravitation   geosynchronous orbit

Figure 10.10   People, such as this camera operator, are an essential resource in communication systems.

Extending   *What other people are involved in making a television program?*

People

Just as in manufacturing, construction, and other types of systems, communication systems make use of the seven types of technological resources. From talking on the phone to broadcasting a television commercial, communication systems use people, capital, time, information, energy, materials, and tools and machines.

People are essential to effective communication systems. They decide on the message, help deliver it, and receive it. Actors and camera operators are needed for television shows and films (Figure 10.10). Writers write books, scripts for radio and television, and newspapers. Artists and photographers produce illustrations. Printers run machines that turn out many copies of books, newspapers, and magazines. People are needed to run and repair all the different kinds of communication systems.

Capital

Capital is needed to set up and run a communication system. Machines and tools must be bought. People must be paid for their work. Work spaces, such as television studios and print shops, must be built or rented. Light and heat must be provided.

Time

Moving a message takes time. It must travel from the transmitter to the receiver. The time needed depends on the length of the message and the rate at which the message is sent.

The time needed to relay messages has changed dramatically as a result of technology. Consider one way people sent messages over long distances just 100 years ago—by handwritten letters that usually traveled by train. Now, messages can be sent anywhere in the world in a fraction of a second. Can the time required for communication continue to change?

Information

Information is at the center of all communication. People use communication to share and receive all types of information. For communication to be useful, information is needed about the audience that will receive the message. Advertisers need to find out what will make people buy their products. What works for one group of people will not work for everybody. For example, teenagers and much older people normally like very different types of music. Information helps advertising companies communicate an effective, appropriate message (Figure 10.11).

Information is also needed to keep communication systems inexpensive and of high quality. Research and good decision making are needed to choose everything from the best writers to the best quality paper. To maintain communication systems, engineers and technicians use information of many kinds. Information technologies make it possible to analyze and interpret data, including text, images, and

Figure 10.11 New materials and printing technologies allow advertisers to communicate messages in dramatic ways.

Applying Concepts *What resources were used to communicate these graphic images?*

sound, in ways that are not possible with human senses alone. These uses may result in positive or negative impacts.

Energy

Energy is used to move a message from place to place. Electricity powers radio and television transmitters and receivers. It also runs computer systems, printing presses, and copiers. In telephones, sound energy is changed into electrical energy and then back to sound energy. Light energy is used to create images on photographic film.

Materials

Materials are constantly being changed and improved to make communication faster, less expensive, and better. New camera film makes it possible to take pictures in very low or very bright light. Paper and inks can give book covers and printing the look of silver or gold. New optical disks store a large amount of information in a small space.

Tools and Machines

Tools and machines are used and constantly updated to send messages quickly, on time, and according to budget. Printers and artists use hand tools such as airbrushes and rulers, pens, and pencils. Machines such as televisions, cameras, printing presses, satellites, and radio transmitters are used to design, send, receive, and store messages.

SECTION 3 Assessment

Recall and Comprehension

1. Describe three ways in which people are important in communications systems.
2. What two factors determine the amount of time it takes for a message to be sent and received?
3. How is information used in advertising? How is it used in other types of communication systems?

Critical Thinking

1. **Applying** Pick one example of a communication system. Explain how it uses each of the seven technological resources.
2. **Applying** Give an example of how each of these resources is used in a communication system:
   (a) material
   (b) tool
   (c) machine

QUICK ACTIVITY

Accurate communication can sometimes be difficult. Stand in front of the chalkboard with your back to the class. Have another student hand you an object. He or she can only describe the object to you, not identify it. As your classmate describes the object, try to draw it accurately on the chalkboard. Now, explain the saying, "A picture is worth 1000 words."

For more related Design Activities, see pages 298–301.

Connecting to STEM

Communications Satellites

If you throw a ball up into the air, it will eventually fall back to the ground. Why doesn't it continue moving along the path in which you threw it? Because of the force of gravity.

The Earth (and anything with mass) has a gravitational force that pulls things toward its center. **Gravitation** is the force of attraction between any two objects with mass. The greater the mass, the greater the gravitational force.

Using Gravity to Stay in Orbit

Of course you've already heard of gravity, but did you know that gravity is also what keeps satellites in orbit? Without the gravitational pull of the Earth, a satellite—just like a thrown ball—would move in a straight line away from the Earth. Gravitational pull and a satellite's initial speed keep the satellite moving in a path around the Earth.

Geosynchronous Orbits

Satellites in **geosynchronous orbit** move around the Earth at the same speed at which the Earth rotates around its axis. You can learn something about geosynchronous orbits by simply looking at their name: geo- means Earth, syn- means same, and chronous means time (literally, same time as the Earth). The Earth rotates around its axis once every day, or 24 hours. A geosynchronous satellite completes its orbit every 24 hours.

A satellite's speed is related to its distance above the Earth. To keep their speed consistent with that of the Earth, all geosynchronous satellites travel at the same distance above the Earth, approximately 22,000 miles (36,000 km). The first geosynchronous satellite was launched in the 1960s. Today, only a few decades later, hundreds of satellites orbit the globe. Imagine the view from 22,000 miles above the Earth!

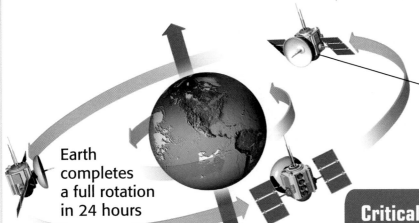

Earth completes a full rotation in 24 hours

Geosynchronous satellite completes an orbit every 24 hours.

If satellites move around the Earth, how can a satellite dish that is located at a fixed place on Earth receive signals at any time of day or night? Because it receives signals from satellites that always keep the same position relative to the Earth. They do this by moving in a geosynchronous orbit.

Critical Thinking

1. **Characterizing** In your own words, describe a geosynchronous orbit.

2. **Inferring** Because everything with mass exerts a gravitational force, a satellite exerts a gravitational force on the Earth. Why is this force unimportant to the way in which satellites work? (*Hint:* Gravitational force is proportional to an object's mass.)

Review and Assessment

Chapter Summary

- Communication systems allow information to be transferred from a source to a destination. Real communication takes place only when the message is sent, received, and understood.

- Humans, animals, and machines communicate among themselves and with each other.

- All communication systems have inputs, processes, and outputs. The communication process includes a message, a source, an encoder, a channel, a decoder, and a receiver.

- We use communication systems to inform, persuade, educate, and entertain.

- The choice of a communication system is based on the type and size of the audience. The design of the message is influenced by the intended audience, the medium, and the purpose.

- Modern communication systems can be divided into two categories: graphic communication and electronic communication. In graphic communication, the channel carries images or printed words. In electronic communication, the channel carries electrical signals.

- Like other technological systems, communication systems make use of the seven types of technological resources.

Building Vocabulary

Your teacher may give you a crossword puzzle. Complete the puzzle using the following words from this chapter. Exchange puzzles with a classmate to check each other's answers.

1. channel
2. communication
3. communication medium
4. decoder
5. electronic communication
6. encoder
7. geosynchronous orbit
8. graphic communication
9. gravitation
10. machine-to-machine communication
11. mass media
12. message
13. person-to-person communication
14. receiver
15. source
16. transmit

See your teacher for the Crosstech puzzle.

Reviewing Content

1. What is machine-to-machine communication? Give two examples of how this type of communication might be used.

2. In your own words, provide definitions for the following terms: source, encoder, channel, decoder, receiver.

3. What is graphic communication? What types of materials does it use?

4. What are mass media? Provide three specific examples of mass media that you have used.

5. How are people used as a resource in communication systems? What skills might be used by these people?

Applying Your Knowledge

1. Draw a systems diagram of a telephone communication system. Identify the input, process, output, and feedback.

2. Which communication system would you use to inform middle-school students about a new school policy? What system would you use to inform the entire community?

3. How might a message be designed for each of these purposes?
   (a) to educate people about the need to protect the environment
   (b) to entertain a group of kindergarten children during a school assembly
   (c) to persuade teenagers that smoking is harmful

4. List five examples each of graphic and electronic communication systems.

5. How can a telephone be used to inform, persuade, educate, and entertain?

Critical Thinking

1. **Extending** Think of a technology that would allow people, animals, and machines to communicate with each other. Describe how it would work.

2. **Characterizing** Decide if each of the following communication systems is graphic, electronic, or a combination of both: email, Internet Web pages, painted pictures, film-based photography, billboards.

3. **Relating Cause and Effect** How does time affect the communication of a telephone message sent by satellite?

4. **Applying** Pick a communication system, and then describe how it uses each of the seven technological resources.

 Connecting to STEM
science • technology • engineering • math

The Speed of Sound

Sound travels at about 66,000 feet per minute. Electrical signals travel at the speed of light, or 186,000 miles per second. If you talk on the telephone, the sound of your voice is converted into electrical signals. Which travels faster—sound or electrical signals? (*Hint:* Try converting the speed of sound to feet per second.)

Design Activity 19
COMMUNICATIONS SATELLITE

Problem Situation

Communications satellites fly in geosynchronous orbit around the Earth. A geosynchronous orbit is one in which the satellite always stays in the same position relative to the Earth's surface. It circles the Earth at the same rate at which the Earth rotates.

Materials

You will need:
- foam board
- laser pointer
- mirrors
- modeling clay
- modeling foam
- small pieces of wood
- stiff wire

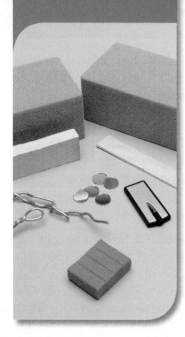

Your Challenge

You and your team members are to create a model showing a communications satellite and the Earth in proportion to each other. Using a laser pointer and mirrors, show how signals are transmitted from one point on Earth to another.

> Go to your **Student Activity Guide, Design Activity 19.** Complete the activity in your Guide, and state the design challenge in your own words.

1 Clarify the Design Specifications and Constraints

To solve the problem, your design must meet the following specifications and constraints:
- The size of the Earth and the satellite's distance from it should be in proportion to each other. The size of the satellite does not have to be in proportion.
- The Earth model should be at least 12 inches in diameter.
- The laser should be inserted in the Earth model so it points upward from it.

> In your Guide, state the design specifications and constraints. Add any others that your team or your teacher included.

2 Research and Investigate

To better complete the design challenge, you need to first gather information to help you build a knowledge base.

> In your Guide, complete Knowledge and Skill Builder 1: Size and Proportion.

In your Guide, complete Knowledge and Skill Builder 2: Transmission and Reception.

In your Guide, complete Knowledge and Skill Builder 3: Communicating Around the Globe.

In your Guide, complete Knowledge and Skill Builder 4: Experimenting with Materials.

③ Generate Alternative Designs

In your Guide, describe two possible solutions that your team has created for the design challenge.

④ Choose and Justify the Optimal Solution

Referring to your Guide, explain why you selected the solution you did, and why it was the better choice.

⑤ Develop a Prototype

Construct your solution. Put a photograph or sketch of your final design in your Guide.

In any technological activity, you will use seven resources: people, capital, time, information, energy, materials, and tools and machines. In your Guide, indicate which resources were most important in this activity, and how you made trade-offs among them.

⑥ Test and Evaluate

How will you test and evaluate your design? In your Guide, describe the testing procedure you will use. Justify how the results will show that the design solves the problem and meets the specifications and constraints.

⑦ Redesign the Solution

Respond to the questions in your Guide about how you would redesign your solution based on the knowledge and information that you gained during the activity.

⑧ Communicate Your Achievements

In your Guide, describe the plan you will use to present your solution to your class. Show what handouts, view-graphs, and/or PowerPoint slides you will use.

Design Activity 20

YOUTUBE IT!

Materials

You will need:

- video camera, or video equipment available to you
- iMovie™ editing, or video editing equipment

Problem Situation

The environmental club is trying to organize students to participate in an international beach clean-up day, sponsored by the Littoral Society. On this day, people around the world pick up debris from beaches and waterways, then catalog and weigh it to develop an international record. The club has not had success previously in motivating students to join, so they have turned to the technology class for help. They want to educate and persuade classmates to join them in the clean-up.

Your Challenge

Select an issue—from joining a club to participating in a community volunteer effort—about which you would like to educate and persuade others. Develop a one-minute YouTube video with students from another school or schools to encourage others to join your effort.

> Go to your **Student Activity Guide, Design Activity 20.** Complete this activity in the Guide, and state what the design challenge is in your own words.

1 Clarify the Design Specifications and Constraints

To solve the problem, your design must meet the following specifications and constraints:

- The video should inform others why becoming involved is a good idea, what they will be doing, and how it benefits them and others.
- You will need at least one team member who is at another school, so you will be working as a virtual team.
- The video should be one minute long.

> In your Guide, state the design specifications and constraints. Add any others that your team or your teacher included.

2 Research and Investigate

To better complete the design challenge, you need to first gather information to help you build a knowledge base.

In your Guide, complete the following Knowledge and Skills Builders:

- Building a Team
- YouTube
- Lights, Camera,…
- Action!

3 Generate Alternative Designs

In your Guide, describe two possible solutions that your team has created for the design challenge. In lieu of a sketch, include a brief overview of the key ideas of your presentation as well as the images you will use. Indicate how the images will be transmitted, and how you anticipate they will be received and interpreted. In your description, indicate what you consider to be each solution's strengths and weaknesses.

4 Choose and Justify the Optimal Solution

Referring to the Student Activity Guide, explain why you selected the solution you did, and why it was the better choice.

5 Develop a Prototype

Develop a script for your chosen design. Shoot and edit the video.

Attach a copy of your script to the folio. Include a still photo of your team in action.

In your Guide, indicate which technological resources were most important in this activity, and how you made trade-offs among them.

6 Test and Evaluate

How will you test and evaluate your design? In your Guide, describe the testing procedures you will use. Justify how the results will show that the design solves the problem and meets the specifications and constraints.

7 Redesign the Solution

Respond to the questions in your Guide about how you would redesign your solution. Your redesign should be based on knowledge and information that you gained during the activity.

8 Communicate Your Achievements

In your Guide, describe the plan you will use to present your solution to your class. Show what handouts, view-graphs, and/or PowerPoint slides you will include.

Electronics and the Computer

In this chapter, you will learn about electronics and the following different parts of a computer system:
- motherboard
- CPU
- input devices
- output devices
- memory

Electricity became available less than 100 years ago, but today it is impossible to imagine life without it.

- Electrical signals carry information quickly over wires or through the air via radio waves.

- Electricity is used to process and store data.

- Electric light has extended our day.

- Electronic entertainment has changed the way we use our free time.

Electronic technology has also given birth to the computer—and to our current Information Age. In the United States today, there is one computer for about every three people, and 80 million people are Internet users. Worldwide, 200 million people use the Internet. These numbers are continually increasing.

Computers have created an information revolution as significant as the Industrial Revolution of the 1800s. In this chapter, you will learn about the technology that made computers possible.

Understanding Electronics

Benchmarks for Learning

- The use of electronics has completely changed our world over the last 100 years.
- All materials are made up of tiny particles called atoms.
- Electrical current is the flow of electrons through a conductor.

Reading Strategy

Listing Make a list of all the electrical units that are defined in this section. Next to each unit, write a definition in your own words.

I. Units in Electricity
- Ampere—number of electrons moving in a second
- Volt—

_____

Vocabulary

atom	neutron	volt
element	electron	Ohm's law
nucleus	current	resistance
proton	ampere	ohm

Electrical Energy

Much of today's technology uses electrical energy. The television, the refrigerator and kitchen stove, all types of lights, and your computer run off the electrical energy supplied to the building. Anything powered by batteries—cell phones, laptop computers, handheld games—also uses a portable form of electricity. Without electricity, much of the technology on which we depend would not be possible.

How does electricity work? How is it stored in batteries? To answer these questions, we need to explore the world of things that cannot be seen.

Atoms—The Smallest Pieces of Our World

The world around you is made up of invisible particles called atoms. An **atom** is the smallest unit of a material that still has all of the properties of that material. Take an iron bar, for example. If you keep separating the iron bar into smaller and smaller pieces, the smallest unit that would still have all the properties of iron is an iron atom. Helium, which is a gas, also can be broken down into many identical atoms (Figure 11.1).

An **element,** such as helium or iron, is a pure substance that consists of only one type of atom. Other familiar elements are oxygen and nitrogen, which make up the air you breathe, and metals such as gold, nickel, and lead.

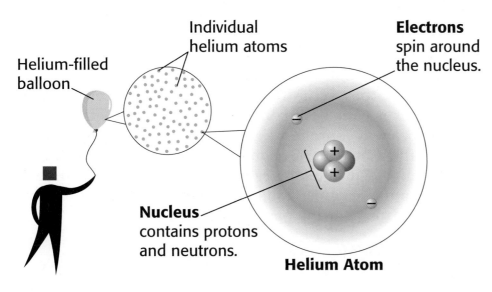

Individual helium atoms

Helium-filled balloon

Electrons spin around the nucleus.

Nucleus contains protons and neutrons.

Helium Atom

Figure 11.1 Helium is made of many identical atoms.

Interpreting *How many protons and neutrons are in every helium atom?*

Protons, Neutrons, and Electrons

As incredibly small as atoms are, they are made up of even smaller particles. Most atoms consist of protons, neutrons, and electrons.

Atoms have a center portion called a **nucleus.** A nucleus is made up of tiny particles called protons and neutrons. A **proton** is a positively charged particle. A **neutron,** which has the same mass as a proton, does not carry any charge.

An **electron** is a tiny particle that carries a negative charge. In an atom, electrons circle the nucleus very rapidly. The path an electron moves along can be called its "orbit." In most natural atoms, the number of protons equals the number of electrons. The positively charged protons equal the negatively charged electrons. The atom, therefore, has no charge.

Movement of Electrons

Particles with opposite charges are attracted to each other (Figure 11.2). Particles with the same charges repel, or push each other away. In an atom, electrons are attracted to the nucleus because of the protons' positive charge. Their rapid motion around the nucleus keeps them from falling into it.

In some types of atoms, the electrons are held tightly to the nucleus. These types of materials are called insulators. In an insulator, electrons cannot move freely among atoms. Foam and plastic are examples of insulators.

In other types of materials, electrons can be pulled away from an atom and easily move from one atom to another. Materials whose atoms give up electrons easily are called conductors—i.e., they are able to pass on heat and electricity. Copper wire is an example of a good conductor.

Figure 11.2 Objects with different charges attract each other, whereas objects with similar charges repel each other.

Inferring *Based on this art, what type of interaction would you expect between a proton and an electron?*

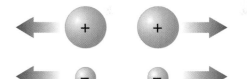

(A) Like charges repel each other.

(B) Opposite charges attract each other.

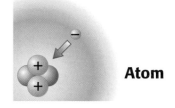

Atom

Producing Electrical Current

The movement of electrons is what creates electricity. In copper wire, for example, electrons are so free to move from atom to atom that they actually flow through the wire. The flow of electrons is called **current.** Electrons can flow through a thin copper wire, through a piece of solid material, or even through the atmosphere, as in a lightning bolt (Figure 11.3).

The unit used to measure the flow of electrical current is an **ampere** or amp. One amp is equal to about 6 billion *billion* electrons flowing past a point in 1 second.

Applying a Force

When you want to move a chair, you have to push on it. You have to exert a force. In the same way, to get a current to flow, a force has to be applied. The force that produces a current is called an electromotive force. The unit used to measure electromotive force is called a **volt.** Just as pushing harder on a chair gets it to move farther, applying a greater voltage produces a larger electrical current.

A battery is one source of electricity (Figure 11.4). A battery has a positive terminal and a negative terminal. When the terminals are connected by a good conductor, such as a wire, electrons start moving through the wire. They are attracted to the positive terminal and repelled from the negative one. This behavior of electrons sets up a current, or flow of electrons.

Figure 11.3 Lightning is produced by electrons moving through the air, whereas the electricity we use is produced by electrons flowing through wire.

Comparing *How is lightning similar to household electricity?*

Figure 11.4 In a battery, electrons flow from the negative terminal to the positive terminal.

Interpreting *In this battery, what material is used as the positive terminal? The negative terminal?*

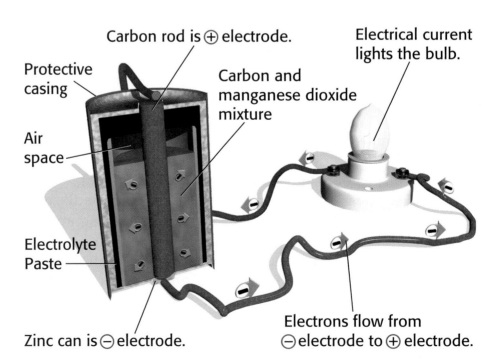

Carbon rod is ⊕ electrode.

Electrical current lights the bulb.

Protective casing

Carbon and manganese dioxide mixture

Air space

Electrolyte Paste

Zinc can is ⊖ electrode.

Electrons flow from ⊖ electrode to ⊕ electrode.

Figure 11.5 Power lines provide a constant supply of electrical energy.

Inferring *Does the protective insulation that covers these wires have a high or low resistance?*

Ohm's Law

An equation called **Ohm's law** describes the relationship between current, voltage, and resistance:

$$\text{Current (amps)} = \frac{\text{Voltage (volts)}}{\text{Resistance (ohms)}}$$

For a given voltage (force), more current flows through a good conductor than through a poor one. **Resistance** opposes the flow of current through a material. The unit of resistance, an **ohm,** is a measure of how well a material conducts electricity (Figure 11.5). A poor conductor of electricity, such as an insulator, has a high resistance (many ohms). A good conductor has a low resistance (few ohms).

SECTION 1 Assessment

Recall and Comprehension

1. Give an example of how the use of electronics has changed the world since your birth.
2. If the atom of an element has 20 protons, how many electrons does it have?
3. What makes some materials good conductors of electricity and other materials good insulators?

Critical Thinking

1. **Contrasting** Explain the difference between a proton and a neutron.
2. **Calculating** In an electronic circuit powered by a 9-volt battery, 1 ampere of current flows through the circuit. If the 9-volt battery is replaced by a 20-volt battery, does more or less current flow? Use Ohm's law to calculate your answer.

QUICK ACTIVITY

Make your own battery from a grape, a paper towel, two large washers made of different metals, and a sensitive voltmeter. Squash the grape in the paper towel. Sandwich the "juiced" paper towel between the two washers. Measure and record the voltage between the washers. What differences do you find if you use different metals and juices? Explain your findings. **For more related Design Activities, see pages 336–338.**

Connecting to STEM

Electrical Current: Understanding Ohm's Law

Because we can't see electrons, it can be difficult to understand the flow of electricity. Recall Ohm's law, which states that electrical current is equal to the voltage of a circuit divided by its resistance. To understand how these variables work together, picture electricity not as a stream of electrons but as water flowing through a pipe.

Decreasing the Resistance

Increasing the voltage is one way of increasing the electrical current that flows through a wire. Another way is to make the wire bigger. Again, think in terms of water flowing through a pipe. If the pipe is made larger, more water can flow through because there is more room for the water to move.

Like a larger pipe, a thicker wire allows a greater current to flow through. There is less resistance to the electrons.

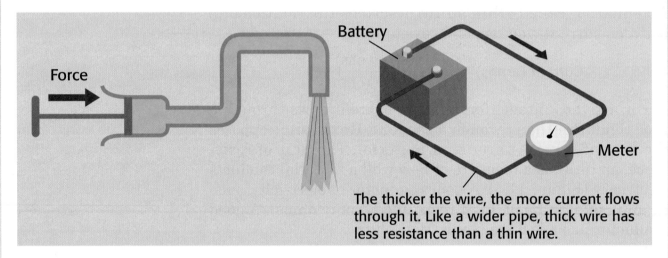

Force

Battery

Meter

The thicker the wire, the more current flows through it. Like a wider pipe, thick wire has less resistance than a thin wire.

Increasing the Voltage

Recall that electrical current is the flow of electrons moving through a wire. The force of this current is its voltage, which is measured in amps. Just as opening a faucet causes more water to flow, turning up the voltage of a circuit increases its electrical current. The greater the voltage, the greater the electrical current.

$$\text{Current (amps)} = \frac{\text{Voltage (volts)}}{\text{Resistance (ohms)}}$$

According to Ohm's law, as voltage gets larger, current gets larger. As resistance gets larger, current gets smaller.

Critical Thinking

1. **Describing** Making a wire thicker decreases which variable—current, voltage, or resistance? Explain your answer.

2. **Calculating** If a light bulb with a resistance of 1,200 ohms is connected to a 120-volt line, how much current flows in the circuit?

SECTION 2
Electronic Circuits

Reading Strategy

Outlining Make an outline to keep track of all the terms defined in this section.

> I. Electronic
>   Components
>   A. Resistor
>   1. reduces
>   current flow
>   B. Transistor

Benchmarks for Learning

- Electronic circuits are made up of components. Each component has a specific function in the circuit.

- An integrated circuit is a small electronic circuit made on semiconductor material.

- The power of a computer depends on the number of components on its circuits.

Vocabulary

semiconductor
diode
thermistor
photoresistor

capacitor
circuit
integrated circuit

analog circuit
digital circuit
bit

Electronic Components

An **electronic component** is a device that controls the flow of electricity. Electronic components may make small signals larger, they may limit the amount of current that passes through a device, or they may control the direction of the current. Certain components can also adjust the flow of current depending on the light or temperature. Common electronic components include resistors, transistors, and semiconductors.

Resistors

One of the simplest electronic components is the resistor. A **resistor** controls the flow of current through a circuit. The greater a component's resistance, the less electricity will pass through. Resistors come in a wide range of values, from less than 1 ohm to tens of millions of ohms. One of the most common types of resistors is a light bulb (Figure 11.6).

Transistors

One of the most important electronic components today is the transistor, which was invented in 1947. A **transistor** is a type of resistor that lets a small amount of current control the flow of a much larger amount of current. Transistors are used to control electric motors. They can also be used to control the storage of a small amount of electric charge used to represent information, as in a computer. The transistor is very small. It is a square wafer a few thousandths of

Figure 11.6 Resistors such as this light bulb are used in many different types of circuits.

Summarizing *What does a resistor do?*

Comparing and Contrasting
How are LEDs similar to light bulbs? How are they different?

an inch on a side. It is packaged in a larger metal or plastic container to make it easy to handle.

Semiconductors

A **semiconductor** is a material that is neither a good insulator nor a good conductor. The most common semiconductor material is silicon, a nonmetallic element.

One type of component made using semiconductors is the diode. A **diode** lets current flow in one direction but not the other. It is created when two different materials meet at a junction. A common diode is the LED (**l**ight-**e**mitting **d**iode). LEDs produce red, green, blue, or white light (Figure 11.7). LEDs are different from ordinary light bulbs. They do not have a wire filament that can break or burn out.

Other Electronic Components

A **thermistor** has a resistance that changes with the temperature. Thermistors can be used to make electronic thermometers. They can also be used as part of the control system for a refrigerator or oven. A **photoresistor** has a resistance that changes with the amount of light that hits it. Photoresistors can be used to turn on lights when it gets dark. They also can be used to measure light.

Other types of components are used in building circuits. **Capacitors** are components that store electrical charge. Coils store energy in the form of a magnetic field. A battery, explained in Section 1, produces electrical current from a chemical paste that fills the battery. Switches and motors are other examples of capacitors.

Circuits

These different electronic components are connected together and arranged into a circuit. A **circuit** is a group of components connected together to do a specific job. Any task that is accomplished using electricity is controlled by a circuit.

Early circuits were made of large components connected by several wires. The wires were connected to each other by soldering. Solder is a metal that melts easily and makes a good connection. It has a low resistance. Care had to be taken to make sure that wires in the circuit did not accidentally touch. If they did, this would set up an unintended flow of current, called a short circuit. To prevent a short circuit, wires often had a covering of insulation.

Connecting Components to Form Circuits

Circuits can be made to perform very specific functions. A series circuit connects components in one continuous path. If the circuit is broken at any point, current stops flowing. A flashlight that includes a battery, a bulb, and a switch is an example of a series circuit (Figure 11.8).

Parallel circuits have more than one path in which electricity flows. If the circuit breaks in one path (for example, if one bulb burns out), the current continues to flow through the other paths. The connection of some Christmas tree lights is an example of a parallel circuit.

More complex circuits use more components. Some components are connected in series and some in parallel with other components. For example, a circuit might be used to record telephone conversations. The device connects to the phone line and to your tape recorder. When you answer the phone, the device begins recording.

Figure 11.8 In a series circuit (A and B), all components are located along the same wire. A parallel circuit (below) connects lights along different paths.

Interpreting *If L1 goes out in both circuits, what will happen to the remaining light(s)?*

(A)

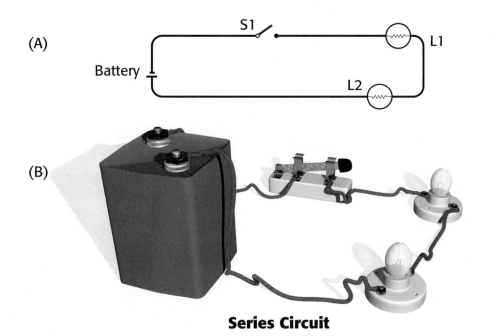

Series Circuit

(B)

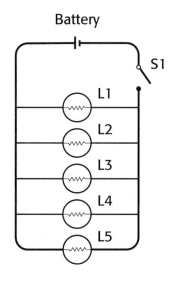

Parallel Circuit

For: Computer Activity
Visit: www.mytechedkit.com

Hardware Basics: Go Inside the Box Online

A computer is a machine, just like a blender, an electric drill, or a dishwasher. Every computer is made up of physical parts that work together to help it run properly. Go online to learn all the parts that go into a computer and what each part does.

Printed Circuits

Over the years, circuit components have become smaller. As a result, individual components have become harder to solder. The demand for circuits has also increased dramatically. The printed circuit board provided a method in which a small circuit could be reproduced over and over again, without introducing mistakes (Figure 11.9).

A printed circuit board is a thin board made of an insulating material, such as fiberglass. On one or both sides of the board, a thin layer of a good conductor—often copper—is bonded onto the board. Patterns etched in the copper form paths in which electricity can flow. Holes for mounting components are drilled into the board. Components are then soldered to the conducting paths on the board.

The conducting paths are placed photographically on the board, so that many boards can be made with exactly the same circuit. Once the components are mounted on the board, they are all soldered at once by an automatic soldering machine.

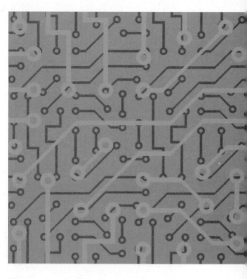

Figure 11.9 Early printed circuit boards were made using a photograph that showed the layout of the components.

Inferring *How does the use of a photograph help to make many identical copies of a circuit?*

Integrated Circuits

One of the most important inventions of the twentieth century was the integrated circuit. An **integrated circuit** provides a complete circuit on a tiny piece of semiconductor. This type of circuit, which is often less than one-tenth of an inch long by one-tenth of an inch wide, is commonly referred to as a chip. Each chip contains many components and conducting paths.

A chip is designed by an engineer. The engineer makes a drawing of the chip several hundred times larger than actual size. The drawing is reduced photographically, forming a mask. The mask is then used to put patterns on a wafer of semiconductor material, usually silicon. Many identical circuits are made at once on a round wafer that is several inches across.

Computers are built using many large integrated circuits. Because of the reduced size of integrated circuits, computers that took up several rooms of space 20 years ago now fit on a desktop. Chips have replaced large, bulky circuits in many other systems as well.

The power of a computer is largely dependent on the number of components that can be placed on circuits. The first integrated circuit was composed of a single transistor and a few other components. Currently, a single integrated circuit may contain several billion transistors.

Analog Circuits

Information can be represented by electricity in several ways. One way is through a voltage change based on the information that is being represented. For example, a voltage representing a person's speech would get higher as the person spoke louder. An analog is something that is similar to something else, so the voltage is considered an analog of the person's speech.

An electronic circuit that works with analog signals is called an **analog circuit.** Voltages in such a circuit change smoothly, as do the things they represent, such as a person's voice. Analog circuits are used in devices such as clocks with minute and hour hands. Another example is an analog thermometer, in which the height of the liquid rises and falls smoothly with the temperature.

Digital Circuits

Sometimes information must be very accurate or must be sent over long distances. Under these conditions, analog circuits do not carry the information well. Instead, digital technology is used. In a **digital circuit,** information is coded into a series of 0s and 1s.

Digital circuits are based on the binary number system, which uses only 0s and 1s. A voltage below a certain value is coded as a 0, and a voltage above that value is coded as a 1. Each 0 or 1 is called a **bit,** short for **b**inary dig**it.** Digital circuits, the basis of how computers work, are explained in more detail in the next section.

SECTION 2 Assessment

Recall and Comprehension

1. Explain how a diode controls current in an electronic circuit, and give an example of how it is used.
2. What is a printed circuit? How does it connect electronic components?
3. Name at least three different types of circuits. Briefly describe each one.
4. Explain what integrated circuits are and how they work.

Critical Thinking

1. **Comparing** Compare the functions of these electronic components: a resistor, a transistor, a thermistor, and a photoresistor.
2. **Summarizing** How have circuits changed from the original circuits that were soldered together by hand to the integrated circuits of today?

QUICK ACTIVITY

Make a capacitor by placing a 12" square of plastic wrap between two 10" square pieces of aluminum foil. Make a 1" border and fold it over. Do not let the two foil sheets come in contact with each other. Briefly connect the positive terminal of a 9-volt battery to one of the foil pieces and the negative terminal to the other. Disconnect the battery, and use a voltmeter to measure the stored charge.
For more related Design Activities, see pages 336–338.

Technology in the Real World

Integrated Circuits—"Chips"

The integrated circuit, abbreviated as IC and also known as a chip, has created a revolution in redefining small. Amazing changes have taken place in electronics because of the shrinking size (and cost) of an electronic circuit.

Consider this comparison: If the same type of technological breakthrough as the chip had happened in the automobile industry, a Rolls-Royce would now cost about $500 and would get 1,500 miles per gallon! Quite an innovation!

What Is an IC?

An IC is simply a smaller version of an electronic circuit. It contains the same electronic components—resistors, semiconductors, and transistors. Instead of being attached to a circuit board with wires and soldering, however, the electronic components are etched onto a single crystal, or chip. The chip is made of a semiconductor material such as silicon.

The circuit wires etched into the silicon chip are about 4,000 times thinner than a human hair. Today's fastest microprocessors, which are ICs containing the CPU of a computer, contain several billion transistors on their chips and operate at speeds of over 5 gigahertz. Microprocessors 10 years from now may contain more than 100 million transistors and operate at speeds of over 20 gigahertz.

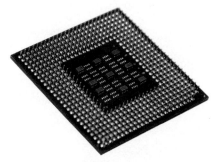

The Pentium 4 was the first commercially available chip that could attain a speed of 2 GHz.

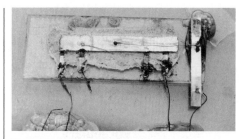

Texas Instruments' first IC.

Creating a Revolution in Electronics

Chips being made today use less material, take up less space, and are cheaper to manufacture than individually made circuit boards. They have greatly reduced the cost of computers and all types of electronic products.

The impact of ICs surprised even its inventors. In 1958, two separate inventors—Jack Kilby at Texas Instruments and Robert Noyce at Fairchild Semiconductor Corporation—both received patents for slightly different versions of the chip. Kilby's version used germanium for the semiconductor, whereas Noyce's used silicon.

According to Kilby, "What we didn't realize then was that the integrated circuit would reduce the cost of electronic functions by a factor of a million to one—nothing had ever done that for anything before."

Critical Thinking

1. **Summarizing** What are the most significant advantages of ICs over chips with separate components?

2. **Extending** After doing research, decide for yourself if you think there are limits to how small a chip can be made. Explain your reasoning.

3. **Inferring** Silicon Valley is a region in California where many computer industries are located. Based on what you know about ICs, explain the origin of the term Silicon Valley.

Computers

Benchmarks for Learning

- Computers are general-purpose tools of technology.
- Computers use binary data—1s and 0s—to represent information.
- Computers can be used as systems or can be small parts of larger systems.

Reading Strategy

Listing As you read about the history of computers, make a list of all the changes that were brought about by small integrated circuits, or chips.

Vocabulary

byte	microcomputer	mainframe computer
computer	personal computer	supercomputer

Early Machines for Computing

For thousands of years, people have used machines to help with calculations. The Chinese abacus, which uses a system of beads to perform arithmetic, is still in use today. Napier's bones, invented in 1617, was used to multiply. The first mechanical adding machine was invented by Blaise Pascal in 1645. It used sets of wheels, moved by a needle, to add numbers and give a sum (Figure 11.10).

From Calculators to Census Machines

The first all-purpose calculator was developed in the mid-1800s by Charles Babbage. It could be instructed,

Figure 11.10 The Pascaline was a mechanical adding machine invented by Blaise Pascal in 1645.

Comparing *How are early computing machines similar to the computers of today?*

or programmed, to do sets of calculations. Ada Lovelace, Babbage's co-worker, wrote simple programs that instructed the calculator to carry out these calculations. Ada Lovelace was the world's first computer programmer. A modern computer programming language, ADA, is named after her.

In 1890, Herman Hollerith devised a way of automating the process of compiling information for the U.S. Census. Information was coded as holes punched into cards. The information could be tabulated by a machine. Hollerith's method was a complete system. It had a machine to punch the cards (input), a tabulator for sorting the cards (process), and a counter to record the results (output). A sorting box rearranged the cards for reprocessing (feedback). In 1911, Hollerith's Tabulating Machine Company became part of a company that later became International Business Machines (IBM) Corporation. IBM is the largest computer company in the world today.

The World's First Computer

The ENIAC (Electronic Numerical Integrator And Computer) was the world's first electronic digital computer (Figure 11.11). The ENIAC was developed by the U.S. Army during World War II. It used 18,000 vacuum tubes. The size of the tubes made the computer enormous. The ENIAC was 10 feet tall, 3 feet deep, and 100 feet long.

The vacuum tubes of the ENIAC gave off light and heat, which attracted moths. The moths became trapped in the wires and moving parts of the computer. People had to clean out the moths, a process they called "debugging." Modern computers don't have problems with moths, but people still use the term *debugging* when they find and fix problems in a computer.

Better Chips, Faster Computers

With the technology of integrated circuits, computers became a different sort of machine. They still carried out basic functions, such as calculations and storing data. With smaller and cheaper circuits, however, computers also became smaller, portable, and affordable for individuals.

Computers also became much, much faster. Despite its size and cost, the ENIAC would be considered an extremely slow computer by today's standards. Modern computers have reached the nanosecond range for processing data.

Figure 11.11 The ENIAC was the world's first computer.

Clarifying *What component of the ENIAC made this computer so large?*

Figure 11.12 Since the first chips were made, the number of transistors per chip has increased dramatically.

Summarizing *What is the relationship between processing speed and the number of components per chip?*

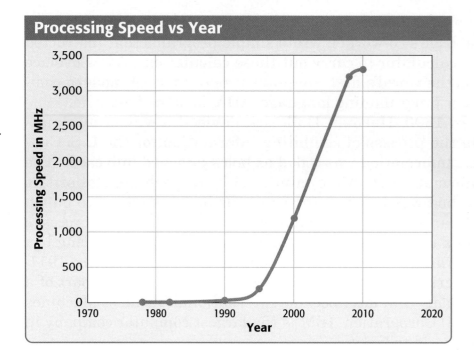

Processing Speed vs Year

Figure 11.13 The number of transistors on a CPU has increased dramatically.

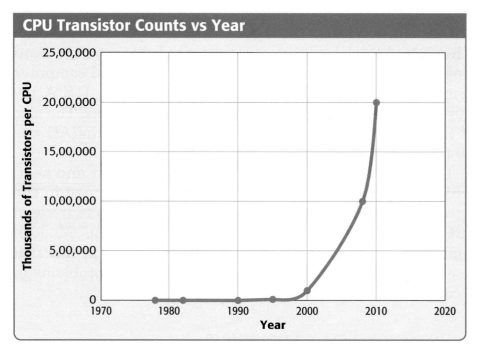

CPU Transistor Counts vs Year

Figure 11.12 shows the increase in the speed of computers (processing speed) over the last 25 years alone. This trend toward more power in a smaller package will continue. The processing speed increase is a result of the CPU containing more transistors. Figure 11.13 indicates the change of transistors per CPU versus time.

Computers Today

Today, a **computer** can be defined as any electronic machine that stores, processes, and retrieves information. Because

computers can be programmed to carry out so many functions, they are one of the most useful and widespread tools of technology.

Customized for Different Tasks

Computers have different capabilities. Some are faster than others, some have more memory. They come in all sizes. They range from a chip less than 1-inch square to a room full of equipment. They are sometimes used as stand-alone systems, but often are part of a larger system. Computers are often used to help people form a feedback loop. For example, a computer system can function as a central control center in a transportation system. It can help switch trains from track to track or bring planes in to land safely (Figure 11.14).

Bits and Bytes

Like the first computers, today's computers are digital, using the binary system of 1s and 0s to represent information. Any number can be represented by a binary number. For example, instead of representing the number 359 using numbers from 0 through 9, a computer would represent this same value using only the digits 0 and 1.

Binary digits, or bits, are organized into groups of eight to make them easier to work with. Each group of eight bits is called a **byte** (pronounced "bite"). Each byte can represent one of 256 different characters (numbers, letters, punctuation, or other information). A grouping of 1,024 bytes is called a kilobyte (KB). One thousand KB, or 1,024,000 bytes, equals a megabyte (MB). Other units are the gigabyte and terabyte (Figure 11.15).

Figure 11.15 Computers work in units called bytes. In this table, each unit shown is 1,000 times greater than the one above it.

Inferring *Based on the table above, what do you think the prefix giga- means?*

Computer Units	
Unit	**Number of Bytes**
Kilobyte (KB)	1,024
Megabyte (MB)	1,024,000
Gigabyte (GB)	1,024,000,000
Terabyte (TB)	1,024,000,000,000

Figure 11.16 Because they can be made so small, microcomputers are embedded in all types of machines.

Hypothesizing *Why are microcomputers not usually used to store large amounts of data?*

Figure 11.17 Laptop computers are very powerful and have excellent graphics. Most students in college prefer using them.

Types of Computers

Computers fall into one of four categories: microcomputers, personal computers, mainframe computers, and supercomputers. These different types of computers serve different purposes.

Microcomputers are extremely small computers that are built into electronic games, automatic bank teller machines (ATMs), cash registers, home appliances, cars, and personal computers (Figure 11.16). They can be as small as a single chip.

A **personal computer** (PC) is a general-purpose single-user computer designed to be operated by one person at a time. PCs can be desktops, laptop models, or small notebook-sized computers called *netbooks* that primarily are used for general computing and accessing Web-based applications (Figure 11.17). They may include many microcomputer chips. PCs are more powerful today than the fastest supercomputers were just 10 years ago.

A **mainframe computer** is a large computer used by large companies, government agencies, and universities. Mainframes are used to make out payroll checks, keep personnel records, keep track of orders, or keep lists of warehoused items. They may have very large secondary devices containing storage, such as hard disks and tapes.

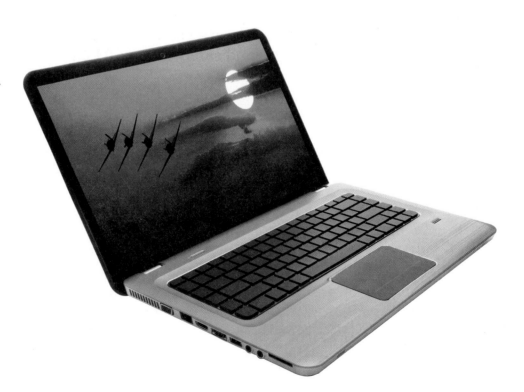

A **supercomputer** is the fastest and largest type of computer made today (Figure 11.18). Supercomputers are most often used for research, for analyzing huge amounts of data, or for other very large jobs. Supercomputer speed is measured by the number of multiplications or divisions the computer can carry out per second. The largest ones can carry out over a quadrillion operations per second.

Figure 11.18
Supercomputers, such as the CRAY 2.4 supercomputer shown here, often have very large capacities for storage.

Contrasting *How are supercomputers different from personal computers?*

SECTION 3 Assessment

Recall and Comprehension

1. Give three examples in which computers are used as small parts of larger systems.
2. What is a gigabyte? How much bigger is it than a megabyte?
3. Describe the situations in which a mainframe computer is used. Is it likely to be faster or slower than a supercomputer?

Critical Thinking

1. **Analyzing** Explain why computers are considered general-purpose tools of technology.
2. **Inferring** Why are computers sometimes referred to as digital computers?
3. **Summarizing** How have computers changed since they were first invented?

QUICK ACTIVITY

Conduct an interview with at least five adults to learn how computer use has changed. Some questions you might ask are: When you were my age, did you have a computer? Do you have a computer now? Do you use a computer for work? What types of software do you use? Do you use email? Share your survey results with the class. **For more related Design Activities, see pages 336–338.**

SECTION 4
The Computer as a System

Benchmarks for Learning

- Computer systems are made up of hardware, peripheral devices, and software.
- Computers operate under a set of instructions called a program. Computer programs are written for many different tasks.

Reading Strategy

Listing Make a list of all the different components of a computer system. Next to each component, write a brief description of what that component does.

I. Hardware
- Motherboard—controls all the system hardware
- CPU—

Vocabulary

motherboard
central processing
  unit (CPU)

read-only memory (ROM)
random-access
  memory (RAM)

operating system
application program
computer virus

Figure 11.19 The motherboard is the main circuit board to which components are attached.

Summarizing *What does a motherboard do?*

Computer Systems

Like much of the technology you have already read about, a computer is a type of system—with inputs, a process, and outputs. Its input devices provide both command inputs and resource inputs. For example, a user at a terminal might type in a program (command input). He or she then types in data (resource input). The computer processor acts on these resources in response to the command input. It is carrying out the system's process. The output of the computer, such as a completed calculation or an altered photograph, is the output of the system. Feedback may come from a person, or through hardware or software. Computer systems are made up of hardware, peripheral devices, and software.

Computer Hardware

Computer hardware refers to the physical components of a computer system. This includes a motherboard, a microprocessor with the central processing unit (CPU), memory, different types of secondary storage, and a power supply unit. It may also include other hardware, such as a modem or a network card.

The Motherboard

Everything in your computer is hooked up through the motherboard (Figure 11.19). A **motherboard** is a printed circuit board that controls all of the system's hardware. It includes the CPU, the computer memory, and often a modem or network card. A motherboard controls the power and function of the entire computer.

LIVING GREEN
Managing Electronic Waste

What do you do when your computer crashes and there's no hope of repair? When you get a new cell phone, what do you do with your old one? Every year, approximately 1.8 million tons of obsolete electronics, including computer CPUs, monitors, keyboards, and all other components, end up in landfills. By comparison, only about 18% of e-waste (including all types of electronics) is recycled in the United States each year. Electronic waste contains hazardous materials such as lead, cadmium, and mercury, which pose serious risks to the environment as well as to human health.

MAKE A CHANGE
The first step to ending this problem is simple: never, ever throw any type of electronic device into the trash bin or dumpster. Any item that is reusable should be donated to a school or other community-service organization. Any broken or unusable product should be disposed of responsibly. However, the emerging business of e-cycling has problems of its own. Many recyclers do not actually recycle your e-waste at all; rather they export old CPUs, motherboards, and other materials to less developed countries. Here, the materials are burned to collect the trace amounts of gold and copper, and personal information can be retrieved. For more information on the global trade in e-waste, visit http://www.pbs.org/frontlineworld/stories/ghana804/.

TRY THIS
Locate the certified Responsible Recyclers in your area and spread the word. Visit http://www.electronicstakeback.com/recycling/recyclers_chart.htm or the Environmental Protection Agency's Web site on e-cycling (http://www.epa.gov/osw/conserve/rrr/recycle.htm) and identify the agencies in your area that accept donated electronics, or are certified to recycle these products in a fair and environmentally sound way. Then, organize a community drive at your school to collect old and unusable electronics and make sure they are donated or given to your local responsible recycler.

The Central Processing Unit

The "brain" of the computer is the central processing unit, or CPU (Figure 11.20). The **central processing unit** of a computer reads programs and converts each program's instructions into actions. In a modern desktop computer, the CPU is a single chip about 1 square inch in size.

The power of a CPU refers to how fast it is. CPUs are rated in hertz, which is a measure of the frequency at which the microprocessor runs. The higher the number of hertz, the more instructions the chip can carry out at a given time. Powerful PCs run at frequencies of more than 2 billion hertz (2 gigahertz). They can carry out hundreds of millions of instructions per second. Large mainframe computers can handle billions of instructions per second.

Memory

The place where programs are stored is called a computer's memory. The memory also stores the information being used actively by the computer. Modern computers use memory chips that can store more than 1 billion characters. Most computers use two types of memory: read-only memory and random-access memory. Each type of memory is optimized for its intended use.

Read-only memory, or ROM, is fixed memory that provides instructions needed to start the computer. It is built

Figure 11.20 Like any technological system, a computer has inputs, a process, and outputs.

Extending *How might a computer system be customized for a particular task, such as recording and editing music?*

Exploring Computer Hardware A computer system has several basic physical parts that are used to enter data into the computer, process the data, and retrieve information out of the computer.

CPU The CPU is the control center of the computer. The CPU processes and stores information, and coordinates the functions of the other parts of the computer.

Scanner A scanner is an input device by which the computer gathers data. It can transfer input from the printed page directly into the computer.

into chips on the computer's motherboard. These chips don't lose their contents when the power is switched off.

Random-access memory, or RAM, is the main type of memory in a computer. RAM is contained in chips that are wired onto the computer's motherboard. It is used to run programs that you load into the computer. Data is stored in RAM only when the computer is powered up and you are working with a program. When you are finished working with that program, the data disappears from RAM and the memory can be used by some other program.

RAM is measured in bytes. A computer with 1 megabyte of RAM can store 1 million bytes of data. Typical RAM in a PC ranges from 128 megabytes (laptop portables) to more than 2 gigabytes (desktop computers). Mainframes and supercomputers may have many gigabytes of RAM.

Secondary Storage

In addition to RAM and ROM, a computer must have another way of storing data. The hard drive may not provide sufficient room for the information that is being used. Also, an extra backup copy ensures that information will not be lost if the computer breaks down. Secondary storage is provided by external devices that store information for later use. Secondary storage includes zip disks, magnetic tapes, and compact discs. On all of these formats except compact discs, data is stored magnetically.

Digital Camera   A digital camera is an input device that records images on an electronic image sensor instead of using film.

Webcam   A webcam is a camera designed to take digital photographs and transmit them over the Internet or other network.

Inkjet printer   An inkjet printer is an output device that squirts ink from a print cartridge that moves from side to side as the paper is fed through the printer.

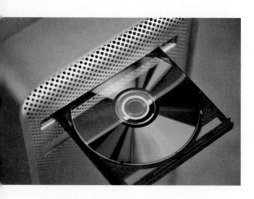

Figure 11.21 CD-ROMs can store up to 650 MB of data. Data is stored as pits in the grooves of the CD.

Summarizing *Why is it useful to have different types of storage devices?*

Compact Discs and DVDs

A CD-ROM (compact disc—read-only memory) is a commonly used secondary storage device (Figure 11.21). CD-ROMs are removable plastic discs on which optical data has been encoded by lasers. Most computers now come with a CD-ROM drive as a standard feature. Unlike magnetic storage media, CD-ROMs store data as tiny reflective or nonreflective spots in grooves on the CD. A CD-ROM drive reads the data by shining a laser onto the surface and sensing how the light is reflected by the spots in the grooves.

Standard CD-ROMs cannot be written to by home users. CD-Rs (*R* for "recordable") can be written to only once. In other words, after you have saved information to the disc, you cannot erase it or write over it. CD-RWs (*RW* for "rewritable") can be written to over and over again. Information is written to CD-Rs and CD-RWs using a CD burner.

A standard CD-ROM holds about 700 MB of data. The 20 million books and other print materials of the Library of Congress would equal about 10 terabytes (TB) of data. A terabyte is 1,000 gigabytes, so 10 TB is 10,000,000,000,000 or 10^{13} bytes. That means the entire collection of the Library of Congress could fit on about 15,000 CD-ROMs. About 240 CDs take up one foot of bookshelf space. So it would only require about 60 feet of bookshelf space to house the entire Library of Congress collection on CDs.

A Digital Versatile Disc (DVD) is like a CD-ROM except that it can also be used to store high-quality movies in addition to sound and data. DVDs can store considerably more data than CDs. They are sometimes mistakenly referred to as "Digital Video Discs," a misnomer because they can be used for much more than just storage of video.

Current DVD optical disc technologies use a red laser to read and write data. *Blu-ray* is a new optical disc format developed by a group of electronics companies including Apple, Dell, HP, Sony, and others. It uses a blue-violet laser instead (hence the name blu-ray). A blu-ray disc can store about five times the amount of data as traditional DVDs.

USB Flash Drives

Flash memory is a special kind of computer memory that can be electrically erased and reprogrammed. It is used in ROM chips within the computer itself to store basic information about the computer's configuration. It's also used in memory cards and memory sticks for digital cameras that require removable, reusable storage and in USB flash drives

(also called jump, thumb, pen, or key drives). Most removable flash memory devices include a chip that stores data and a microcontroller that permits the operating system to communicate with the chip (Figure 11.22). As the technology of flash memory improves, the capacity of flash devices will increase significantly. Early flash devices held 32 MB to 256 MB, but capacities of several hundred GB are now available. The small size, increasing capacity, and ease of connection of these removable devices makes them widely used as replacements for hard drives, Zip, and Jaz drives.

Common Magnetic Storage Devices, Zip® and Jaz® drives

Other forms of magnetic storage devices include a variety of USB-connected hard drives, which can hold up to as much as a terabyte of data. A terabyte (about 10^{12} bytes) equals 1,000 gigabytes or 1,000,000 megabytes of data. These include Zip, Jaz, and Rev (removable) drives from Iomega Corporation. In addition, devices such as Apple's iPod, Creative Labs' Zen, and Microsoft's Zune function as both MP3 players (meaning they can play the popular MP3 format) and as high-capacity, transportable storage devices.

Capacities of Common Storage Devices	
Device	**Capacity**
Internal hard drive	60 GB–1.5 TB and more
External hard drive (USB connection)	40 GB to 2 TB and more
CD-ROM	650 MB–700 MB
MP3 player (iPod, Zen, Zune)	2 GB–30 GB and more
Jaz and Zip drives	100 MB–2 GB
Flash memory cards and drives	32 MB–4 GB and more
DVD (Digital Versatile Disks)	8.5 GB

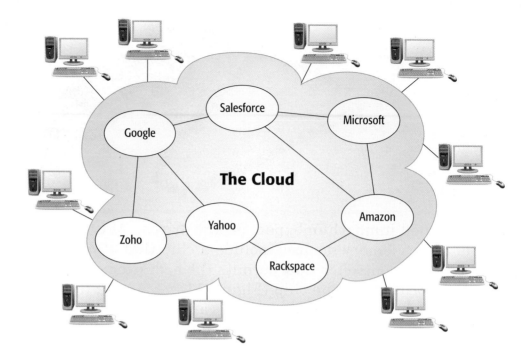

Figure 11.23 A conceptual diagram illustrating cloud computing and possible connections between entities.

Cloud Computing

A new approach to computing is using servers on the Internet to provide software and other resources to individuals and businesses who might not wish to incur the expense of storing information locally. This Internet-based computing is called **cloud computing** because the computer network that serves users is like a cloud of computers, as shown in Figure 11.23. An example of cloud computing is users accessing Google Apps for a variety of tasks.

Hard Disk Drives

In most computers, the **hard drive** provides most of the computer's memory. A hard disk drive is usually built into the computer. The surface of the disk (called the medium) is coated with a thin layer of iron oxide, a magnetic material. A tiny electromagnet, or head, is placed near the disk as it moves. A voltage applied to the head magnetizes tiny bits of iron oxide on the disk.

Information is stored on disks as magnetic fields. This storage process is called "writing to disk." Information can also be taken back from the disk by the head. The magnetic fields in the disk are changed back into electrical impulses. This is called "reading from disk." Floppy disks are usually 3.5 inches square and store up to 1.44 megabytes of data.

Depending on the computer, a hard drive can store up to several terabytes of data, or about 50,000,000 pages of typed text. Additional information is kept in secondary storage devices.

Tapes

Tape storage is often used by businesses to back up data after each day's work. This method of storing data is much the same as for floppy disks, except that the storage medium is magnetic tape, similar to audio recording tape. The tape is often held in cartridges for ease of loading, unloading, and storage. Tapes are available that can store over a terabyte of data. Backup tape drives are very fast. They transfer information at 30 megabytes per second and can read a 300-page novel in about 3 seconds.

Peripheral Devices

Peripheral devices either provide input to a computer (input devices) or display output from a computer (output devices). They are connected to the computer using individual ports. Peripheral devices include the keyboard, mouse, monitor, video cameras, speakers, and printer.

Input Devices

CD-ROMs, tapes, hard disks, floppy disks, and various types of cameras all provide input to a computer (Figure 11.24). They can hold a program or data that the computer can move, or load, into its main memory.

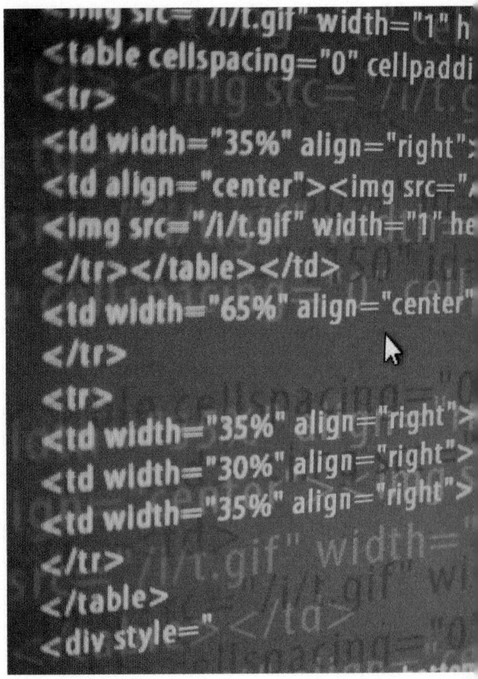

The most familiar input device is a computer keyboard. People use a keyboard to provide data to a computer. A keyboard can also be used to program a computer. Communicating with a computer by means of a keyboard is an interactive process. Input through the keyboard is followed quickly by output from the computer.

Another common input device is the mouse. A computer mouse is a pointing device that provides a way for computer users to interact directly with graphics on the screen. It translates hand motions into electrical signals that are sensed by the computer. Touch pads, joysticks, trackballs, and game pads are similar types of input devices.

In today's world, a **touch screen** is the newest method of input. It is normally a piece of glass with a touch-responsive surface. Generally, these surfaces have a minute electrical current flowing through them, and touching them creates an irregularity in the voltage. This change is used to determine the location of the touch. Most new cell phones, computer monitors, and GPS systems have touch screens.

Touch screens are composed of several different parts. The physical surface that the user interacts with is called the **touch sensor.** The **controller** is a small card that translates the interaction into information that devices such as

Figure 11.24 We often are not aware of the computer programming that underpins the digital devices we use.

Comparing *How is a digital camera similar to other input devices, such as a keyboard?*

a cell phone, iPhone, or GPS can understand. The **software driver** allows the touch screen and the PC to work together by telling the PC's operating system how to interpret the information sent from the controller.

There are several different types of touch screens: A **resistive** touch screen is generally made up from several layers. When touched, two of the metallic layers create a circuit that allows a current to flow through at that point. A **capacitive** touch screen is usually made from an insulator (usually glass), coated by a solution that conducts electricity. Because the human body can act like a capacitor (it can store electrical charge), this screen was developed so that it responds when it senses the body's capacitance (the body's ability to store electrical energy). A **scanner** transfers data from a printed page directly to the computer. Using special software, a scanner converts letters, numbers, and images into a code of bytes. In this way, many pages of text can be put into a computer's secondary storage for later use.

Digital and video cameras allow users to capture images, create Web pages, add photographs to email, and make video calls over the Internet. Web cams (video cameras attached to computers, which transmit images across the Internet every few seconds) have become very popular.

Computer inputs can also be in the form of human speech. Special software can recognize spoken words and change them into bytes. Some software can recognize more than 100,000 different words. This technology is becoming an increasingly important part of computer systems.

Output Devices

Computer output appears in many forms. The most common, the video monitor, has a cathode ray tube (CRT) screen, which changes electrical signals to light images. CRTs are used in televisions to produce the screen image. Video monitors can be used to display text, graphics (pictures), or both at the same time. Audio output from a computer takes the form of tones, beeps, music, and voices. These are heard through speakers (Figure 11.25).

A printer is another common output device. A printer records output on paper. A record of data or text on paper is called a hard copy. Printers are used to output drawings and photographs, as well as letters, and spreadsheets. Two common kinds of computer printers are laser printers and inkjet printers.

In a laser printer, a laser beam changes the electrical charge on a printing drum. The final printed copy is made from toner that has been heat-fixed onto the paper. An inkjet

Figure 11.25 Physicist Stephen Hawking communicates by selecting words on his computer screen. The computer converts the words to speech using a voice synthesizer.

Contrasting *How is a voice synthesizer different from voice recognition technology?*

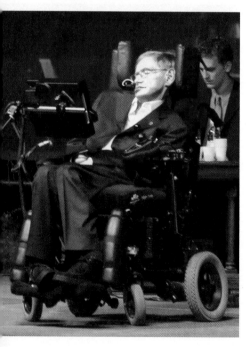

printer squirts ink from a print cartridge, which moves from side to side as the paper is fed through the printer. You will learn more about these printers in Chapter 13.

A specialized printer, a plotter, also outputs data to paper. Plotters are devices that use one or more pens to make detailed drawings. Plotters are often used in systems that make, change, and store drawings, such as computer-aided design (CAD) systems.

Computer Software

A computer does its work according to a list of instructions, called a software program (Figure 11.26). A program is meant to be followed mechanically by a computer. Computers are built to carry out instructions written in a very simple type of language called machine language. Each type of computer has its own machine language. It can execute a program only if the program is expressed in that language.

Because software programs can be rewritten to do different tasks, a computer is a general-purpose tool. A programmer provides the computer with instructions for a specific task, but a completely different task can be carried out using a different program.

Types of Software

Computers use two important types of software: operating system programs and application programs. A computer's **operating system** controls the computer and its components. It also makes the components available to other kinds of software. An operating system is chosen to fit the computer and the job it must do. Examples of operating systems are Windows ME, Windows 2000, Mac OS X, Linux, UNIX, and MS-DOS. Sometimes more than one operating system is available for a particular computer.

Figure 11.26 Software programs are used to create documents, graphics, videos, sound recordings, and many other types of output.

Summarizing *How is operating system software different from application software?*

Figure 11.27 Computer programmers create and update all types of software programs.

Hypothesizing *What types of skills do you think computer programmers use?*

An **application program** gives the computer instructions for carrying out a specific task. Thousands of new computer programs are written every year. Many of these application programs are purchased for use. They include computer games, word processors, and drawing programs. Application programs are typically stored on disks or CDs.

Programming Languages

Complex applications available for purchase may have taken a team of programmers years to write. Sometimes, however, no program exists to do a job. In this case, a new program must be written (or an existing program customized) by the user or a trained computer programmer (Figure 11.27). Programming a computer may be easy and quick, or it may require long, hard effort. Computer programs are written in one of many programming languages. Each language has features that make it best for writing a certain kind of program.

One common programming language is BASIC (Beginner's All-purpose Symbolic Instruction Code). It is used by students, businesspeople, and hobbyists because it is fairly easy to learn and use. It allows beginners to write a program, test it, and debug it. BASIC can be used on most small personal computers as well as on large mainframes.

Another common language is HTML, which stands for Hypertext Markup Language. HTML is used in Web page design. Another language, called Java, is also used to program Web pages. Unlike HTML and many other languages, Java was designed to run on any operating system. Because Java has wider compatibility than do other languages, it can be used for many different kinds of computers and consumer gadgets. Programs written in Java are used commonly to animate smaller programs called applets.

Adobe Flash provides programming for multimedia, allowing animations and simulations that can be part of Web sites. Flash programming relies on vector graphics, wherein geometric shapes are created by using points, lines, and polygons to represent the graphic image, in contrast to the pixels you might obtain from a photographic image. The Flash Player can be downloaded quickly. In addition, applications are being developed for a variety of handheld devices.

Computer Viruses

Just as infections can attack the human body, spread, and cause damage to its systems, a computer virus can infect a computer program and spread to other software on the computer. A **computer virus** is a software program that attaches itself to other programs. When a virus infects a program, it can alter or delete files, or consume computer memory (Figure 11.28). Viruses are a real threat to our information-based society.

Programs have been developed to combat viruses. Once installed on a computer, these programs detect viruses by comparing actual program length to expected program length or by checking for the presence of known programs that are actually viruses.

Most antiviral programs keep checking your system's activity, watching for and reporting suspicious virus-like actions. But no antivirus program can detect every virus, so they need to be frequently revised.

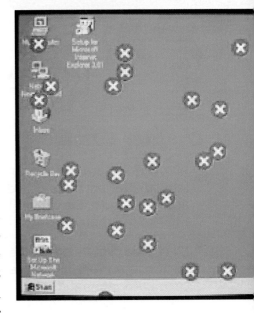

Figure 11.28 The Marburg virus creates serious error messages on a user's screen.

Summarizing *How are computer viruses spread?*

SECTION 4 Assessment

Recall and Comprehension

1. Identify three different ways of storing data on a computer.
2. What is an operating system? What functions does it carry out in a computer?
3. Name three different types of input devices. What is the function of each?
4. What two computer languages are commonly used to program Web pages?

Critical Thinking

1. **Applying** Draw a system diagram of the computer that you use most regularly. Show the inputs, process, outputs, and feedback.
2. **Comparing and Contrasting** How are ROM and RAM similar? How are they different?
3. **Inferring** Why is secondary storage critical to a computer system?

QUICK ACTIVITY

A barcode reader is an input device that is used to read a special code called the Universal Product Code (UPC). A UPC is a series of vertical lines of different thicknesses that identifies a specific product, such as a box of cereal. Do research on UPC symbols. Find out when they were first used. How is information scanned in at checkout? **For more related Design Activities, see pages 336–338.**

Chapter Summary

- All materials are made from atoms. Atoms contain smaller particles called protons, neutrons, and electrons. Electricity is produced by the movement of electrons.

- Electronic circuits are made up of components. Each component has a specific function in the circuit. An integrated circuit is a complete electronic circuit made on a single piece of semiconductor material.

- The power of a computer is largely dependent on the number of components on its circuits. Because of improvements in these circuits, computing power has greatly increased during the past 50 years, and the cost of computer systems has fallen.

- A computer follows the general system model. It has inputs, processes, outputs, and feedback. Computer systems are made up of hardware, peripheral devices, and software.

- Computers have different capabilities and functions. They include microcomputers, personal computers, mainframe computers, and supercomputers. The classification relates to how powerful and fast each computer is.

- Computers can be programmed, or instructed, to do many different jobs. They operate under a set of instructions called a software program. Computer software is divided into operating systems and application programs.

Building Vocabulary

Your teacher may give you a crossword puzzle. Complete the puzzle using the following words from this chapter. Exchange puzzles with a classmate to check each other's answers.

1. ampere
2. analog circuit
3. application program
4. atom
5. binary
6. bit
7. byte
8. central processing unit
9. circuit
10. computer
11. computer virus
12. current
13. digital circuit
14. diode
15. electromotive force
16. electron
17. electronic component
18. element
19. floppy disk
20. hard copy
21. hard disk
22. hardware
23. inkjet printer
24. integrated circuit
25. laser printer
26. mainframe
27. memory
28. microcomputer
29. motherboard
30. mouse
31. neutron
32. nucleus
33. ohm
34. Ohm's law
35. operating system
36. personal computer
37. photoresistor
38. printed circuit
39. proton
40. RAM
41. resistance
42. resistor
43. ROM
44. scanner
45. secondary storage
46. semiconductor
47. short circuit
48. software program
49. solder
50. supercomputer
51. thermistor
52. transistor
53. volt

See your teacher for the Crosstech puzzle.

Reviewing Content

1. Name the three particles found inside atoms. What is the charge (if any) on each type of particle?

2. Describe five electronic components that can be used to build circuits.

3. List the four categories of computers. Which type do most people use?

4. Identify four parts of a computer.

5. What are three types of memory used by computers? How is each type used?

Applying Your Knowledge

1. Research to find three examples of good conductors, and explain why they are good.

2. Based on your own experience and observation, give one example each of how the use of electronics has changed manufacturing, transportation, communication, and healthcare technologies.

3. Find three products that use an LED, and explain how the LED is used in each.

4. Describe the hardware, peripheral devices, and software used in the computer system you use most often.

5. A computer can be used for mailing letters to thousands of people. List the major components that you would expect to find in such a computer system. Draw the system using a general system model.

Critical Thinking

1. **Contrasting** How does the movement of electrons differ between a conductor and an insulator?

2. **Drawing Conclusions** Why was the invention of the integrated circuit important in the history of technology?

3. **Evaluating** Should a telephone system use analog or digital technology? Why?

4. **Contrasting** Describe the difference between operating system software and applications software.

5. **Taking a Position** Do you think it is a good idea to have totally automated systems that require no involvement by people? Why or why not? Give an example to support your answer.

 Connecting to STEM
science • technology • engineering • math

The Binary System

All computer functions are based on the binary number system. The binary system uses only 0s and 1s. Do research on the Internet to see how it compares with our system of numbering, which is a base-10 system. How would you write the numbers 3, 15, and 100 using the binary system?

Design Activity 21

BLINKING IN THE DARK

Problem Situation

Several students like to run after school, but as autumn and winter approach, it gets darker and darker in the afternoons. From dusk to dawn is a dangerous time for motorists to see runners and cyclists because of limited visibility. Design a blinking safety light system for use by cyclists and runners as a headband or an armband that will be highly visible and last a long time.

Materials

You will need:

- 3 AA batteries
- battery holders
- headband or armband (stretchable or closed with VELCRO®)
- heat-shrink wire wrap
- LEDs
- RadioShack Blinking LED (276-036C)
- switch
- wire

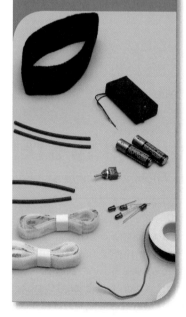

Your Challenge

You and your teammates are to design and construct a blinking light headband or armband.

> Go to your **Student Activity Guide, Design Activity 21.** Complete the activity in the Guide, and state the design challenge in your own words.

① Clarify the Design Specifications and Constraints

To solve the problem, your design must meet the following specifications and constraints:
- The blinking light system must be attached to a headband or an armband.
- The blinking light system must be highly visible.
- The device must last a long time.
- The device must use three AA batteries.
- The device must use LEDs, including a blinking LED.

> In your Guide, state the specifications and constraints. Add any others that your team or your teacher included.

② Research and Investigate

To better complete the design challenge, you first need to gather information to help you build a knowledge base.

> In your Guide, complete Knowledge and Skill Builder 1: Visibility.
>
> In your Guide, complete Knowledge and Skill Builder 2: Series or Parallel?

In your Guide, complete Knowledge and Skill Builder 3: Headband and Armbands.

3 Generate Alternative Designs

In your Guide, describe two possible solutions that your team has created for the problem.

4 Choose and Justify the Optimal Solution

Refer to your Guide. Explain why you selected the solution you did, and why it was the better choice.

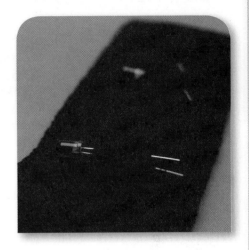

5 Develop a Prototype

Construct your blinking light headband or armband. Include a drawing or a photograph of your final design in your Guide.

In any technological activity, you will use seven resources: people, capital, time, information, energy, materials, and tools and machines. In your Guide, indicate which resources were most important in this activity, and how you made trade-offs among them.

6 Test and Evaluate

How will you test and evaluate your design? In your Guide, describe the testing procedure you will use. Justify how the results will show that the design solves the problem and meets the specifications and constraints.

7 Redesign the Solution

Respond to the questions in your Guide about how you would redesign your solution. Your design should be based on knowledge and information that you gained during the activity.

8 Communicate Your Achievements

In your Guide, describe the plan you will use to present your solution to your class. Show any handouts or PowerPoint slides you will use.

Design Activity 22
DESIGN YOUR OWN PC SYSTEM

Problem Situation

Determining what type of computer to purchase from the many choices available is an important decision. Different people use computers in different ways, such as to play a lot of computer games, to write, to send and receive email, or to surf the Internet. Computer companies have made it easy to buy a computer online, and they have provided a variety of options at different prices.

Imagine that you (or a relative) are given the opportunity to buy a new computer. Decide for whom you will be designing a computer system, and how the computer will be used. You will have $1,500 to purchase the computer system and necessary software (not including gaming software). Design the computer system that will have the best capabilities for the intended user.

Materials

You will need:
- computer with Internet access

Your Challenge

You and your teammates are to design a computer system.

Go to your **Student Activity Guide, Design Activity 22.** Complete the activity in the Guide, and state the design challenge in your own words.

① Clarify the Design Specifications and Constraints

To solve the problem, your design must meet the following specifications and constraints:
- The computer system must cost no more than $1,500.
- The computer system must have the best capabilities for the intended user.
- You must use online PC sites or documentation from a computer store.

In your Guide, state the specifications and constraints. Add any others that your team or your teacher included.

② Research and Investigate

To better complete the design challenge, you first need to gather information to help you build a knowledge base.

In your Guide, complete Knowledge and Skill Builder 1: Developing Your Specifications.

In your Guide, complete Knowledge and Skill Builder 2: Benchmarking and Multi-Criteria Decision Analysis.

In your Guide, complete Knowledge and Skill Builder 3: CPUs, Monitors, and Other Devices.

3 Generate Alternative Designs

In your Guide, describe two possible solutions that your team has created for the problem.

4 Choose and Justify the Optimal Solution

Refer to your Guide. Explain why you selected the solution you did, and why it was the better choice.

5 Develop a Prototype

Design your computer system. Include a schematic of your final design in your Guide.

In any technological activity, you will use seven resources: people, capital, time, information, energy, materials, and tools and machines. In your Guide, indicate which resources were most important in this activity, and how you made trade-offs among them.

6 Test and Evaluate

How will you test and evaluate your design? In your Guide, describe the testing procedure you will use. Justify how the results will show that the design solves the problem and meets the specifications and constraints.

7 Redesign the Solution

Respond to the questions in your Guide about how you would redesign your solution. Your redesign should be based on knowledge and information that you gained during the activity.

8 Communicate Your Achievements

In your Guide, describe the plan you will use to present your solution to the class. Show any handouts and/or PowerPoint slides you will use.

Information Technology

In this chapter, you will learn about the following technologies, which allow us to send signals around the world:
- radio waves
- antennas
- satellites
- computer networks
- World Wide Web

At 1,245 feet high, the Oriental Pearl Television Tower in Shanghai, China, is one of the tallest structures in Asia. The tower sends electronic information to residents throughout the city. These electronic signals travel through the air as radio and television waves.

The Shanghai tower is just one type of technology that allows us to communicate information. Today, most of the information that we use comes to us in an electronic form.

- Television stations broadcast live news coverage to different countries on different continents.

- The Internet links thousands of computers worldwide, bringing every type of information to people in any country.

- Automated teller machines (ATMs) allow us to carry out banking transactions any time of day or night.

- Cell phones allow individuals to engage in telephone conversations from almost anywhere.

Radio, Telephone, and Television

Benchmarks for Learning

- Information technologies are developing rapidly. As a result, the amount of data that can be stored and made widely accessible is growing at a faster rate each year.
- Most of the information we use comes to us in electronic form. In an electronic communication system, the channel carries an electronic signal.
- In radio communication, a message is sent through the air from a transmitting antenna to a receiving antenna.
- In a television communication system, an image is changed into electrical signals.

Reading Strategy

Outlining As you read the section, create an outline of the information that you learn. Use the headings and your own notes to organize the information.

> I. Information Revolution
>   A. Radio
>   1.
>   2.
>   3.

Vocabulary

frequency
radio broadcasting
point-to-point transmission

landline telephone
telephone switching
Internet service provider

carrier frequency
uplink
downlink

The Information Revolution

Much as machine power brought about the Industrial Revolution, electronic technology brought about the Information Revolution. Technology has made possible a world in which people can exchange information and ideas instantaneously. Every day, information is sent and received using different types of devices.

In the past, information was distributed in limited ways (Figure 12.1). Now, information technologies include text messaging, Facebook, instant messaging, email, data communication, radio and television broadcasting, electronic imaging, videos, CDs and DVDs, motion pictures, multimedia presentations, and Internet Web pages. Each of these technologies can create, store, exchange, or use information in electronic form.

Unlike other technologies, which process core materials such as steel or concrete, electronic technology processes knowledge and information. Information is unique because it becomes more valuable the more it is used. It is the commodity we deal with the most today. The world produces between 1 and 2 billion gigabytes of new information every year. This is equivalent to a 150,000-page book for every single person on Earth! Information technologies are developing rapidly so that the amount of data that can be stored and made widely accessible is growing exponentially (at a faster rate each year).

How is all this information stored? Printed documents, such as books, account for only three thousandths of one percent (0.003%) of the new information stored. In our Information Age, most information is "born digital." It is recorded on a computer as bits of information. Digital information is inexpensive to copy and distribute, and it can be searched easily.

Recall that the process for every communication system includes a transmitter, a receiver, and a channel. The channel carries the message from the transmitter to the receiver. In electronic communication systems, the channel carries electronic signals. The channel can be the air or cables made of copper wire or glass. Technologies that send messages through the air include radio, telephone, and television.

Radio

During the mid-1800s, James Maxwell, a Scottish physicist, showed on paper that it was possible to send signals through the air. Not until the 1900s, however, were devices built that could actually send and receive radio waves.

The first truly long-distance radio transmission took place on December 12, 1901, in a deserted hospital building in Newfoundland, Canada. Guglielmo Marconi heard the three short signals of the Morse code for the letter S. The code had been sent across the Atlantic Ocean over radio waves and then picked up by his receiver. Wires were no longer needed to send messages! This new invention was called the wireless, and later the radio (Figure 12.2).

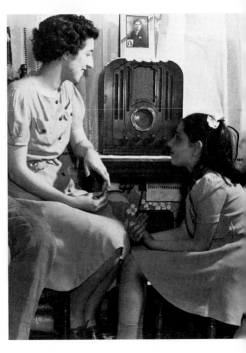

Figure 12.1 Radio was the first mass broadcast medium.

Interpreting *How has technology influenced the way people use radios today?*

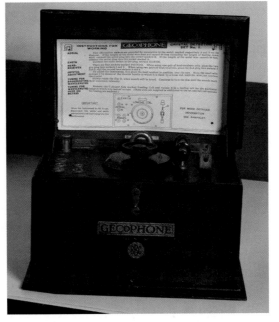

Figure 12.2 This Gecophone, built in 1925, was the first radio that amplified sound without using headphones.

Identifying *What other features of radios have changed since their invention?*

Radio was first used on ships. Because no wires were needed for ship-to-ship messages nor for ship-to-shore messages, people on ships could send routine calls for help.

As technology improved, new inventions such as the vacuum tube allowed voices and music to be transmitted over radio waves. Regular radio broadcasts began in the 1920s at station KDKA in Pittsburgh, Pennsylvania. Today, there are dozens of radio stations in almost every part of the country.

How Does Radio Work?

Most radios are built in the same way. Radio transmitters create an electric voltage that changes direction, or alternates, from positive to negative many times each second. The number of times an electronic signal alternates each second is called its **frequency.** Frequency is measured in cycles per second, and 1 cycle per second is equal to 1 hertz.

When alternating voltage from the transmitter is sent to an antenna, an electromagnetic wave is launched into the air. A radio signal is one type of electromagnetic wave. Radio waves travel outward in all directions away from the antenna.

The message—traveling as a radio signal—goes through the air from a transmitting antenna to a receiving antenna. Because each transmitter uses a different frequency, many transmitters in the same area can send messages at the same time. That is why you can choose from among many different stations. Once the radio wave reaches the antenna on your radio, the radio converts the radio signal into sound waves that you can hear (Figure 12.3).

Signals of different wavelengths travel differently through the atmosphere. Radio signals with low frequencies have long wavelengths. Radio signals with high frequencies have short wavelengths. The right wavelength must be chosen for each kind of radio communication. High-frequency radio waves bounce off the atmosphere's upper layer, the ionosphere. They are used for communication between two very distant points on the Earth. Super high-frequency radio waves are used for satellite communication. They travel straight through the ionosphere.

Using Radio Today

Today, radios broadcast music, news, and sporting events. They provide for two-way communication between automobiles, boats, airplanes, and homes. They enable communication with people in space.

Radio communication systems can be used in different ways. In **radio broadcasting,** one transmitter sends

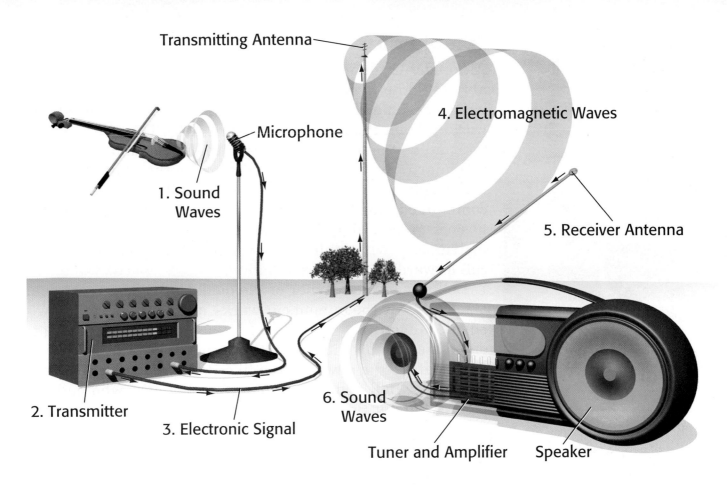

Transmitting Antenna

Microphone

4. Electromagnetic Waves

1. Sound Waves

5. Receiver Antenna

2. Transmitter

3. Electronic Signal

6. Sound Waves

Tuner and Amplifier Speaker

a signal out to many listeners. Signals are broadcast by AM and FM radio stations. If a message is sent from a single transmitter to a single receiver, the service is called **point-to-point transmission.** Such point-to-point transmissions are useful for communicating with remote places, such as fire-ranger towers, or for mobile users, such as an appliance-repair truck.

Figure 12.3 A radio communication system contains a transmitter, a receiver, an antenna for each, and a channel.

Summarize *How does a radio communication system work?*

Digital Radio

Radio is poised to undergo a revolution. For years, listeners have heard static on radio signals. Using digital radio, the radio communication industry is now able to deliver CD-quality, interference-free sound.

Digital radio employs a smart receiver, which uses a microcomputer to eliminate noise that accompanies analog radio signals. The microcomputers in the digital radio transmitters process sounds into patterns of numbers. This allows for a solid signal without distortions, static, or feedback.

The cost of digital technology will add about $100 to an existing car stereo system. Some digital car stereos will include a screen that can display pictures of singers and musicians or news. Unlike services offered by satellite radio programmers, the digital broadcasts will be free.

Telephone

Alexander Graham Bell, the inventor of the telephone, probably could not have imagined how successful his invention would become. Today, about 3 billion people around the world talk on the telephone (Figure 12.4). There are almost 950 million telephone numbers in the world, and more than one quarter of these are cell phones. In the United States, the telephone is the most commonly used form of electronic communication. The crank-box telephone of the late 1800s has been replaced by telephones as advanced as pocket-sized instruments with built-in answering machines, call waiting, call forwarding, and speed dialing.

Bell was also a teacher of the deaf. He knew a great deal about the way the human ear works, and he used this knowledge to invent the telephone. A few hours after Bell patented his new invention, another American inventor, Elisha Gray, patented a device much like Bell's telephone. For several years, Bell and Gray fought in court over who had the patent. Finally, Bell won.

How Does a Telephone Work?

A telephone mouthpiece contains tiny carbon granules (Figure 12.5). When you speak into it, the sound of your voice presses on these granules. When you speak loudly, the pressure is greater, causing the granules to become tightly packed. More electric current flows through the telephone circuit. When you speak softly, the granules are more loosely packed, and less electric current flows. The current flowing through the telephone line changes as you change the way you speak.

This smoothly varying electrical current, called analog signal, is changed into digital pulses. These pulses are similar to computer data. The amplified pulses—a type of electrical signal—are then sent through the phone line to the central office switching computer.

On the receiving end of the phone line, the digital pulses are changed back into analog electrical current. Then, they are sent to the earpiece of the receiving telephone. The electrical signal makes a thin piece of metal vibrate, producing a sound that you can hear. Because these vibrations change in strength as your speech changes, your speech is reproduced exactly.

This type of telephone is called a **landline telephone** because signals are sent through wires or cables on land. With landline phones, digital technology makes it possible to use services such as call forwarding and call waiting. It also makes the use of voice synthesizers possible. Voice synthesizers

Figure 12.4 Every day, the sight of people talking into thin air is becoming less science fiction, more reality.

Describing *How has cell phone technology affected the way people communicate?*

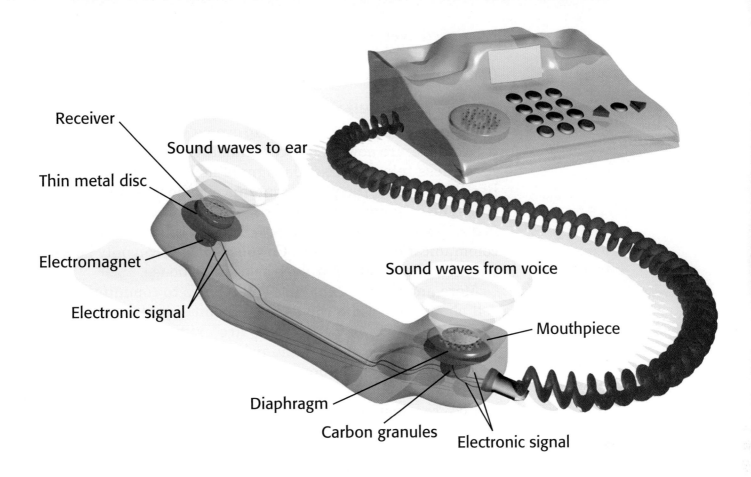

Receiver

Sound waves to ear

Thin metal disc

Electromagnet

Sound waves from voice

Electronic signal

Mouthpiece

Diaphragm

Carbon granules　Electronic signal

are electronic circuits that create sounds similar to human speech. These machines provide information such as telephone numbers and the time of day over the phone.

Telephone Switching

Every time you make a call, the message goes to a central office, where a computer sends the signal to the number you dialed. The connection of your telephone to another telephone is called **telephone switching** (Figure 12.6).

The first telephone switch was invented in the late 1800s by Almon Strowger, a funeral home director in a small town in the Midwest. The town's telephone operator, the wife of the owner of another funeral home, always connected people asking for a funeral home with her husband's business. Mr. Strowger saw his business failing. To save it, he invented a device that would connect telephones without an operator. The switch he invented was the first automatic system to be used in a public exchange.

Today, telephone switching is extremely important. At any given time, millions of telephones must be able to be connected with any other telephone in the world. Switches are connected using one of several channels. Microwave relay stations send signals through the air. More than 1,000 conversations can

Figure 12.5 A telephone converts sound to electrical signals and electrical signals back to sound.

Clarifying *How is sound transmitted by a telephone?*

Figure 12.6 Phone switching used to be done by people working at a switchboard.

Inferring *How did technology affect these women's jobs?*

be carried over a single microwave signal. Fiber-optic cables carry other signals underground. More than 10,000 conversations can be carried over a single optical fiber.

Cordless and Cell Phones

Telephone technology continues to evolve. Cordless telephones, for example, are now considered standard in many households. A cordless telephone sends signals using radio waves. The handheld part of the phone contains a radio transmitter. This transmitter sends a signal to the nearby base station, which is a radio receiver. The receiver picks up the signal, allowing communication to continue. Cordless phones can only be used within a very small area, such as a single house.

More and more, people use cell phones to communicate. A cell phone uses radio waves to transmit signals over a large area (Figure 12.7). Cell phones are extremely popular because they can be carried and used anywhere, including in a moving vehicle. Cell phones are powered by small batteries that can be recharged when they lose energy.

Like cordless telephones, cell phones send radio signals. As a person using a cell phone moves, the cell phone signal is picked up by an antenna in a particular area, called a *cell*. When a person leaves one cell, the signal is picked up by antennas in the next cell along the way. Once the signal is picked up, it is amplified and retransmitted to a switching office, where the call is routed to the destination. Cell phones are becoming so popular that some people no longer purchase landline service and use cell phones instead.

Figure 12.7 A typical large city can have hundreds of towers that contain radio transmitters and receivers.

Inferring *How do you think the population of a town affects the number of towers that are built?*

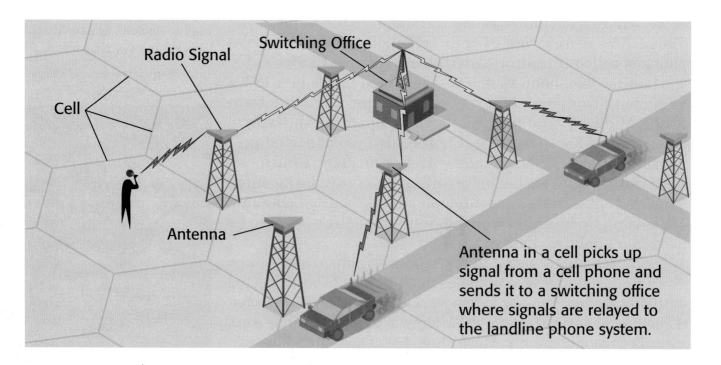

Radio Signal

Switching Office

Cell

Antenna

Antenna in a cell picks up signal from a cell phone and sends it to a switching office where signals are relayed to the landline phone system.

Figure 12.8 Many cell phones now have multiple capabilities.

In addition, many technologies are being combined in cell phones. Many cell phones now look more like personal digital assistants with the capabilities of cameras, calendars, email, credit card charging, and video-conferencing all in one small device (Figure 12.8). The newest technology to be added to cell phones is the ability to play MP3 files. Apple's iPhone combines an iPod MP3 player with a cell phone, and offers email, Web browsing, and maps as well.

Cell Phones in the Future

Cell phones will make use of faster connection speed technology. The 4G ("G" means "generation") technology will increase data transfer speed and will also reduce dropped calls (which occur when a user transitions from one cell area to another). Increased data transfer speeds will work better for playing games and will enable cell phone users to communicate with video, not just voice or text.

One problem with current cell phones used for games or photos is the small screen size. It's likely that in the future, cell phones may include a projector built into the phone that will enable users to project an image onto a screen or nearby wall.

Some companies, such as Cisco Systems, are designing technologies that would allow users to transfer files on the phones by aiming their phone at another digital device (another phone or a computer) and using a sweeping motion to transfer data.

There is even talk of designing solar-powered phones that would use collectors worn on the wrist to capture solar energy.

One of the most exciting developments is the ability to download cell phone apps (applications) that allow the cell

Technology in the Real World

Free Telephone Calling over the Internet

How soon will it be before home telephone calling is altogether free of cost? The days of sending and receiving telephone signals over copper wires may be coming to an end. Those wires are being replaced by telephone services provided over the Internet. These services are known as voice over Internet protocol, or VoIP. Though there is a cost for Internet service, using VoIP over the Internet may not involve any extra charges, so people can make their calls for free. Every week, about 150,000 traditional telephone lines disappear. In 2005 VoIP entered the mainstream residential telephony market in the United States, and it is rapidly becoming a billion dollar industry.

Some companies that provide VoIP service depend on a central service provider, but the most interesting companies are ones that provide peer-to-peer (P2P) technology, using special software to allow people to use home computers or Internet devices to communicate directly. One P2P company, Skype™, includes free voice and video conferencing. Skype™ claims to have over 100 million users.

"I knew it was over when I downloaded Skype," Michael Powell, the former chairman of the Federal Communications Commission, explained. "When the inventors of KaZaA [a file sharing application commonly used to exchange MP3 music files over the Internet] are distributing for free a little program that you can use to talk to anybody else, and the quality is fantastic, and it's free. It's over. The world will change now inevitably."

Each Skype™ user must have the software running on his or her computer. This software is available free of charge and can be downloaded from the company Web site, www .skype.com. Analysts say that millions of people in the United States and Europe will abandon traditional phone lines in favor of wireless and broadband telephony.

Using a USB telephone adapter and a standard telephone, you can make Internet calls and also place normal calls over telephone lines.

phone to perform functions that are completely unrelated to making phone calls. There are thousands of apps that users can download, including ones that allow users to access their Facebook pages or surf the Web. Other apps permit you to scan bar codes on a product, and then check prices of comparable products on the Web. Downloading songs, playing music, watching movies, fitness training, counting calories, getting weather and sports news, and finding traffic information are other popular apps.

Internet Telephones

As the Internet grows in popularity, more and more people are using Internet telephone services. With an Internet telephone, instead of using a telephone number, you contact another user through that person's email address. Internet telephones are cheaper to use than cell phones or landlines, but the quality is not as good.

When you speak into the computer microphone, a sound card converts your voice into a digital signal. The signal is then sent to the computer modem. The modem connects to an Internet service provider, which receives the data on its computer (called a server). An **Internet service provider** (ISP) is a company that provides its user's access to the Internet. The ISP sends the data over telephone lines to a computer on the other end. The data is then converted by the second computer's sound card and sent to speakers or headphones. This all happens very quickly, even if the person you are talking to is very far away.

Online Communities

One factor that helps explain the popularity of the World Wide Web is the way the Web makes it so easy for people to create communities and networks of friends all over the world. Three Web tools that contribute to that are wikis, blogs, and social networking sites. **Wikis** are Web sites that allow users to add and change information on the site. Information on wiki pages can be linked to other Web sites and wiki pages. Lots of information can be gathered and kept up-to-date when many people contribute their expertise.

Blogs, or "web logs," are Web sites where individuals can post text and photos, like a journal. Some blogs are about a single subject such as cooking or politics, but many others are more personal, almost like a public diary about daily life. Blogs can also contain links to other sites on the Web. Readers can leave comments and responses to blog entries. Some very popular blogging sites are *LiveJournal, Blogger,* and *Xanga*. The sites provide the software and support, and users create accounts and then simply enter information and photos into their blogs.

Social Networking

In the last few years, internet social networking has become quite a trend; teens and seniors alike have become enamored, if not obsessed, with Facebook, MySpace, and LinkedIn. These sites allow people to share pictures and events, and communicate across the world. Over 175 million people log into Facebook daily. What's the attraction?

Facebook

In 2005, Facebook had over 5.5 million users. In 2007, there were over 50 million. In February of 2009, Facebook had over 400 million active users. Facebook is the most highly populated social networking site.

Facebook was founded in 2004, and is a social utility to connect people around the world. From friends, to family, to co-workers, Facebook allows people to communicate, share pictures, and discuss their present activities. Facebook has become internationally used.

MySpace

Another social networking site is MySpace. MySpace was launched in August 2003, and became the most popular social network in June 2006. MySpace is very much like Facebook, except it allows for more personalization. It allows users to choose a background for their page, and choose emoticons to display their "mood." Because MySpace was written in HTML, it is easily changed by the user, who has only to insert new text into the script that is already present to change the look of the page.

Other Networking Sites

Whereas MySpace and Facebook are sites primarily designed for friends to keep in touch, **LinkedIn** is more for professional use. LinkedIn is a business-oriented networking site launched in 2003 that now has 60 million users worldwide. The purpose of LinkedIn is to allow users to reach people they have met, and trust, in business. These people are known as *connections*.

Another social networking site is **Twitter.** Twitter allows users to post updates on what they're doing and how they are feeling at the moment. The update is called a *tweet* and is limited to a length of 140 characters. Users can list their

Connecting to STEM

The Electromagnetic Spectrum

Radio, television, and telephone communication are alike in one way. They all use electromagnetic waves to send a message from a transmitter to a receiver. Radio waves, microwaves, and light waves are all types of electromagnetic waves. These and other different types of waves make up the electromagnetic spectrum.

Different Types of Waves

Waves along the electromagnetic spectrum are different because they occur at different frequencies. As the frequency of a wave gets higher, its wavelength gets shorter. Microwave radios, for example, use higher frequencies than FM radios or television broadcasts. Because their frequencies are higher and their wavelengths shorter, these waves are called microwaves, or "little waves." Other types of waves in the electromagnetic spectrum include X-rays, visible light, and ultraviolet light.

Wavelength and Frequency

All electromagnetic waves travel through space at the speed of light—about 186,000 miles each second. One second after a wave is sent out from a radio transmitter, it is 186,000 miles (or about 300,000,000 meters) away, in every direction. This is true no matter what the frequency of the wave.

The relationship between frequency and wavelength is given by the equation

$$\lambda = c/f$$

where λ is the wavelength, c is the speed of light, and f is the frequency.

During that same second, the radio transmitter generates wave cycles. The number of cycles generated is defined by the frequency of the signal. Recall that a hertz is equal to 1 cycle per second. Thus, a 10-megahertz transmitter generates 10 million cycles in 1 second. We already know that 186,000 miles is equal to 300,000,000 meters. Using this information, we can calculate the wavelength, or the distance covered by one cycle, as follows:

$$\text{Wavelength} = \frac{300{,}000{,}000 \text{ meters/second}}{10{,}000{,}000 \text{ cycles per second}}$$

$$= \textbf{30 meters}$$

A wavelength is the distance from one peak to the next peak on the waveform. This distance is 30 meters when the frequency is 10 MHz.

Long   Wavelength   Short

Low   Frequency   High

Radio Waves   Infrared Rays

Critical Thinking

1. **Summarizing** What are electromagnetic waves? How are they produced?

2. **Calculating** Use the formula above to answer the following question: If an antenna receives a signal at a frequency of 14 MHz, how long must the wavelength be?

tweets in categories. As opposed to having "friends," or "connections," Twitter allows the user to follow people. Once a user follows another, they can read tweets. Twitter is used internationally, and, although not as popular as Facebook, it is steadily growing and becoming an important social networking community.

Television

Television, or TV, is a broadcast communication system in which electrical signals are converted into visual images. It is much like radio—with a transmitter, channel, and receiver—except that the information being carried is seen as well as heard. Electrical signals are transmitted through the air. When they reach an antenna (the receiver), a television receiver (the monitor or television set) changes the video into images on a screen.

In a television communication system, the transmitter is the television camera. The camera encodes the visual scene into electrical signals. These electrical signals make up a video. Video signals may be stored on videotape or on digital videodiscs (DVDs). Videotaped programs can be played back by a videocassette recorder (VCR). Digital video programs are played back through DVD players or through the videocamera. Video may also be sent directly by cable (as in cable TV). It may also be sent over distances through the air using TV transmitters.

Broadcast and Cable Television

In broadcast television, each transmitter has its own frequency, called a **carrier frequency,** on which it sends the video. Each television channel is assigned a carrier frequency. A television receiver, or television set, is simply a monitor that contains a tuner. The tuner is used to choose a channel.

Cable television systems use very high antennas or relay stations to receive distant signals. At a subscriber's home, all the information from different channels is carried on a single cable. Because of this extra programming, there is cable television in most large cities, even where broadcast reception is good.

In a closed-circuit system, a different kind of cable television, a cable connects the camera directly to the monitor. This type of television system allows people to watch a concert or lecture when there is not enough room to seat everyone. It also allows security guards to watch several areas of a building at the same time.

High-Definition Television

High-definition television (HDTV) is the most significant improvement to television since color replaced black and white. HDTV provides clearer images and better sound than conventional television. There are about three times as many scanning lines in a HDTV as in a regular television. The picture is also wider, providing a larger viewing area similar to a movie screen.

A separate HDTV tuner is required in HDTV sets to decode digital television signals. HDTVs can receive broadcasts from land-based or satellite sources.

Using Satellites for Communication

A satellite is an object that orbits, or revolves around, a larger object. The moon is a natural satellite that orbits the Earth. Artificial satellites have been in use for decades. They are placed in orbit around the Earth to relay communication signals, survey the Earth's surface and weather patterns, and gather images for military purposes.

Communication satellites send and receive radio and television signals (Figure 12.9). They make it possible to communicate globally by telephone, fax, or computer. The signals are sent from a station on the Earth's surface. The satellite receives signals and then retransmits them to other places on the Earth. With the right number of satellites in space, telephone conversations, radio programs, and television programs can be sent anywhere in the world.

A transmitting station on the Earth sends signals to a satellite on one frequency. This is called the **uplink.** The satellite receives the signal and changes it to a different frequency. The satellite then sends this changed signal back toward the Earth to defined areas. This is called the **downlink.** Anyone with a special antenna, called a satellite dish,

Figure 12.9 This communication satellite receives microwave signals from the Earth (uplink), amplifies the signals, and then retransmits them back to the Earth at a different frequency (downlink).

Applying *Why are satellites used so extensively in communication systems?*

in the covered area can pick up the transmission. Satellite dishes can vary in size depending on the intended use.

Satellite Radio and Telephone

Satellite radios are being installed in some new cars. This system provides about 100 channels of radio programming beamed to the Earth by satellites. The digital signals provide high-quality reception without static. Satellite radio providers charge a monthly fee to users, so many of the stations are commercial-free. Satellite radios can also be used at home. They require a separate small antenna that receives satellite signals.

Satellites are sometimes used for telephone communication from remote locations where landline or cell phone systems have not been installed. The signals are beamed to a geosynchronous satellite that orbits 22,000 miles above the Earth. Because of this distance, it takes the signal about half a second to arrive at the satellite, be retransmitted, and then received back on Earth.

Satellite Television

Because a TV antenna must be in the line of sight of the transmitting antenna, broadcast television has a limited range. The line of sight is limited by large objects, as well as by the curve of the Earth. People in remote areas often cannot receive any broadcast television stations, but they can easily receive signals via satellite.

There are two major types of satellite television systems today: conventional satellite TV and the newer direct satellite system (DSS), shown in Figure 12.10. Conventional satellite TV requires a large satellite dish to receive signals correctly. These dishes can be rotated to receive signals from about 35 different satellites. These systems provide coverage of news without an anchorperson or reporter. Viewers see news as it happens, before live cameras. They can also watch shows uncut, without commercials. Although some channels can be received without cost, most are coded, and program providers sell an electronic "key" to unlock the signals. Over 1,000 audio channels, including every conceivable type of music, are available. Every game in every sport is available on at least one satellite channel, as is a wide range of international programs.

DSS satellite TV units operate at a higher frequency than conventional units. As a result, DSS dishes are smaller than conventional satellite dishes. A DSS dish is stationary, but it can pick up two satellite locations at the same time. Normally, viewers pay monthly for a programming package that provides more than 150 television and

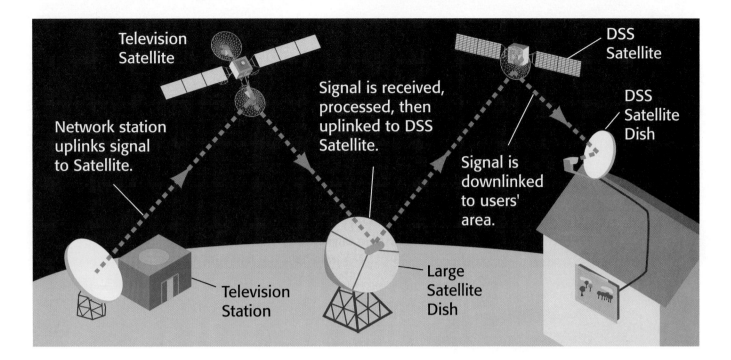

Television Satellite

Network station uplinks signal to Satellite.

Signal is received, processed, then uplinked to DSS Satellite.

DSS Satellite

DSS Satellite Dish

Signal is downlinked to users' area.

Television Station

Large Satellite Dish

music channels. The monthly costs are somewhat higher for DSS than for conventional satellite service.

DSS services actually receive signals from the larger conventional satellite dishes. Signals are received on large dishes from the networks. They are processed, put into digital form, and then sent (uplinked) to DSS satellites. From these satellites, the signals are sent back down (downlinked) to users' DSS satellites. Each time the signal is processed, some quality is lost, so DSS signals are not quite as good as signals received directly by large dishes.

Figure 12.10 DSS services use small dishes that can be mounted easily on the roof or side of a building.

Inferring *Which type of satellite dish provides a clearer television image, conventional or DSS?*

SECTION 1 Assessment

Recall and Comprehension

1. What is meant by electronic information technology? What technologies are included?
2. Identify the most commonly used electronic information technology in the United States.
3. Why is fiber-optic cable better than copper wire for large numbers of telephone circuits?

Critical Thinking

1. **Taking a Position** Cell phones are useful, but they are also associated with a higher incidence of driving accidents. What type of law should be used to govern cell phone use in cars?
2. **Making Judgments** Would you choose satellite radio for your car or home? Explain why or why not.

QUICK ACTIVITY

A simple broadcasting system can be made using two small loudspeakers and a roll of two-conductor wire. Connect one end of the wire to the terminals of one of the speakers. In another room, do the same with the other speaker. In a loud voice, speak directly into one of the speakers while another student listens quietly to the other speaker. What happened? Why? **For more related Design Activities, see pages 378–381.**

For: Lost Dog Activity
Visit: www.mytechedkit.com

GPS Receivers: Help Find a Lost Dog Online

Did you know technology exists that can pinpoint your location any place on Earth? Go online to learn how global positioning system (GPS) devices can even help you find a missing pet.

11,000 miles above Earth, 24 GPS satellites are in orbit, sending out radio signals to GPS receivers. The GPS receivers use those signals to determine precise locations with regard to latitude, longitude, and altitude.

Computer Networks and the Internet

Reading Strategy

Mapping Create a map for each type of computer network. Use Figure 12.14 to help you get started.

Vocabulary

computer network	wire cable	wide area network
server	fiber-optic cable	email
modem	peer-to-peer network	browser
network interface card	local area network	

Benchmarks for Learning

- A computer network includes a server, a network operating system, network interface cards, and cables.
- The three types of networks are peer-to-peer networks, local area networks, and wide area networks.
- The Internet has revolutionized our society by providing worldwide information access at very low cost.
- Computers use the Internet for electronic mail, telnet, file transfer, and access to the World Wide Web.
- The large range of personal and professional information technologies and communication devices allows for remote collaboration and rapid sharing of ideas unrestricted by geographic location.

Computer Networks

As the demand for information has grown, so has the need to organize, store, and archive information as data. As companies have become more dependent on computers, single computers have become less capable of meeting their storage and processing needs.

Computer networks have provided one solution to this problem. A **computer network** is made up of interconnected computers and often includes peripheral devices such as printers, scanners, and storage hardware.

Computer networks enable computers to share data, software, and different peripheral devices. Networks are used to make airline reservations quickly from anywhere in the world. They are used to operate automatic teller machines for banking. With the help of cables, computer networks now extend around the globe (Figure 12.11). The largest computer network is the Internet. A basic network includes a server, a network operating system, network interface cards, and cables.

Servers

A **server** is a computer or a bank of computers that provides information or services to other computers in a network. A Web server is a computer that hosts Web pages on the Internet. Depending on its size, a Web server can host from dozens to hundreds of Web sites.

A network operating system is software that manages all the other programs on a server and access to files. The operating system determines the order in which other programs should run and manages how the various programs share computer memory. It also acts like a traffic cop in controlling the flow of information input and output to and from attached hardware devices.

Modems

When a computer connects to the Internet using a telephone line, a modem provides the connection. A **modem** is a device that converts data signals into analog sounds that a telephone line can carry.

Digital equipment cannot use telephone lines directly because data signals are different from telephone voice signals. The word *modem*, which is short for *mo*dulator-*dem*odulator, is derived from this process. At one end of an Internet connection, a modem converts data signals from a computer into analog signals that the telephone line can carry. This process is called modulation. At the other end of the connection, another modem converts the sound back into data signals. This is called demodulation.

Network Interface Cards

A **network interface card** (NIC) is a computer circuit board or card that is installed in a networked computer or server

Figure 12.11 Communication systems use satellite and fiber optic networks to transmit information.

Describing *What areas have the most cables?*

(Figure 12.12). A NIC allows continuous connection to a computer network. At any time, data can be transmitted to and from a server or other workstations equipped with a NIC.

Cables

Wire cables are copper wires used to connect the computers in a network. A single cable includes hundreds of wires and can be many miles long. The more wires, the more data that can be carried at the same time. Cables that use concentric wire conductors are called coaxial cables. These cables are made of copper, a costly metal.

More and more, data networks are using light rather than electricity to carry information. **Fiber-optic cables** are cables made of very thin strands of coated glass fibers (Figure 12.13). They can carry much more information than copper wire. Information is sent from a transmitter which, for fiber-optic cables, is a very bright light that can be controlled. Two types of fiber-optic transmitters are lasers and light-emitting diodes (LEDs). Information is put onto the light wave by changing the voltage powering the laser or LED. As the voltage changes, the light becomes brighter or darker, creating a code that represents the information.

Light is high-frequency electromagnetic radiation. It has a frequency far above radio waves in the spectrum, so it has a much shorter wavelength. Because light has a large bandwidth, it can carry large amounts of information. A single fiber in a multiple-fiber cable can carry more than 10,000 telephone circuits, and a single cable may have more than 144 such fibers. Most major cities now use fiber-optic cables

Figure 12.12 A network interface card allows a computer to have a dedicated connection to a computer network.

Comparing *How are NICs similar to modems?*

Figure 12.13 Each optical fiber carries digital signals that are used in television and computers.

Inferring *Why is the cladding designed to reflect light?*

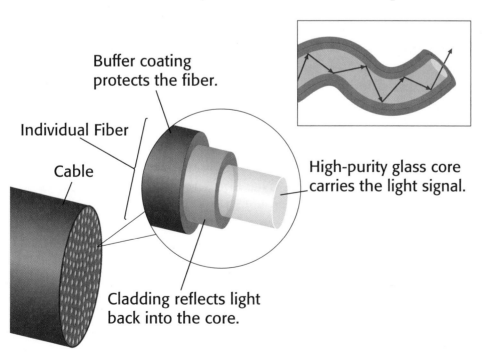

Buffer coating protects the fiber.

Individual Fiber

Cable

High-purity glass core carries the light signal.

Cladding reflects light back into the core.

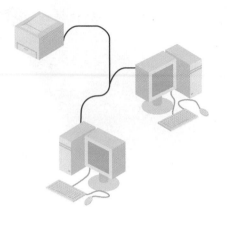
Peer-to-peer network

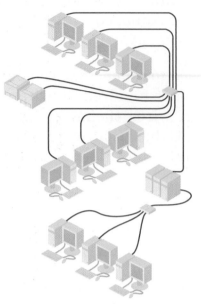
Local area network

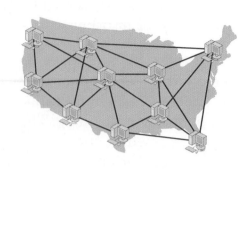
Wide area network

Figure 12.14 Many office functions are now automated through local area networks.

Inferring *Which type of network is the Internet?*

in their data communication and telephone systems. Fiber-optic cables span both the Atlantic and Pacific Oceans.

Types of Computer Networks

Three common types of computer networks are peer-to-peer networks, local area networks, and wide area networks (Figure 12.14). A **peer-to-peer network** is normally small, connecting from two to five stand-alone computers to share information, or peripheral devices such as drives or printers. Each computer may act as a server, sharing files with all other users.

A **local area network** (LAN) is a data network that allows computers to share data throughout a department, an office, or a building. The computers can also share files, a printer, a modem, or memory. LANs can include several computers or hundreds of computers, as might be the case in a medium-sized business. The computers of a LAN are connected to one or more servers.

A **wide area network** (WAN) connects many hundreds of computers that are far apart. They can be connected by telephone lines or by using satellites. A WAN is normally made up of numerous LANs. The Internet, which spans the globe, is the largest wide area network in the world.

The Internet

Over the last few years, the Internet and the World Wide Web have had a tremendous impact on the way we communicate and the way we live our daily lives. Anyone with a computer can do research from home and obtain information from all over the world in seconds. Small businesses can have access to millions of customers. Today, students can access more information than is stored in the Library of Congress (Figure 12.15).

The Internet has truly revolutionized our society. It is the fastest growing instrument of communication in the history of civilization. It took radio broadcasters 38 years to reach an audience of 50 million. It took television 13 years. It took the Internet only 4. The Internet combines computing and communication technology in such a way as to provide users with worldwide information access at very low cost. Today, hundreds of millions of people are Internet users in more than 200 countries around the world.

Simple Beginnings in the Military

You are probably familiar with the Internet, but what is it exactly? The **Internet** is a worldwide network of computer networks. It began in the late 1960s when the Pentagon's Advanced Research Projects Agency (ARPA) needed to distribute its computers away from one central location. This was a precaution so that in case of war, the destruction of one or more segments of the network would not result in a total communication collapse. The original network was called ARPANET.

A few years later, the National Science Foundation funded computing centers at five major universities. These centers formed the beginning of what has become known as the Internet backbone in the United States. A backbone is a set of large, high-speed transmission lines that carry data from smaller lines with which it interconnects. The connection points are known as network nodes.

The World Wide Web

Originally, the Internet was only a text-based system. Simple text messages were sent as email (electronic mail). Soon people needed a way to carry graphic information as well. This need led to the development of the World Wide Web (WWW). Once the Web was created, it grew exponentially. By November 1996, the number of servers connected to the Internet had grown to about 28,000. Today, more than 100 million households in the United States (about 95 percent) have telephones, and 54 percent have access to the Internet. Analysts estimate that thousands of new Web servers go online daily.

Fair Use

Many people use the Internet for purposes of conducting and sharing research. Students seek information for papers they

Figure 12.15 This is one artist's vision of the world of information.

Extending *How would you describe—with words or pictures—all the world's information?*

are assigned to write, scientists share their research with colleagues, authors collect interesting examples and use them to illustrate their writing.

When people use information that they collect over the Web, or from other sources, they must make sure that they abide by **fair use policies** when using copyrighted materials. A copyright is the legal right to the material that is granted to the original author or publisher and protects the copyright holder from people using the work without permission.

Fair use policies, however, give people the right to limit use of copyrighted material without permission if there are social benefits to the use. However, the amount of a copyrighted piece that you can quote is limited and you must identify the source. When authors write books, the publisher provides a style guide that provides detailed examples for how to give appropriate credit to others when incorporating their ideas, text, or images in one's own work. This is correctly done by adding an appropriate citation. Here is an example:

> In their book *Engineering and Technology Education: Learning by Design*, Hacker and Burghardt state "When people use information that they collect over the Web, or from other sources, they must make sure that they abide by fair use policies when using copyrighted materials."[1] The source of the quotation is identified as shown below:

[1]Hacker, M. and Burghardt, D. (2010). Engineering and Technology Education: Learning by Design. Pearson Prentice Hall Publishing. Boston, MA.

Internet Service Providers and Email

Internet service providers (ISPs) are companies that provide users with access to the Internet. Some of the well-known ISPs are America Online (AOL), Earthlink, and CompuServe. ISPs charge users a monthly fee to establish their connection. ISPs use banks of computers to route communication to and from subscribers, over their own computer network to the Internet backbone. ISPs provide full Internet and Web access, including email accounts and mailboxes. ISPs establish connections for subscribers over telephone lines (dial-up connections), fiber-optic cables (cable connections), or wireless networks.

Email is an Internet service by which a text message is sent using electronic signals. (The *e* in email stands for electronic.) Email allows users to send and receive messages to and from one another, and to participate in discussion groups, called Listservs®. Email messages can be text only or can include graphics. Messages can also include attachments, which are copies of other files (such as documents or photos) stored on your computer. An "electronic mailbox" stores all the messages for a single user. Replies can be made instantly via email and users can save the messages they receive. No paper or printer ever has to be used.

Telnet and FTP

Terminal access protocol, or telnet, is an Internet service that allows a computer user remote access to all the files on another computer. Files can be changed and moved around as if you were using the remote computer directly. To transfer files from another computer to your own, you use a service called File Transfer Protocol (FTP). FTP allows your computer to quickly access files from another computer and download them to your own.

Instant Messaging

Instant messaging is an Internet service that establishes a private email connection between people that you include on a buddy (or contact) list (Figure 12.16). Instead of reading messages that have been placed in your electronic mailbox at some prior time, instant messaging permits real-time communication. A separate window opens when you use instant messaging. When two people on each other's lists are online at the same time, they can both see the window and have continuous electronic conversations.

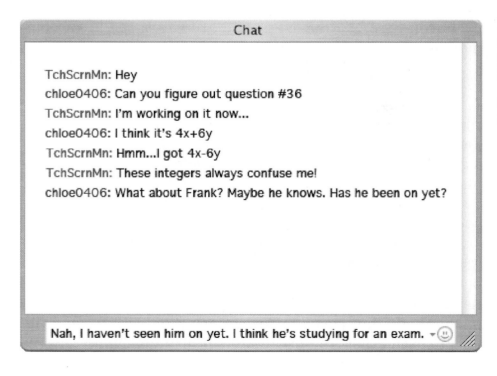

Chat

TchScrnMn: Hey

chloe0406: Can you figure out question #36

TchScrnMn: I'm working on it now...

chloe0406: I think it's 4x+6y

TchScrnMn: Hmm...I got 4x-6y

TchScrnMn: These integers always confuse me!

chloe0406: What about Frank? Maybe he knows. Has he been on yet?

Nah, I haven't seen him on yet. I think he's studying for an exam.

Figure 12.16 Instant messaging allows people to have real-time chats over the Internet.

Summarizing *How is instant messaging different from e-mail?*

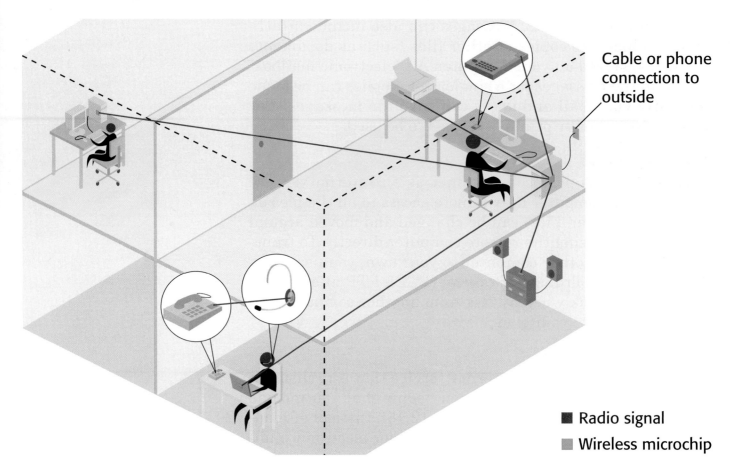

Cable or phone connection to outside

■ Radio signal
■ Wireless microchip

Figure 12.17 Wireless personal area networks use radar waves to transmit signals over short distances.

Extending *What are the advantages of wireless over traditional technology?*

Figure 12.18 Bluetooth technology enables small devices to communicate over short distances.

Web Browsers

The World Wide Web is the most advanced information system on the Internet. It is a system of linked resources in a graphical format used to find information. It has become the heart of the Internet. To access Web pages, computers use a software program called a browser. Internet Explorer, Firefox, and Netscape are the three most common browsers.

Wireless Personal Area Networks

A **wireless personal area network** (WPAN) is a short-range, wireless network that links personal devices together without the use of cables (Figure 12.17). WPANs use a globally available radio signal for devices such as cordless phones, wireless headsets, remote controls, and garage door openers. Different WPAN technologies that have been developed include Bluetooth, Wi-Fi, and HomeRF.

Bluetooth is a new technology that provides short-range wireless connections among desktop and laptop computers, personal digital assistants, cellular phones, printers, scanners, digital cameras, and even home appliances. Many cell phones and other portable devices use Bluetooth to send signals to each other (Figure 12.18). For example, many people use Bluetooth to send photos from their cell phones to their computers. The technology will let the devices share information within a 30-foot range without having to be physically connected. With increased power, the range can be extended to 300 feet. To enable a device to use Bluetooth, an inexpensive chip must be added.

Wi-Fi, which stands for wireless fidelity, is an increasingly popular way of creating WPANs at high data speeds (Figure 12.19). With a special plug-in circuit card, Wi-Fi provides Internet access to computers within 300 feet of a Wi-Fi transmitter. These locations are known as hot spots and are found in homes, coffee shops, Internet cafes (known as cyber cafes), hotels, airport lounges, and libraries.

HomeRF, or home radio frequency, permits computers, peripherals, cordless telephones, and other devices to communicate voice and data without using cables over a range of 150 feet. HomeRF is used mainly with cordless phones. It can support up to eight toll-quality voice connections, eight prioritized streaming media sessions, and multiple Internet and shared resource connections at broadband speeds.

Figure 12.19 Cyber cafes typically use a Wi-Fi network to link their computers to each other and to the Internet.

Comparing *Which WPAN transmits over a greater area—Wi-Fi or HomeRF?*

SECTION 2 Assessment

Recall and Comprehension

1. What is a computer network, and what are its basic components?
2. How is light used to carry information over fiber-optic cables?
3. Name four services available on the Internet.
4. What is a browser? Describe what a browser does when you use the Internet.

Critical Thinking

1. **Contrasting** Describe the differences between peer-to-peer networks, local area networks, and wide area networks.
2. **Applying** How do you use the Internet? How has its use changed the way you live your life?

QUICK ACTIVITY

Fiber-optic glass is so clear that light will "bounce" down a single fiber as far as 2 miles before it needs to be amplified. Window glass, on the other hand, becomes opaque after about 16 inches. Using library and Internet sources, research the history of fiber optics. Be sure to include where and when fiber-optic glass was first developed, as well as the different uses of fiber optics. **For more related Design Activities, see pages 378–381.**

People in Technology

Former CEO of eBay
Meg Whitman

"The Internet . . . is one of those revolutionary technologies that has the ability to change so much of how we live, how we communicate, how we plan, how we get information . . ."

—Meg Whitman

In 1998, when Meg Whitman began as manager of eBay, the online auction and trading company, it resembled the Classifieds section of a local newspaper. It was mostly a site for trading odd specialty items such as Pez toys and Beanie Babies. eBay has survived many crises by constantly innovating and improving their services.

Whitman regularly gives talks about e-commerce and the digital economy.

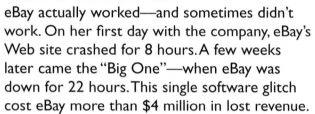

Understanding eBay's Technology

What did Whitman do differently than managers at other companies? For one thing, she realized that she needed to understand how eBay actually worked—and sometimes didn't work. On her first day with the company, eBay's Web site crashed for 8 hours. A few weeks later came the "Big One"—when eBay was down for 22 hours. This single software glitch cost eBay more than $4 million in lost revenue.

Whitman decided that she needed to understand Web technology. She learned about servers, Web software, and large-scale networks. She also began to see how e-businesses are different from other businesses. Decisions must be made much, much faster on the Internet, and people expect information much faster. In 2002, eBay installed its new "V3" next-generation shopping technology. With V3, users can get real-time (instead of hourly or daily) updates for the data warehouse and for items that are being auctioned.

Why has eBay thrived when so many other e-businesses have failed? According to Whitman, it's because eBay offered something new—a truly global community. Even though buyers and sellers may not live close to each other, they can meet online.

Critical Thinking

1. **Summarizing** How is e-business different from a land-based business?

2. **Applying** How might eBay's V3 technology make the site more appealing to users?

Electronic Games and Recording Systems

Reading Strategy

Listing As you read the section, make a list of all the different types of electronic technologies that exist today. Next to each item on the list, write a brief description of what it is and how it is used.

1. Electronic Games
- Game Boy – handheld

2. Playback
- CDs and DVDs

Vocabulary

compact disc	digital videodisc	MP4
MiniDisc	MP3	

Electronic Games

Electronic games are among the most popular applications of electronic technology. Eighty-four percent of teenagers play video and electronic games. Many of these games can be purchased on CDs or downloaded from the Internet. Other games use separate game consoles such as Sony Playstation 3 and Microsoft's X-Box (Figure 12.20). Some of these consoles also play CDs and DVDs.

There are also many types of handheld games. These range from individual games such as chess and Monopoly

Benchmarks for Learning

- Electronic games and music are among the most popular consumer applications of electronic information.
- Recorded information is stored on magnetic and optical media and on integrated circuits.
- CDs and MiniDiscs are replacing tape as the preferred music medium.
- Technologies such as MP3 and MP4 permit downloading music and other types of media directly from the Internet.

Figure 12.20 Games on consoles such as the Microsoft Xbox 360 can be played in either single player or multiplayer (on line) mode.

Predicting *How do you think game technology will change over the next few years?*

to units like Sony's PlayStation Portable (PSP), which offer separate games on Universal Media Discs (UMDs). Writing programs for games has become an industry unto itself. The electronic game industry is so big that more money is spent on computer games than on movies.

The first video game, called Spacewar, was developed in 1962. Each player used keyboard controls or a joystick to move a small ship, which could fire a stream of torpedoes as it moved across the screen. Spacewar was developed by a group of computer programmers at M.I.T. The game became so addictive that play was banned during working hours at the lab where it was designed.

Today, thousands of new computer games are written every year. The programs can take as long as two years to develop. Each game is set in a complex, realistic world with computer-controlled characters that have human abilities. Action and adventure games allow players to control characters on screen, often in a fight against evil or some other civilization. Sports games allow players to play against each other on opposing teams or even play against the computer. Games such as SimCity give players godlike control over the characters and their environment (Figure 12.21).

Figure 12.21 In SimCity 3000, players build whole communities and give the inhabitants certain characteristics.

Hypothesizing *What types of skills might be important in playing this type of game?*

Educational video games have existed for a long time, but recently gaming developers have started creating gaming platforms that are fun, challenging, and give the player the ability to make decisions about sustainability. Sustainability means ensuring the long-term well-being of life on Earth and taking responsibility for its natural resources. Available games, such as the IBM Corporation's *Power Up*, focuses on saving an imaginary planet from destruction through the development of renewable energy sources. Others, like the Cloud Institute's *Fish* game allow the player, a professional fisherman, to catch as many fish as possible while learning about the concepts of species preservation. The game can be played for free at http://www.cloudinstitute.org/games/. If you're already a fan of games such as SimCity, check out *City Rain*, and practice your skills at restructuring a city for green living.

MAKE A CHANGE

While many of you are probably very fond of playing video games, have you ever given any thought to designing your own video game? Not sure where to start? Many of the newest environmental games are available free on the Internet. Start by looking for and playing what's available. Next, list some environmental issues that you find important and see if you could brainstorm a way to explore one of these problems using a game.

TRY THIS

Choose one of the ideas you've brainstormed and a gaming platform that you enjoy, such as SimCity or Civilization IV, and develop a new set of game rules and goals that use the same game to attack your environmental concern. Start by identifying your audience and the problem to solve. Many video games are written much like movie or television scripts and begin with a storyboard, or picture explanation of what each screen would look like. Develop your storyboard, a list of rules and goals, and a script of an example game that solves your environmental problem.

Recording and Playback Systems

Electronic information systems that store and play back information are called recording and playback systems. Magnetic recording tape (cassette tape) is one medium on

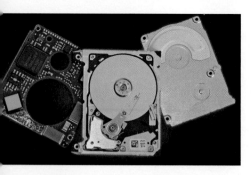

Figure 12.22 Storage media, such as this Microdrive compact flash card, are becoming more compact and durable.

Extending *What type of device is used to "read" these media?*

which information is stored. Compact discs (CDs), Mini-Discs (MDs), and Digital VideoDiscs (DVDs) are examples of storage media (Figure 12.22). Other devices, such as personal data assistants (PDAs) and MP3 players, store information on integrated circuits called flash memory cards, such as Compact Flash or SD and XD cards. Flash memory is a type of memory that can be electronically erased and reprogrammed.

Whatever the medium is, it can be taken from one place to another. The message can then be replayed many times. Recording systems can store audio, video, or data.

Magnetic Tape

A popular way to record sound, video, and data is on magnetic tape. Magnetic tape is a long, thin piece of Mylar plastic that has been coated with a metal oxide. The oxide coating can be magnetized by being pulled past a tape head. The head creates a changing magnetic field when the voltage going to it changes.

In a recording studio, sound is changed into a varying electric voltage by a microphone. An amplifier boosts, or increases, the small electric voltage from the microphone. The electrical signal is then sent to the recording head. The tape moves past the varying magnetic field in the tape head and the oxide coating on the tape is magnetized to a varying degree.

On playback, the process is reversed. The tape is moved past a playback head. The playback head picks up the changing magnetic field on the tape and changes it to a voltage. An amplifer boosts the voltage so that it is large enough to drive a speaker. The speaker then changes the voltage back to sound waves.

Compact Discs

Compact discs (CDs) got their name because, at just under 5 inches in diameter, they are smaller than the 12-inch phonograph records they replaced. A **compact disc** is a plastic disc that contains digital information encoded as a pattern of pits. A standard CD is 1.2 mm (0.05 inches) thick and is composed of a plastic base, one or more thin reflective metal (usually aluminum) layers, and a clear lacquer coating. An audio CD can hold up to 74 minutes of recorded sound and up to 99 separate tracks.

Compact discs combine optical and digital technologies to produce superior-quality audio and video recordings. In a recording studio, sounds or images are changed into binary

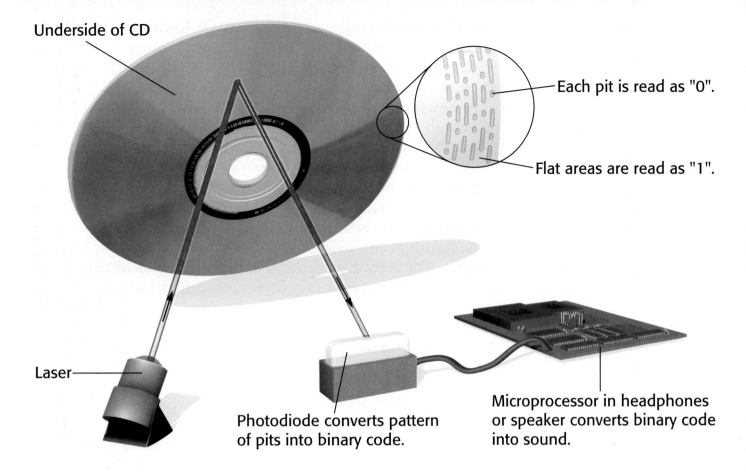

Underside of CD

Each pit is read as "0".

Flat areas are read as "1".

Laser

Photodiode converts pattern of pits into binary code.

Microprocessor in headphones or speaker converts binary code into sound.

code. This binary code is then transferred onto the surface of a master disc. A laser beam does this by burning millions of tiny pits into the surface of the CD. When a CD is played, a drive motor in the CD player spins the disc and a laser beam shines light on it. A light-sensing device "reads" the light reflected from the pits. A deep pit reflects a little bit of light. A shallow pit reflects more light. The reflected light is changed into digital pulses of electricity and the pulses are turned into sound (Figure 12.23).

CDs are used for recording, storing, and playing back audio, video, text, and other information in digital form. A single CD has the storage capacity of about 700 floppy disks—enough to store 300,000 pages of text. A CD-ROM (compact disc, read-only memory) is used to store computer data as text, graphics, and sound. Newer technology allows users to record onto special CDs using a laser CD burner.

Figure 12.23 A CD player must find and read the data stored as pits on the CD.

Identifying *What is the optical technology used by CD players?*

MiniDiscs

By the late 1980s, sales of cassette tapes were down and the music industry realized that tapes had reached the end of their life cycle. Recordable CDs are fairly large to be used as portable devices, so a new system called the MiniDisc system was developed. A **MiniDisc** (MD) is a small, 2-inch-wide disc that holds audio recordings. MDs have the same high quality as regular CDs, but in a smaller, more convenient package. They are housed in a plastic cartridge that protects against scratches and other types of damage.

A new technology called ATRAC allows 80 minutes of stereo music to be stored on this small disc. ATRAC (Adaptive TRansform Acoustic Coding) technology removes sounds that are masked to people's ears. For example, if loud cymbals are heard along with a soft drumbeat, the softer sound can be removed without the human ear noticing, saving data storage space on the MD.

Recordings can be made on MDs using digital or analog sources. The recording process uses a laser to heat tiny spots on the disc, whereas a magnetic recording head writes data using magnetic fields. Tracks can be numbered, played in different sequences, and given titles. Manufacturers claim that MDs can be re-recorded a million times without losing quality.

Digital Versatile Discs

A **digital versatile disc** (DVD) is a plastic disc that carries visual and audio information. DVDs are replacing CDs and VHS tapes as the most common medium for games and movies. DVDs provide excellent picture and sound quality. They can store about 28 times the amount of information that can be stored on a CD. Double-sided DVDs store the equivalent of 30 hours of VHS-quality video. They can be instantly rewound, and are quite durable.

DVD video is encoded in a way similar to the compression in ATRAC MiniDisc technology. To save storage space, areas of the picture that do not change, and information that is not easily perceived by the human eye, are removed.

CDs and DVDs have different formats. You cannot use a CD drive to play DVDs, but some DVD recorders can play both DVDs and CDs.

MP3

Instead of buying CDs or tapes at a music store, you can download good quality music files from the Internet. **MP3** is a digital recording technology that reduces the size of computer files containing music so that they can be more easily moved or stored.

Many MP3 files can be downloaded for free, but these are not usually performed by well-known musicians. Because most music is copyrighted (belongs to the artist or record company), record companies do not like music to be downloaded without their permission.

MP3 files are played with an MP3 player, or more accurately, a digital audio player, because they can play a variety of audio file formats. MP3 players can be downloaded to your computer as software. Portable MP3 players work like portable stereos. Apple's iPod is a popular MP3 player. It works with iTunes, the MP3 player software that you download to your computer. Using iTunes, you can purchase music, convert music from your CDs into MP3 files, and download video and podcasts, then move the files onto your iPod. You can also take MP3 files from iTunes and burn them to CDs that can be played on any CD player. Other MP3 players such as Creative's Zen and Microsoft's Zune offer comparable features. An advantage of MP3 players is that they use electronic memory to store audio so there is no skipping, which can happen with CDs, and you don't need to worry about damaging or storing CDs. One possible drawback is that not every player can play every music file format. For instance, WMA files are Microsoft audio files that cannot be played on iPods, and files purchased from iTunes can only be played on iPods.

MP4

MP3 files are primarily used for audio. MP4 is a format that can also be used for video files for players such as Apple iPods and Microsoft Zunes. MP4 is also the format used to upload YouTube videos.

SECTION 3 Assessment

Recall and Comprehension

1. What is the most popular consumer application for electronic information technology?
2. Describe the types of media that are used to store recorded information.
3. Name three types of recording or playback systems. How does each one work?

Critical Thinking

1. **Drawing Conclusions** Explain why CDs and Mini-Discs are replacing recording tape as the preferred music medium.
2. **Inferring** Why are record companies reluctant to support MP3 technology?

QUICK ACTIVITY

Optical storage devices rely on laser technology for recording, storing, and playing back data. Without the development of the laser, CDs would not exist. Make a list of some other inventions, materials, and discoveries that have contributed to the development of CD technology. See who can make the longest list in your class.
For more related Design Activities, see pages 378–381.

Chapter Summary

- A communication system is called an electronic communication system if the channel uses electricity to carry information. The use of electricity and electronics in communication has created the current Information Age.

- Radios send information through the air from a transmitting antenna to a receiving antenna.

- The telephone is the most commonly used electronic communication system today. Conventional telephone technology has been combined with radio technology to produce cell phones and cordless telephones. Conventional telephone technology has been combined with radio technology to produce new devices such as cell phones and cordless telephones.

- In television, images are changed into electrical signals. These signals are sent through the air to a receiver that changes them back into images.

- Satellite communication uses radio waves that are relayed through satellites. Satellite technology is now used for telephone, radio, and television.

- Computer networks enable several computers to be joined together to share data. A basic network includes a server, a network operating system, network interface cards, and cables.

- Computers use the Internet for telnet, file transfer, email, and access to the World Wide Web.

- Sound is recorded on magnetic tape and optical discs (often called compact discs or CDs). Video is recorded on magnetic tape and digital videodiscs (DVDs).

Building Vocabulary

Your teacher may give you a crossword puzzle. Complete the puzzle using the following words from this chapter. Exchange puzzles with a classmate to check each other's answers.

1. browser
2. carrier frequency
3. cell phone
4. compact disc
5. computer network
6. demodulation
7. digital videodisc
8. downlink
9. email
10. fiber-optic cable
11. frequency
12. instant messaging
13. Internet
14. Internet service provider
15. Internet telephone
16. landline telephone
17. local area network
18. MiniDisc
19. modem
20. modulation
21. MP3
22. network interface card
23. network operating system
24. peer-to-peer network
25. point-to-point transmission
26. radio broadcasting
27. satellite dish
28. server
29. telephone switching
30. television
31. telnet
32. uplink
33. wide area network
34. wire cable
35. World Wide Web

See your teacher for the Crosstech puzzle.

Reviewing Content

1. What two new products were the result of combining radio and telephone technology?

2. In a television system, which component changes the image into electrical signals?

3. Name three tasks accomplished by satellites.

4. Explain the advantages of a computer network.

5. Identify the roles that computer game characters assume.

Applying Your Knowledge

1. Choose three electronic communication systems. In each, identify the channel that carries the electronic signal.

2. Do a survey of 20 people. Find out how many own cell phones and how many owned cell phones five years ago. What do you think accounts for this change in the way people communicate?

3. Do research to determine, by percentage, the principal uses of the Internet. Your categories might include sending email, shopping, doing research, playing games, instant messaging, and so forth.

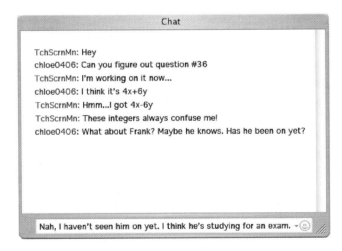

Chat

TchScrnMn: Hey
chloe0406: Can you figure out question #36
TchScrnMn: I'm working on it now...
chloe0406: I think it's 4x+6y
TchScrnMn: Hmm...I got 4x-6y
TchScrnMn: These integers always confuse me!
chloe0406: What about Frank? Maybe he knows. Has he been on yet?

Nah, I haven't seen him on yet. I think he's studying for an exam.

4. Cell phones are used widely across the globe. Some people say that cell phones emit harmful radiation and can cause cancer. What is your hypothesis about the relationship between cell phone use and cancer? Search the Web to find evidence that supports or refutes your hypothesis.

5. Identify or provide examples of fair use practices that apply appropriate citation of sources when using information from books or digital resources. Do so by showing an example of how appropriate credit is given to someone when an author used

his or her ideas. You can find sample citations at the end of most research papers and can search for "style guide" to find preferred formats.

6. Find out about the computer network in your school or in a local business. What type of network is used? How many computers are connected? What peripheral devices are connected? What kind of cables are used?

7. Find two examples of electronic games. For each, identify the category of game it fits into and the roles of the characters.

Critical Thinking

1. **Making a Graph** Create a graph showing how electronic information technology has grown over the last 20 years. Show the technologies on the x-axis and the timeline on the y-axis.

2. **Drawing Conclusions** How has the Internet revolutionized our society?

3. **Stating an Opinion** Some computer games have been criticized for being too violent and having a negative effect on young people. Do you agree with this opinion? Why or why not?

4. **Comparing and Contrasting** Describe the similarities and differences between music CDs and MiniDiscs. Explain the advantages of each.

5. **Evaluating** What makes DVDs such a popular alternative to VHS videotapes?

Connecting to STEM
science · technology · engineering · math

CD Burners

CD burners are becoming more and more popular as a way of storing information and also making copies of music files. In this chapter, you learned how a professionally recorded CD is made and played. Now do research to find out how a CD burner works. How is information transferred from the computer to the surface of the CD? Is this process different from the way professional CDs are made?

Design Activity 23

PowerPoint® DESIGN FOLIO

Problem Situation

You and your classmates have been using the hard-copy version of the design activity in your Student Activity Guide. However, this is the twenty-first century. An electronic version is needed that can be customized to serve your needs. It will help you make presentations, because much of the material needed will already be available.

Materials

You will need:

- Microsoft PowerPoint® or another presentation software package

Your Challenge

You and your teammates are to create an electronic version of the design activity to replace the hard-copy version in your Student Activity Guide.

> Go to your **Student Activity Guide, Design Activity 23.** Complete the activity in your Guide, and state the design challenge in your own words.

① Clarify the Design Specifications and Constraints

To solve the problem, your design must meet the following specifications and constraints:

- The electronic design activity should allow for pictures, videos, and drawings.
- All the steps of the informed design process should be included.
- PowerPoint or another presentation software package should be used.

> In your Guide, state the specifications and constraints. Add any others that your team or your teacher included.

② Research and Investigate

To better complete the design challenge, you need to first gather information to help you build a knowledge base.

> In your Guide, complete Knowledge and Skill Builder 1: The Informed Design Process.

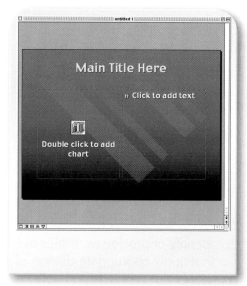

In your Guide, complete Knowledge and Skill Builder 2: Investigating PowerPoint Slides.

In your Guide, complete Knowledge and Skill Builder 3: Inserting Special Features.

3 Generate Alternative Designs

In your Guide, describe two possible solutions that your team has created for the problem.

4 Choose and Justify the Optimal Solution

Refer to your Guide. Explain why you selected the solution you did, and why it was the better one.

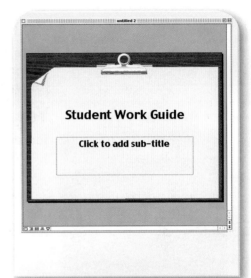

5 Develop a Prototype

Create your electronic design activity. You may use information from an earlier design to illustrate the various elements of the design activity. Print out the slides and attach them to your Guide, or provide a file for your instructor to view.

In any technological activity, you will use seven resources: people, capital, time, information, energy, materials, and tools and machines. In your Guide, indicate which resources were most important in this activity and how you made trade-offs among them.

6 Test and Evaluate

How will you test and evaluate your design? In your Guide, describe the testing procedure you will use. Justify how the results show that the design solves the problem and meets the specifications and constraints.

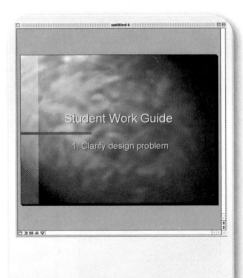

7 Redesign the Solution

Respond to the questions in your Guide about how you would redesign your solution. Your redesign should be based on the knowledge and information that you gained during the activity.

8 Communicate Your Achievements

In your Guide, describe the plan you will use to present your solution to your class. Include the PowerPoint slides you will use, or provide a file for your instructor.

Design Activity 24

WEB SITE DESIGN

Problem Situation

There are all types of Web sites. Many people have their own Web site, and many organizations and schools have theirs. Now, you need to design one for your technology class.

Your Challenge

You and your teammates are to design and create a Web site for your class.

Go to your **Student Activity Guide, Design Activity 24.** Complete the activity in your Guide, and state the design challenge in your own words.

① Clarify the Design Specifications and Constraints

To solve the problem, your design must meet the following specifications and constraints:

- The Web site should reflect the subject matter, activities, and material that students are learning.
- The Web site should be easy to navigate and should contain accurate information.
- Size requirement: (to be filled in in your Guide after completing Knowledge and Skill Builder 4).
- Complexity requirement: (to be filled in in your Guide after completing Knowledge and Skill Builder 4).
- Use of Web design software (such as Microsoft Word and Netscape Page Composer).

In your Guide, state the specifications and constraints. Add any others that your team or your teacher included.

② Research and Investigate

To better complete the design challenge, you need to first gather information to help you build a knowledge base.

In your Guide, complete Knowledge and Skill Builder 1: Developing a Plan.

Materials

You will need:

- computer
- Web design software

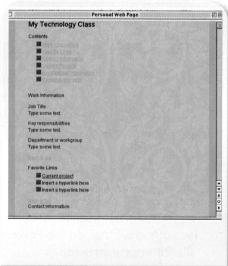

In your Guide, complete Knowledge and Skill Builder 2: Researching Web Pages.

In your Guide, complete Knowledge and Skill Builder 3: Making a Simple Storyboard.

In your Guide, complete Knowledge and Skill Builder 4: Practice Programming.

③ Generate Alternative Designs

In your Guide, describe two possible solutions that your team has created for the problem. Include such considerations as how to include text, pictures, linkages, and email on the Web site.

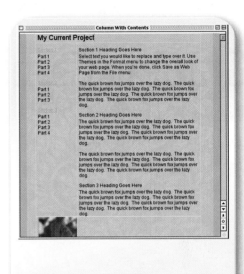

④ Choose and Justify the Optimal Solution

Refer to your Guide. Explain why you selected the solution you did and why it was the better one.

⑤ Develop a Prototype

Design and construct your Web site. Include a final storyboard in your Guide. Attach a printout of the Web pages, or provide a file for your instructor to view.

In any technological activity, you will use seven resources: people, capital, time, information, energy, materials, and tools and machines. In your Guide, indicate which resources were most important in this activity and how you made trade-offs among them.

⑥ Test and Evaluate

How will you test and evaluate your design? In your Guide, describe the testing procedure you will use. Justify how the results show that the design solves the problem and meets the specifications and constraints.

⑦ Redesign the Solution

Respond to the questions in your Guide about how you would redesign your solution. Your redesign should be based on the knowledge and information that you gained during the activity.

⑧ Communicate Your Achievements

In your Guide, describe the plan you will use to present your solution to your class. Show what handouts and/or PowerPoint slides you will use.

CHAPTER 13

Graphic Communication

SECTION 1 Printing
SECTION 2 Photography
SECTION 3 Electronic Imaging

In this chapter, you will learn about the following technologies used in many kinds of graphic communication systems:

- relief printing
- lithography
- laser printing
- film-based photography
- digital photography
- photocopying
- scanning

The word *graph* means "to draw" or "to write." In graphic communication systems, messages that are formed by letters, symbols, drawings, or pictures are transferred to many different types of surfaces, such as a ceramic mug, the exterior of a bus, a CD label, or a computer screen.

In each of these systems, ink or some other medium (paint, pencil, glaze, toner, or even light) is deposited onto a substance to create an image.

Graphic communication systems are used to make many different products, such as

- books
- magazines
- newspapers
- Web pages

- photographs
- tickets
- stamps
- billboards

Printing

Reading Strategy

Listing Make a list of the different types of printing described here. Next to each item on the list, write a quick definition in your own words.

1. Movable Type
- metal letters coated with ink

2. Relief Printing

Vocabulary

movable type	screen printing	desktop publishing
relief printing	lithography	computer-to-plate
gravure printing	dye-sublimation printer	printing

What Is Graphic Communication?

As in all communication systems, the graphic communication channel carries a message from its source to its destination. In graphic communication, the channel carries symbols, words, or pictorial images to convey information, ideas, and feelings.

Graphic communication systems use different types of channels. In writing, the channel is ink or pencil lead (graphite). In printing, it is the printing ink. In conventional photography, it is the film. When artists produce a painting, the channel is the paint. In photocopying, it is the toner.

From Papyrus to Paper

Paper is still the primary medium, or channel, for presenting graphic messages. Paper was first made by the Egyptians more than 4,000 years ago (about 2500 B.C.). They made paper from fibers of the papyrus plant. The fibers were soaked in water, mashed together, and then matted to form thin sheets.

Paper similar to what we use today was invented by the Chinese about 2,000 years later (Figure 13.1). It wasn't until about 600 years ago (A.D. 1400), however, that high-quality paper was developed.

The availability of lightweight, low-cost paper made it easier for people to record their ideas and share them with others. With this simple development, technology boomed. More people could read about what others had done. They could understand and then build on and improve the ideas of others.

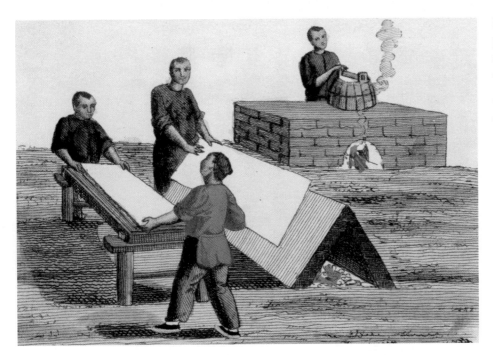

Figure 13.1 Early Chinese paper was made by dipping a screen made of cloth into a vat of pulp and letting it dry.

Clarifying *Why did paper cause advances in technology?*

How Printing Began

At about the time that high-quality paper was developed, one of the most important advances in graphic communication took place. In about A.D. 1450 in Germany, Johannes Gutenberg invented a way of making separate letters, called **movable type.** Individual letters of type were molded out of metal, assembled into the text for a single page, and then coated with ink. Multiple pages were printed on a printing press.

The First Printed Books

Before Gutenberg's invention, an entire page of text was carved out of wood or soft metal and then printed. Because a new template had to be carved for every page, this process was slow and costly. With Gutenberg's type, books could be printed faster and less expensively than before. Once a page was printed, the same letter molds were used to set the next page of type. Gutenberg's printing press was first used to print copies of the Bible. It was soon used to print other books and all types of printed media.

Early Newspapers

Newspapers were the first medium for mass communication. The first colonial newspaper, the *Boston News-Letter,* was published beginning in 1704. For the next hundred or more years, most early newspapers were printed on hand-operated presses. With new technologies, however, companies gradually changed over to machine-run presses.

Figure 13.2 Newspapers are printed and distributed from many different locations.

Predicting *How will Internet communication affect printed newspapers?*

Figure 13.3 In relief printing, raised letters are coated with ink and then pressed to paper.

Clarifying *Why are the letters reversed on the printing block?*

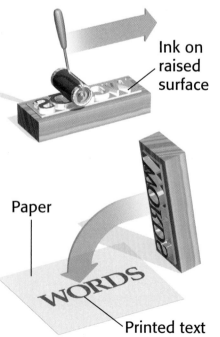

Ink on raised surface

Paper

Printed text

Printing Technologies Improve

Toward the end of the 1800s, newspapers became very popular. By that time, machines had been developed that could print large numbers of newspapers very quickly. Newspapers brought people closer together. People could read about events in the rest of the world soon after the events took place.

The methods of producing newspapers have changed with advances in technology. In 1840, the largest paper in the country, the *New York Sun*, printed only 40,000 newspapers a day. Today, the *New York Times* prints 1.5 million Sunday papers. News stories are written by reporters using word-processing software and emailed directly to their editors. The pages are optically scanned by lasers and sent to printing plants by satellite. Huge printing presses use enormous spools of paper (Figure 13.2). They not only print the pages but also collate, fold, and trim them.

Commercial Printing Today

Many types of printing exist today. Different technologies are used to print large signs, books, or T-shirts. Important printing technologies include relief printing, offset lithography, gravure printing, and screen printing.

Relief Printing

Gutenberg's printing method used raised surfaces, in a method called relief printing. In **relief printing,** only the raised surfaces of the letters are inked. When the letter is pressed against a piece of paper, only the inked surface prints (Figure 13.3).

Relief printing, which is also known as letterpress printing, is still used today. All types of rubber stamps use relief printing. Newspapers and greeting cards are also printed using this method.

Offset Lithography

Today, most printing is done using offset lithography. **Lithography** literally means writing (*graphy*) on stone (*litho*). It takes advantage of the fact that oil and water do not mix. Lithography was invented by the German artist Alois Senefelder. Senefelder drew a line with a waxy crayon on a smooth, flat piece of limestone, wet the entire surface, then applied an oil-based ink. The ink stuck only to the crayon lines. When the limestone surface was pressed against a piece of paper, only the inked crayon lines printed.

Today's offset printing method uses metal sheets rather than pieces of limestone (Figure 13.4), but the principle is the same. The image to be printed is exposed onto film, creating a film negative. The negative is then exposed onto an aluminum plate, creating a positive image on the surface of the metal.

The aluminum plate is wrapped around a cylindrical drum. When a page is to be printed, water is first applied to the aluminum drum. It is attracted to areas of the image that will be white or light in color. An oil-based ink is then applied to the drum and is attracted to all the areas that are not already wet. These areas will be dark in the final image.

Figure 13.4 Offset lithography is based on the principle that oil (the ink) and water do not mix.

Interpreting *Which cylinder transfers the ink to paper?*

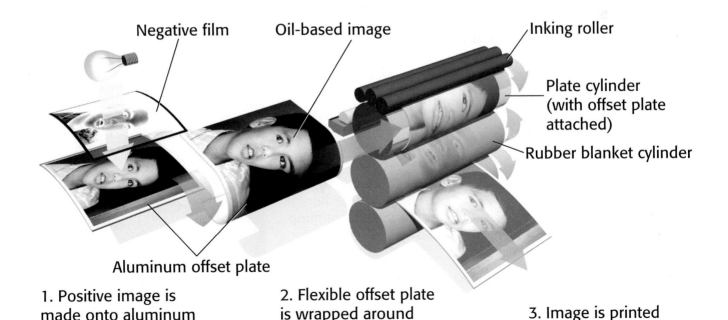

Negative film   Oil-based image   Inking roller

Plate cylinder (with offset plate attached)

Rubber blanket cylinder

Aluminum offset plate

1. Positive image is made onto aluminum offset plate.

2. Flexible offset plate is wrapped around plate cylinder of press.

3. Image is printed onto paper.

The pattern of ink is then transferred (offset) to a rubber cylinder. The image is now reversed. The rubber blanket turns, and paper goes through the press. The image is offset again, this time to the paper. Every time a page is printed, water and ink are reapplied using the same process.

Gravure Printing

In **gravure printing,** ink is printed from a lowered surface. First, a line is scratched into the surface of a piece of metal. The surface is inked and then wiped dry. The ink stays in the scratch. When a piece of paper is pressed against the metal surface, the paper pulls the ink out. This type of printing is called gravure, or intaglio printing. Gravure is used to print the magazine section of most Sunday newspapers. It is also used to print large volumes of materials, such as brochures, catalogs, and company annual reports.

Screen Printing

In **screen printing** (originally known as silk screening), a stencil is used to print designs using ink. Screen printing is used for posters, T-shirts, and the labels on CDs and DVDs. The stencil lets ink pass through some areas and keeps ink from passing through others.

A stencil, often made from plastic, is attached to a silk screen. The silk screen is stretched tightly on a wooden frame. The screen touches the paper or fabric to be printed, transferring the ink.

SECTION 1 Assessment

Recall and Comprehension

1. What is the role of the channel in a graphic communication system? Name three types of channels used in graphic communication.
2. What was the first mass communication medium?
3. Describe how ink is transferred to paper using relief, gravure, screen, and offset printing.

Critical Thinking

1. **Comparing and Contrasting** How is gravure printing similar to offset lithography? How is it different?
2. **Relating Cause and Effect** How have new technologies affected the printing industry?

QUICK ACTIVITY

The use of color in design can sometimes trick your eyes. Use a computer to draw an American flag—but with a few changes. Make the white stars and white stripes black, the blue field orange, and the red stripes green. Print out your flag on a color printer. Stare intensely at the image for at least 60 seconds. Immediately shift your stare to a white sheet of paper. What do you see? **For more related Design Activities, see pages 412–415.**

People in Technology

Johannes Gutenberg

Inventor of the Printing Press
Johannes Gutenberg

In Germany, about 50 years before Columbus discovered America, Johannes Gutenberg developed a technique that completely changed how people communicated. He invented a new way of printing books.

Studying from Handwritten Books

As a young man, Gutenberg had studied from handwritten books. He had watched monks as they worked for years to make copies of the Bible. He felt there should be a way of making books more quickly. He began to think about mechanical printing.

Gutenberg knew that wooden blocks were used to print playing cards. He also knew of metal stamps that were used to press markings onto metal coins. Gutenberg reasoned that printing would be more efficient if separate, movable pieces of type could be used for the letters. He made these pieces out of metal because it could be molded easily and would hold up during heavy use.

The First Printing Press

From these simple one-letter molds, Gutenberg developed the first printing system. To hold paper in place during printing, he used a screw press similar to the kind used to squeeze grapes. Letters for the text were arranged on the press—a process known as setting the type. Gutenberg found an ink that would stick to the metal type. Many copies of a single page were printed, and then the press was set for the next page.

With Gutenberg's press, people could print books faster and less expensively than before. Once books became cheaper, the average person could own them. Soon, many other people were using Gutenberg's invention to print books. The availability of more books allowed knowledge to spread to all corners of the Earth.

Gutenberg used his new printing techniques to print many copies of the Bible.

Critical Thinking

1. **Inferring** Why do you think Gutenberg's type of printing is also called movable type?

2. **Summarizing** How did Gutenberg use existing technologies to develop his printing system?

Benchmarks for Learning

- The five elements needed for film-based photography are light, film, a camera, chemicals, and a darkroom.

- All film cameras have a dark chamber, a lens through which light enters, a shutter, a view-finder, and a place for film.

- Digital cameras do not use film. Images are recorded on an electronic image sensor.

Reading Strategy

Listing As you read about film-based photography, list the steps involved in the process. Include every step, from exposing the film in the camera to fixing the paper using chemicals.

Film Processing
1. Camera lens focuses light on film.
2. Film exposed onto light-sensitive paper.
3.

Vocabulary

film	fixer	digital camera
developer	darkroom	pixel
stop bath	single-lens reflex camera	

Film-Based Photography

Artists use their eyes, minds, and hands in creating a picture. In film-based photography, light and chemicals put the picture on the paper. The word *photography* means "to write with light." Film-based photography uses light and chemicals to record an image on film and then paper.

Making Photography Portable

The earliest camera-like device was first used in Italy in the 1500s. It was called a camera obscura, which, in Italian, means "dark chamber." The camera obscura was just

Figure 13.5 A portable camera obscura from the early 19th century.

Inferring *Based on the picture, where does light enter the box?*

that—a room with no windows (Figure 13.5). It had a tiny lens set in one wall, where light came through. This beam of light cast onto the wall an upside-down image of whatever was outside. An artist traced and painted over the image onto a canvas.

In 1839, a Frenchman named Louis Daguerre presented a new process he had invented. It was the first true photography. His film was a copper sheet coated with silver and silver iodide. Silver iodide is sensitive to light. A chemical change takes place when light strikes it. This change formed an image on Daguerre's film, which was called a daguerreotype.

In the late 1800s, George Eastman introduced the Kodak camera. Until then, photography had been available only to professionals who had the costly equipment and studios needed for the job. The Kodak camera changed all that. Suddenly, anybody could take pictures. The company advertised, "You press the button, we do the rest." Soon, every family owned a camera and a collection of snapshots.

The Elements of Film-Based Photography

All film-based cameras have at least five parts in common: a dark chamber, a lens, a shutter, a viewfinder, and a place for film. The shutter lets light in for a controlled amount of time, and the viewfinder allows the person to view the photo before taking it. Certain elements are necessary for film-based photography (Figure 13.6).

Early photographs were generally taken indoors using a source of very bright light. For this light, magnesium powder

	1. Light exposes the film. Light comes from the sun, a light bulb, or a flash.
	2. Film has light-sensitive chemicals on a plastic base. Film can be color or black and white.
	3. A camera houses the film and exposes the film for a specific time.
	4. Chemicals are used to develop film and print pictures.
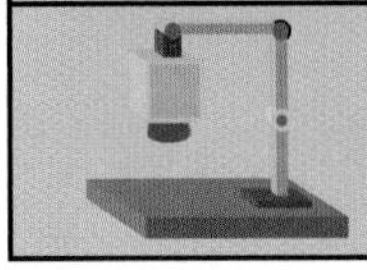	5. A darkroom, which has no natural light, is used to develop film and print photographs.

Figure 13.6 These elements are necessary for developing film-based photos.

Identifying *Which elements of film-based photography are not included in digital photography?*

was burned. It gave off a bright light, but it was smoky and dangerous.

Today, **film** is made from an acetate (plastic) base. Photographic film is coated with tiny grains of silver bromide, which is sensitive to light. Different kinds of film are used to take different kinds of pictures. Color film is used to make prints or slides. Infrared film gives an image of the heat coming from objects. Other film can be used to take pictures at night or in places where there is almost no light.

Photography and Chemistry

When a camera shutter opens, its lens focuses light on the film. The grains are exposed to light. When they are developed, the exposed grains turn parts of the film black. The parts that were not exposed to light are washed away during fixing. They leave clear areas on the film. In this way, a negative is made.

Chemicals are used to develop film and to print photographs (Figure 13.7). The three kinds of chemicals used in black-and-white photography are developer, stop bath, and fixer. **Developer** turns the exposed silver grains black. **Stop bath** stops the developing process. **Fixer** washes away all the unexposed grains. It clears those parts of the film that the light did not touch.

Figure 13.7 In a darkroom, film is exposed onto light-sensitive paper, which is then passed through different chemicals.

Inferring *If you print a negative that has mostly clear areas, will the final photograph be mostly white or mostly black?*

Photographic Enlarger

1. Light passes through a film negative, exposing light-sensitive paper. The clearest areas on the negative allow the most light to strike the paper.

Film negative

Light-sensitive paper

Developer

Stop Bath

Fixer

2. Developer turns exposed silver grains black.

3. Stop bath stops the developing process.

4. Fixer washes away the unexposed grains. These areas will be white on the final image.

Because photographic paper and film are very sensitive to light, they are developed in the dark. The area used can be a corner of a room that is darkened completely. It can even be a closet. A room used for developing is called a **darkroom.** A home darkroom can be very small and simple.

Film-Based Cameras

All film-based cameras have several elements in common. They have a light-tight box, a place for film, an opening through which light can enter, and a shutter to control the length of time that light can enter the camera.

The view camera is the simplest type of film-based camera. It has a lens at the front and film at the back. A shutter just behind the lens controls the amount of time that light comes in. Behind that, a dark chamber called a bellows is stretched out or made smaller to bring the image into focus. The camera is very large and must be supported by a three-legged stand called a tripod. A view camera produces very large negatives of high quality.

In a view camera, the photographer looks at a glass viewing screen to see the exact image that will be photographed. In a viewfinder camera, the photographer looks through a separate viewfinder to see the approximate image that will be photographed. The viewfinder is separate from the path that light takes through the camera. Most point-and-shoot cameras are viewfinder cameras.

A **single-lens reflex (SLR) camera** has a single lens, which is used for viewing and focusing (Figure 13.8).

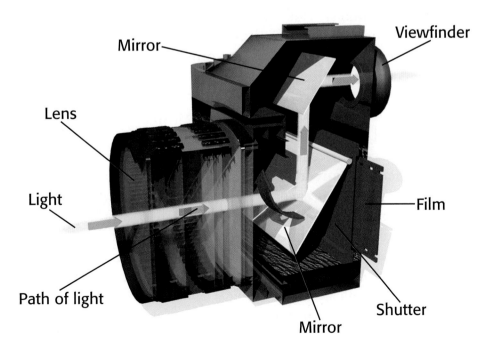

Figure 13.8 An SLR camera allows a person to see exactly what will be photographed.

Interpreting *What do the mirrors inside the camera do?*

An inside mirror reflects the light. In one position, mirrors reflect the light from the lens to the viewing opening. In this position, the photographer can see what will be in the picture. When the picture is taken, one mirror swings out of the way, allowing light to enter the lens and expose the film.

Digital Photography

Digital cameras are similar to film cameras in the way they basically work. The photographer uses a viewfinder to aim the camera, whether it's film or digital. Like film cameras, digital cameras capture an image by opening a shutter and letting light in through a lens.

Instead of using film to record the image, though, a digital camera uses an electronic image sensor. This sensor is called a charge-coupled device, or CCD. The image is then stored in the kind of digital memory that computers use.

In addition to capturing still pictures, many digital cameras are also able to record video and sound. Nearly all include LCD screens that enable the pictures to be previewed right away. Many digital cameras also use this view screen as a viewfinder, so the pictures can be seen even before the photo is taken.

The first digital cameras were sold to consumers in 1995 by Kodak and Apple Computer. Before that date, digital camera technology had been used by organizations such as NASA, which used digital signals to relay images from space back to Earth.

Converting an Image to Pixels

A charge-coupled device (CCD) is made up of a grid of hundreds of thousands of tiny light sensors called photodiodes. Each of these photodiodes represents one tiny element of a picture. A picture element is called a pixel, from the words *picture* and *element*. To understand pixels, look at a surface covered with mosaic tiles or glass (Figure 13.9). Each tile, like one pixel, is one color and brightness.

In a CCD, sensors change light energy into electrical energy. The electricity is converted into a series of digital on-off pulses. These pulses are processed by a microprocessor (a small computer) within the camera and stored in digital memory. The individual pixels are assembled to form an image. Like a mosaic, a digital image is seen as a whole, not as a series of pixels.

An important factor in the quality of picture a digital camera produces is the camera's resolution. Resolution is measured in the number of pixels a camera has. This number

Figure 13.9 This wall is decorated with many small tiles of different colors. Like pixels in a digital camera, each tile is a single color.

Comparing *How are the tiles on the side of this house similar to pixels?*

is usually expressed in terms of millions of pixels, or mega-pixels (MP). The higher the resolution, the tinier the pixels are, and the better the picture will look even printed out at large sizes.

Very early cameras didn't even have one megapixel, and their photos only looked good on computer screens, or printed out at very small sizes. Even the most inexpensive digital cameras today offer five megapixels or more, which produces high-quality images even printed out as large as 16x20. The very highest resolutions available right now on professional cameras go up to 16MP or even 20MP on specialized large-format cameras.

The Photo in Digital Photography

With a film-based camera, the pictures are taken and stored on a roll of film that can be removed from the camera. It needs to be processed before the images can be seen. With a digital camera, the images are stored either in the camera's hard drive or on a removable memory card. The camera is

Figure 13.10 This "point and shoot" digital camera is lightweight and simple to use.

Figure 13.11 This digital SLR camera is more complex to use and slightly heavier than the one in Figure 13.10.

connected to a computer, usually with a USB cable, and the images are moved from the camera to the computer's hard drive. Once the images are in the computer, they can be printed out at different sizes, manipulated using graphics software such as Adobe Photoshop, shared over email or the World Wide Web, and stored for indefinite periods of time.

Many printers that people have in their homes are capable of printing high-quality photos. Special paper is available for printing glossy photos that look very much like the prints that come from processing film. Some printers have memory card readers built into them, so that the card can be taken out of the camera and inserted directly into the printer. Photos can be printed off immediately, without requiring the use of a computer.

There are also specialized photo printers that connect directly to the camera and make 4x6 prints, which is the most common size of photo print that people get from processed film. Most picture frames and photo albums are designed for 4x6 prints.

Most digital cameras come with basic software for storing and organizing photos. This software usually allows photographers to edit or digitally change the photos as well. Basic changes include resizing the photo, so that it can be printed at a larger or smaller size, or cropping the photo, which means cutting off the parts that aren't needed and only keeping the best part of the image. Images can also be lightened or have the color adjusted, so if a photo is too dark or someone's face looks too red, the problems can be fixed. Basic photo editing software can also fix red-eye, that red glow that sometimes shows up in people's eyes when the light is wrong in a photo.

Advanced image editing software is also available. The most popular software is Adobe Photoshop. Using Photoshop, photographers and graphic artists can combine elements from different photos. For example, a person from one photo can be placed in front of the background from another picture. Textures and effects can also be added to photos, to make them look like drawings, for example, or as if they were painted on canvas instead of printed on slick paper.

Photoshop has changed the way that people think about photography. A photo used to be an accurate record of how things looked at a particular place and moment. Now, a photo might be a creation from an artist's mind of an image that never existed in reality.

Improving Digital Technology

Digital photography has a few weaknesses that make some people still prefer film cameras. It is estimated that digital cameras would need a resolution of 20 megapixels to truly capture images as well as film does. Current CCDs in handheld cameras are not capable of managing much over 8MP without actually degrading the image. At that point, the pixels are so small that they cannot capture and record light effectively. Specialized professional cameras with higher resolutions use much larger CCDs than handheld cameras.

However, most people can't tell the difference between an 8x10 photo captured by a 4MP camera and printed on a high-quality printer and one that is produced from film. The differences only become a problem when photos are printed out really large, or if a small area of a photo is selected and then expanded to a very large size.

Another problem with digital cameras is that there is a delay between the time the button is pushed and the time the shutter opens and the image is captured. This is due to the electronic processes that the digital camera goes through before it is ready to capture the image. Digital cameras focus themselves, and determine how much light to let in and for how long. They also need to send an electrical charge to the

Figure 13.12 Digital cameras take crisp, clear images without the need for expensive film development.

For: Digital Camera Activity
Visit: www.mytechedkit.com

Picture This: Explore a Digital Camera Online

Pictures taken with digital cameras are generally stored as images on memory cards. These images can then be transferred to computers and modified using image-editing software. Go online to discover exactly what goes on inside a digital camera.

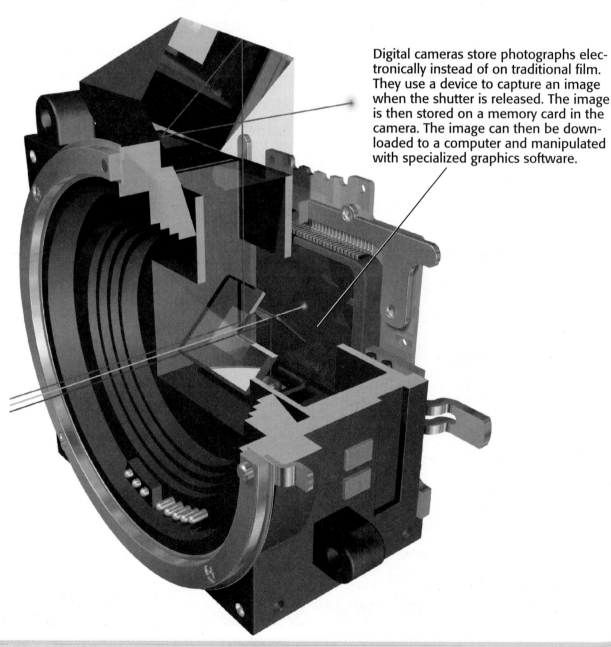

Digital cameras store photographs electronically instead of on traditional film. They use a device to capture an image when the shutter is released. The image is then stored on a memory card in the camera. The image can then be downloaded to a computer and manipulated with specialized graphics software.

CCD. Once a photo has been taken, it must be moved to digital storage before another photo can be taken.

In a film camera, the photographer can determine settings and there is nothing to charge up. Digital cameras are getting faster, however, and there are a few things that the photographer can do to make the camera take the picture faster. On most digital cameras, the button can be pushed part way down, which causes the camera to focus and make light settings but not actually snap the picture. When the button is pushed the rest of the way down, all the camera needs to do is charge the CCD and take the picture.

Some cameras also include what is called burst mode, where several pictures are taken and moved to a very fast, temporary digital storage location. This means that five to ten pictures can be taken in as little as one second before the camera has to pause to move the photos to regular storage. This is very useful for capturing action photos.

A new technology promises to make color capturing in digital images even better. They may even surpass film-based images. The X3 photographic chip takes advantage of the fact that red, blue, and green light penetrate the silicon that the CCD is made of at different rates. The X3 embeds three layers of pixels at different depths in the silicon, basically putting three pixels in the space normally required for one. Using an X3 chip, then, a 3-megapixel camera records three times as much information as a regular digital camera for a given image. Images are clearer and more realistic.

SECTION 2 Assessment

Recall and Comprehension

1. Name the five elements needed for film-based photography.
2. What four parts do all film cameras have in common?
3. Describe the different chemicals that are needed to develop film.
4. What is a single-lens reflex camera? How does the photographer view the image to be photographed?

Critical Thinking

1. **Contrasting** How is a view camera different from an SLR camera?
2. **Evaluating** What are some advantages and disadvantages of a digital camera over a film-based camera?
3. **Predicting** Do you think digital photography will replace film-based photography? Why or why not?

QUICK ACTIVITY

A pinhole camera uses a small hole in a flat surface in place of a lens to magnify an object. Make a very small hole in the center of a 3-in. by 5-in. card. Hold the card close to your eye and look through the pinhole at the text in this book. See if you can change the "focus" by moving the card closer or farther from the text. Retry the activity using different-sized holes. **For more related Design Activities, see pages 412–415.**

SECTION 3
Electronic Imaging

Benchmarks for Learning

- Graphic communication systems use electronics to create digital images that are converted to printed copy.
- Photocopying machines make use of photography and static electricity.
- Digital images can be easily changed, stored, distributed, and reproduced.

Reading Strategy

Listing   Make a list of all the different types of electronic graphic systems. For each item on the list, give an example of how it is used.

Electronic Graphic Systems
1. e-books—download and read a book on screen
2. Desktop Publishing—

Vocabulary

e-book
desktop publishing
computer-to-plate printing

photocopier
facsimile
bitmap

animation
shareware

Electronic Graphic Systems

Today, graphic communication has become a blend of traditional graphics, electronics, and computer technology. Modern graphic communication systems use electronics to create digital images that are converted to printed copy. Examples of electronic graphic systems are desktop publishing systems, photocopiers, facsimile machines, and scanners.

e-Books

One type of electronic graphics has completely sidestepped the printing process. With **e-books,** readers can download text and images directly from the Internet to handheld computers (Figure 13.13). Entire books can be read onscreen. Downloading files from the Internet is easier and cheaper than buying a book.

E-books have evolved significantly from their beginnings, now displaying text and other digital media. One of the most popular e-books is Amazon's Kindle. The Kindle displays text and shades of gray. Apple's iPad is a competitor to Kindle and includes features of a tablet computer and smartphone with touchscreen features and vivid color. The iPad allows users to visit Web sites, see movies, and use email. The Sony Reader and Barnes and Noble Nook compete as well with features similar to those found in the Kindle.

Amazon.com's CEO Jeff Bezos claimed in an announcement on July 19, 2010, that in the prior month, 180 Kindle

Figure 13.13   E-books are actually never printed. The book is downloaded from the Internet and then viewed on screen.

Making Judgments   *What advantages do e-books have over printed books? What advantages do printed books have?*

titles were sold for every 100 hardcover books sold. Bezos called this a "tipping point" because the e-book format has overtaken the hardcover format at the online bookseller.

Desktop Publishing

Desktop publishing is a system for designing, editing, and producing documents using a computer, special software, and a high-quality printer. Desktop publishing lets a person turn out a book or newsletter, page by page. The pages can include text, headlines, and all types of images.

LIVING GREEN
Life in a Paperless World

By its nature, the field of graphic communications typically means some kind of paper is involved. Recently, with concerns over sustainability, many companies are looking for ways to reduce or eliminate paper from many types of businesses and recreational events. E-books are one way to carry multiple books and zero paper. Email has caused a rapid decrease in the amount of letter writing to friends and family. Digital photography saves printing ink and paper by allowing customers to select the pictures to print. Perhaps the largest change in recent history is a large-scale changeover to digital medical records in hospitals. Soon, there will be many fewer forms for patients, nurses, doctors, and other caretakers to fill out, as much of this information will be stored in large digital servers.

MAKE A CHANGE

You've probably conducted a great deal of Internet research during your time in school. Perhaps even more than the research you've conducted in a reference library. Chances are that while you've been working, you've printed out a large amount of information to use in your assignment. The next time you're working on an Internet research project, try bookmarking the pages in a special file saved for research (if you're on a home computer) or emailing the links to yourself if you're at school or in the library to save paper.

TRY THIS

Go paperless in your school for a week. You can complete your homework assignments in a word-processed document and then email them in to your teachers, or bring them in on a portable USB drive. Maybe you can convince your teachers to try it out and email or post assignments to a Web site or blog rather than print them out. How will you handle taking notes in class?

Figure 13.14 With desktop publishing, the finished layout can be viewed on screen before it is printed.

Identifying *How does desktop publishing facilitate the production of books and magazines?*

In a desktop publishing system, the words and pictures appear on the screen just as you want them to appear on paper. The image you see is called a WYSIWYG image. WYSIWYG, which is pronounced "whizzy-wig," stands for "what you see is what you get." The color, font, and layout can be viewed and changed on screen without having to print out the document first.

Desktop publishing allows publishers to print and bind small numbers of books, individual chapters, or collections of specific pages. These books look as professional as commercially printed paperbacks.

Companies use desktop publishing to turn out their own newsletters and advertisements (Figure 13.14). The software is easy to use. Generally, a person chooses the width of the column and the size of the type and its style. Graphics such as pictures, designs, or photographs can then be placed on the page. The page can be rearranged easily. Pictures and text can be moved at the touch of a key.

Pagination

Today, some newspapers use a process called *pagination* to make up their pages. The whole page can be viewed on a screen, or a person can zoom in on one column or sentence. Once the page is ready, it is sent electronically to a photo-typesetter. A phototypesetter works like a computerized printer. It makes a high-quality copy of the newspaper page. From this copy, an offset plate is made and printing begins.

Computer-to-Plate Printing

The newest printing systems are called computer-to-plate systems. These typesetting systems bypass the photographic steps that previously were needed in offset printing.

In **computer-to-plate printing,** data is used to generate text and images onto a screen. The screen produces an image that is transferred directly to an aluminum plate. Printing is then carried out using offset lithography, which was described in Section 1.

Photocopiers

A **photocopier** is a machine that photographically reproduces an original image. Today, photocopiers are common in business offices, but this was not always the case. Prior to the 1960s, making a large number of copies required having them printed commercially.

Photocopiers make use of photography and static electricity. You create static electricity when you walk across a rug on a cold, dry day and then touch a doorknob. You build up an electrical charge on your body. The electricity is discharged when you touch an object that has a different charge. This discharge takes place because, as you learned in earlier chapters, similar charges (two negative charges or two positive charges) repel each other whereas unlike charges (positive and negative) attract each other.

Photocopiers use a metal plate with a coating that is sensitive to light (Figure 13.15). The metal plate gets a positive charge as it passes under a wire in the machine. The paper to

Figure 13.15 A photocopier uses light to copy an image onto paper.

Comparing *How is photocopying similar to photography?*

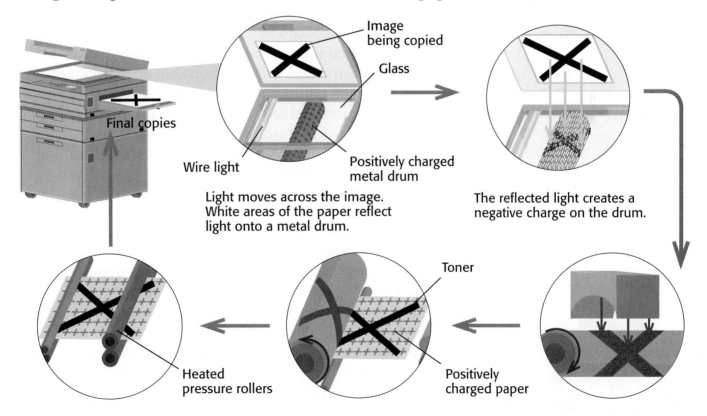

Image being copied

Glass

Wire light

Positively charged metal drum

Light moves across the image. White areas of the paper reflect light onto a metal drum.

The reflected light creates a negative charge on the drum.

Final copies

Toner

Positively charged paper

Heated pressure rollers

Heat fuses the toner to the paper.

Toner moves from the drum to the paper, which has a positive charge.

Toner is attracted to the positively charged areas on the drum.

be copied is exposed to a bright light, which is reflected by the white areas. It is not reflected by the areas that are dark with printing or writing. The light from the white areas wipes out the positive charge on the metal plate. The positive charge remains where the light was not reflected. A negatively charged black powder called toner is dusted over the plate. Toner is also known as dry ink, but it is not actually ink. Toner is a fine, negatively charged, plastic-based powder. It sticks to the positively charged areas. The sheet of paper on which the copy is to be printed is positively charged, so it attracts the toner. When the paper is heated, the heat bonds the toner to the paper, making a finished copy.

Computer Printers

Computer printers have almost entirely replaced mechanical typewriters. Three types of printers are inkjet, laser, and dye-sublimation printers.

An inkjet printer sprays tiny droplets of ink onto paper (Figure 13.16). An ink container is vibrated by a thin crystal connected to an electrical voltage. The vibrations cause ink droplets to be timed exactly and sprayed by a print head onto the page.

The print head of an inkjet printer has a series of nozzles that are fed by ink cartridges. These cartridges contain black or colored ink, in red, blue, and yellow. Using computer software, combinations of these colors can be controlled to make virtually any color. The cost of inkjet printers has dropped over the past few years, and the quality has improved. Today, even photograph-quality prints can be produced with inexpensive inkjet printers.

A laser printer uses a laser beam to quickly turn out high-quality pages of text or images. Instead of using ink, laser printers use toner, a black powder made of iron grains and plastic resin. Laser printers are more expensive than inkjet printers. However, they are cheaper to run because toner costs less per page than the ink required by inkjet printers. They are used when many pages must be printed.

Dye-sublimation (dye-sub) printers use a page-sized printer ribbon, a heated print head, and special coated printer paper. The heated print head, consisting of thousands of heating elements, moves across the ribbon. Panels on the ribbon consist of cyan (greenish blue), magenta, and yellow dye. Heat causes the color on the ribbon to fuse onto the surface of the special paper. Dye-sub printers are quite expensive, producing superior-quality prints for about three to four dollars per page. They are used by professional photographers, graphic artists, and scientists who process satellite images.

Figure 13.16 Inkjet printers are often used with home PCs.

Contrasting *How are inkjet printers different from laser printers?*

Connecting to STEM

The Science and Technology of Laser Printers

Laser printers are gradually replacing other types of printing technologies for quick, inexpensive computer printouts. Like other types of technology you have learned about, the technology used in a laser printer is based on the attraction between positive and negative charges.

Creating an Electrostatic Image

At the heart of a laser printer is a photosensitive drum called a printing drum. The printing drum is made from aluminum and looks a bit

Transferring the Image to Paper

The drum then comes into contact with a black powder called toner. The toner is attracted to the surface of the drum where its electrical charge has been changed. In these places, the toner sticks to the drum. This process is similar to creating a pattern out of glue and then sprinkling flour on top—the flour sticks only where the glue was applied.

The toner is negatively charged. Opposite charges attract each other. The toner sticks to the positively charged areas. Heat fuses the toner to the paper, and then the final print is rolled out of the printer.

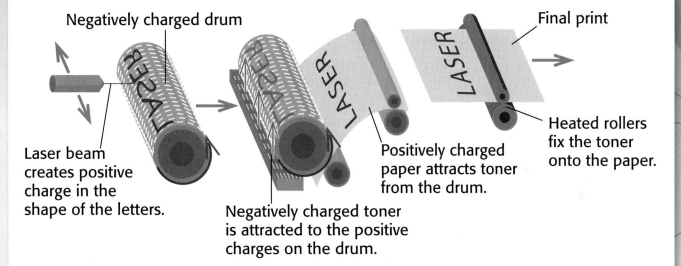

Negatively charged drum

Laser beam creates positive charge in the shape of the letters.

Negatively charged toner is attracted to the positive charges on the drum.

Positively charged paper attracts toner from the drum.

Final print

Heated rollers fix the toner onto the paper.

like a soda can. It is coated with material that is sensitive to light. The drum is passed under an electrical wire that produces a bright flash of light. The light produces a strong negative charge on the drum.

In the printer, a fine laser beam hits the printing drum, changing the electrical charge on the drum. The laser beam hits the drum in the reverse pattern of the text or image to be printed. The drum carries the image to be printed in the form of positive and negative charges. This is called an electrostatic image.

Critical Thinking

1. **Inferring** Laser printers and inkjet printers use different substances to print an image. How are these substances related to the way these printers work?

2. **Comparing** Compare the advantages and disadvantages of laser and inkjet printers. Which would you buy for your own use? Explain your reasoning.

Scanners

A scanner is a computer input device, like a keyboard or mouse. The input to the scanner is a graphical image. The image may be a drawing, a photograph, or a text document.

A scanner operates in much the same way as a digital camera (Figure 13.17). The document to be scanned is placed on a glass bed in the scanner. A strong light scans, or moves across, the document. A mirror reflects the light from the document (more from white areas, less from black areas) onto a CCD with thousands of photodiodes. Each photodiode produces an electrical signal according to the strength of the light that hits it. These signals are converted into digital data (bits) and sent to a computer. Special software reads the data and converts it into information that can be printed as words or pictures.

Fax Machines

A **facsimile** (fax) is the end result of an electronic transmission and reproduction of text or pictures. Newspapers use fax machines to send news photographs from one place to another. Weather maps have been sent this way since the 1920s. Many companies use fax machines to send images and text over long distances (Figure 13.18).

Fax transmission starts by reducing an image to patterns of black and white, or black and different colors. To do this,

Figure 13.17 A scanner converts information from a light beam into digital data.

Drawing Conclusions *Why do you think different-colored filters are needed?*

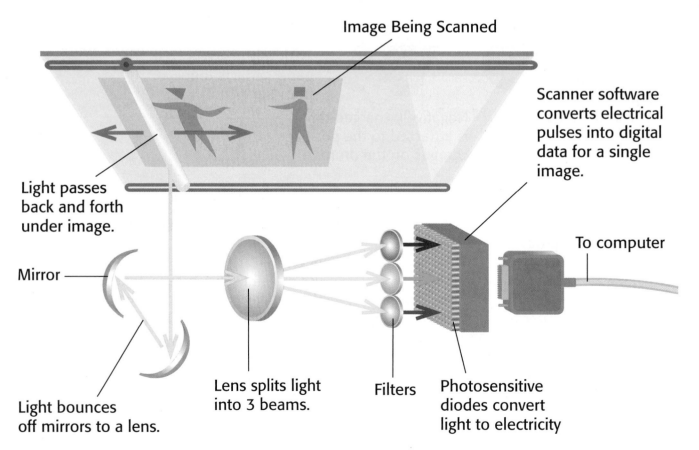

Image Being Scanned

Scanner software converts electrical pulses into digital data for a single image.

To computer

Light passes back and forth under image.

Mirror

Light bounces off mirrors to a lens.

Lens splits light into 3 beams.

Filters

Photosensitive diodes convert light to electricity

Figure 13.18 Fax machines enable people to send replicas of documents and graphics around the world.

Comparing *How are telephones and fax machines similar?*

an optical scanning device moves across the page to be sent. Information from the scanner is then changed to electrical impulses. The impulses are sent over wires to a fax machine in a different location. That machine reverses the process. It turns the impulses into a new, identical image of the page.

Originally, fax machines used telephone lines as the channel to communicate with another fax machine. Now it is also possible to send faxes directly from one computer to another.

Digital Images

Once an image is converted to digital form, it can be changed, stored, distributed, and reproduced at the click of a mouse. A digital image stores visual information as numbers. It can be created using a digital camera or a scanner. Digital images can also be created on a computer using special software such as Adobe Photoshop or CAD. Libraries of digital images can also be purchased on CDs.

Image files can be manipulated in a computer. With software, you can make color adjustments, straighten and crop photographs, fix improper lighting, simulate oil painting techniques, add background textures, and remove scratches, wrinkles, and other flaws. Digital images can be emailed or used as part of a Web page.

Digital Image Files

Digital image files often require large amounts of computer memory. The more pixels used to create the image, the better the quality. However, more pixels require more memory.

Figure 13.19 Images that are used on the Internet are usually JPEGs and GIFs. Both can be compressed so they download easily.

Applying *If you want to send your friend a copy of a digital photograph, would you save the file as a JPEG or a GIF?*

Computers can store digital images in several formats (Figure 13.19). One common format used on PCs is the **bitmap** (BMP) file. BMP files provide good clarity and detail, but they take up considerable amounts of memory. As a result, they take a long time to download from the Internet. Similar PICT (PICTure) files are used on the Apple Macintosh (Mac).

Another image format is the Tagged Image File Format (TIFF). TIFF files can be used on either Macs or PCs. TIFF files are one of the best file types to use for printed images and can be imported into most word-processing, page layout, and drawing programs. They offer high quality but longer download time.

A way to reduce the size of an image file is to use a technique called compression. Compression gets rid of details that are barely seen by the human eye. When data are compressed, the file size is reduced, but so is the quality. As a result, you make a trade-off between image size and quality. Compressed images are used on Internet Web pages because they require less download time. These images are compressed according to standards called Joint Photographic Experts Group (JPEG) or Graphic Interchange Format (GIF). JPEG files can be used to display high-quality photographs. GIF is best used for black-and-white line drawings and pictures with large blocks of solid colors.

BMP, PICT, TIFF, JPEG, and GIF files are raster image formats. A raster image is a pattern of horizontal lines created from rows of pixels.

Figure 13.20 Each animation in this series is quickly displayed one after the other to give the illusion of movement.

Describing *Quickly skim your eyes over these images. What do you see?*

Animating Digital Images

Animation is a series of still images in sequence, much as a movie is a sequence of individual frames, usually at a rate of 24 frames per second. GIF animations are made from individual GIF files (Figure 13.20). These files are seamed together using shareware such as the GIF Construction Set. **Shareware** is software that you use on the honor system: You download it free from the Internet and pay a small registration fee if you use it.

Cartoon characters are commonly animated on Web pages, as are shapes that are made to rotate, and globes that are made to spin.

SECTION 3 Assessment

Recall and Comprehension

1. Name three different graphic communication systems. How does each one produce a final printed or screen image?
2. Describe the process a scanner uses to create a digital image.

Critical Thinking

1. **Describing** Describe how a fax machine is able to send text and pictures electronically.
2. **Comparing** Create a table to compare the computer format, amount of memory, and picture quality of BMP, PICT, TIFF, JPEG, and GIF files.
3. **Evaluating** How are photocopiers and laser printers similar?

QUICK ACTIVITY

Create a printing system from a potato stamp and an inkpad. With a sharp knife, cut a potato in half, and then engrave a design into the cut side of the potato. Remember that the raised area will do the printing and that letters and numbers must be cut in a mirror image to print properly. Use the inkpad to apply ink to the potato, and then stamp your design onto heavy paper. **For more related Design Activities, see pages 412–415.**

13 Review and Assessment

Chapter Summary

- One of the most important advances in graphic communication was the invention of movable type. With movable type, books could be printed more quickly and cheaply.

- Printing from a raised surface is called relief printing. Printing from a lowered surface is called gravure printing. Printing through a stencil is known as screen printing. In offset printing, oil-based ink is applied to a flat sheet of light-sensitive material.

- The word *photography* means "to write with light." All film-based cameras have a dark chamber, a lens, a shutter that lets light into the camera, a viewfinder, and a place for film. Film-based photography requires light, film, a camera, chemicals, and a darkroom.

- Film cameras include the view camera, the single-lens reflex camera, and the APS camera. Digital cameras do not use film. Images are recorded on an electronic image sensor.

- Modern graphic communication systems use electronics to create digital images. These images are then converted to printed copy. Examples of electronic graphic systems are word-processing and desktop-publishing systems, photocopiers, fax machines, and scanners.

- Computers can store digital images in several formats. These include BMP, PICT, TIFF, JPEG, and GIF. All of these formats are raster images that are created from rows of pixels.

Building Vocabulary

Your teacher may give you a crossword puzzle. Complete the puzzle using the following words from this chapter. Exchange puzzles with a classmate to check each other's answers.

1. advanced photo system
2. animation
3. bitmap
4. camera obscura
5. computer-to-plate printing
6. darkroom
7. desktop publishing
8. developer
9. digital camera
10. digital image
11. dye-sublimation printer
12. facsimile
13. film
14. film-based photography
15. fixer
16. gravure printing
17. lithography
18. movable type
19. photocopier
20. pixel
21. relief printing
22. screen printing
23. shareware
24. single-lens reflex camera
25. stop bath
26. toner
27. twin-lens reflex camera
28. view camera
29. word processor

See your teacher for the Crosstech puzzle.

Reviewing Content

1. What is graphic communication? Give five examples of graphic communication technologies.

2. How did Johannes Gutenberg improve communication technology?

3. What was George Eastman's contribution to photography?

4. Name four types of film-based cameras.

5. Describe the use of static electricity in laser printers and photocopiers.

Applying Your Knowledge

1. Using the library or the Internet as a resource, identify five graphic communication careers that interest you. Investigate the advancement requirements for one of these careers.

2. Give an example of some thing printed by each of the following processes:
   (a) relief printing
   (b) gravure printing
   (c) offset printing
   (d) screen printing

3. Describe the type of camera you use and how it serves your needs. If you don't have a camera, describe the camera you would like to have, and why.

4. Interview a few older adults to find out how extra copies of documents were produced before we had photocopying machines. How have these machines changed the way we work and keep records?

5. What type of digital image file would you use for these purposes?
   (a) to send a photo on the Internet
   (b) to print a good-quality color photo
   (c) to put a line drawing on a Web page
   (d) to put an image in a page layout for printing

Critical Thinking

1. **Predicting** How do you think new graphic technologies such as electronic books and Web-based newspapers will affect publishing in the future?

2. **Contrasting** How is computer-to-plate printing different from offset lithography?

3. **Comparing** Compare how a digital camera and a film-based camera work. What functions are common to both types of cameras?

 Connecting to STEM
science • technology • engineering • math

How Many Pixels?

On a display screen, each pixel is made up of a single color. The more pixels, the better the image quality. To determine the total number of pixels in an image, multiply the number of pixels on the horizontal by the number of pixels on the vertical. How many pixels does a 1280 x 1024 image have? How many megapixels does it have? (A megapixel is equal to 1,000,000 pixels.)

Design Activity 25

DEVELOPING A PLOT PLAN

Problem Situation

Your family is planning to build a new house and has to prepare a three-dimensional plot plan for presentation before the architectural review board. To accompany the plan, you will need to prepare a sheet showing your calculations to prove that the house meets the zoning requirements.

Materials

You will need:

- foam block
- foam board or cardboard

Your Challenge

You and your team members must design a building that will satisfy the municipality's zoning requirements. In this community, houses cannot be more than two stories high and must be set back from the street a minimum distance of 30 feet. Houses must also be 15 feet from the rear property line and 10 feet from the side property lines. To guard against buildings being too massive, the municipality requires that the floor area ratio (FAR) be 0.4 or less. The FAR is determined by dividing the floor area on the first two floors (not including any indoor garage space) by the area of the lot. You must then create a 3-D plot plan that shows the house property lines, setbacks, driveways, walkways, swimming pools, and trees and shrubs.

Go to your **Student Activity Guide, Design Activity 25.** Complete the activity in your Guide, and state the design challenge in your own words.

1 Clarify the Design Specifications and Constraints

To solve the problem, your design must meet the following specifications and constraints:

- The plot is a trapezoid.
- The rear property line measures 100 feet.
- The front property line measures 120 feet.
- The left property line, as you look from the street, is perpendicular to the rear property line.
- The right property line is at an angle.
- Use a rigid base such as foam board or cardboard. You can draw directly on this or attach drawing paper. Use a foam block, or join foam board, cut to scale, for the structures.

In your Guide, state the specifications and constraints. Add any others that your team or your teacher included.

② Research and Investigate

To better complete the design challenge, you need to first gather information to help you build a knowledge base.

In your Guide, complete the following Knowledge and Skill Builder 1:

1: Calculating Areas of Geometric Shapes
2: Ratio and Proportion
3: Floor Area Ratio

③ Generate Alternative Designs

In your Guide, describe two possible solutions that your team has created for the problem.

④ Choose and Justify the Optimal Solution

Refer to your Guide. Explain why you selected the solution you did, and why it was the better choice.

⑤ Develop a Prototype

Construct your plot plan. Include a photograph or sketch of your final design in your Guide.

Indicate which resources were most important in this activity, and how you made trade-offs among them.

⑥ Test and Evaluate

How will you test and evaluate your design? In your Guide, describe the testing procedure you will use. Explain how the results show that the design solves the problem and meets the specifications and constraints.

⑦ Redesign the Solution

Respond to the questions in your Guide about how you would redesign your solution. Your redesign should be based on the knowledge and information that you gained during the activity.

⑧ Communicate Your Achievements

Describe the plan you will use to present your solution to your class. Show any handouts and/or PowerPoint slides you will use.

Design Activity 26

T-SHIRT DESIGN

Problem Situation

It is the first day of track practice, everyone is tired, the uniforms are several years old, and the team needs a morale-building experience. Katie suggests that they design a T-shirt for themselves and perhaps to sell to others. Andrew says he would like a mouse pad to remind him of the team when he is using his computer.

Materials

You will need:

- graphics software
- heat press or iron
- inkjet or color laser printer
- transfer paper
- T-shirt, mouse pad, or other medium on which to print

Your Challenge

You and your team members are to create a graphic design that depicts an image for a team or group. The image should deliver a message about the group and its members. It should not be just the team name or photos of members.

> Go to your **Student Activity Guide, Design Activity 26.** Complete this activity in the Guide, and state the design challenge in your own words.

1 Clarify the Design Specifications and Constraints

To solve the problem, your design must meet the following specifications and constraints:

- The image will be printed on a T-shirt, mouse pad, or other medium of your teacher's choosing.
- The design must be no larger than an 8 1/2" x 11" piece of paper.

> In your Guide, state the specifications and constraints. Add any others that your team or your teacher included.

2 Research and Investigate

To better complete the design challenge, you need to first gather information to help you build a knowledge base.

> In your Guide, complete Knowledge and Skill Builder 1: Graphic Images.

> In your Guide, complete Knowledge and Skill Builder 2: Mirror Images.

In your Guide, complete Knowledge and Skill Builder 3: Designing an Image.

③ Generate Alternative Designs

In your Guide, describe two possible solutions that your team has created for the problem.

④ Choose and Justify the Optimal Solution

Refer to your Guide. Explain why you selected the solution you did, and why it was the better choice.

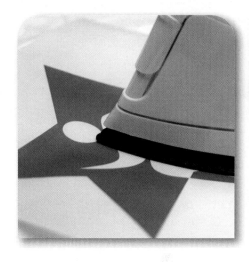

⑤ Develop a Prototype

Make your final design. Include a printout of the design and a photograph of how it looks on the final medium that you have chosen.

Indicate which technological resources were most important in this activity, and how you made trade-offs among them.

⑥ Test and Evaluate

How will you test and evaluate your design? In your Guide, describe the testing procedure you will use. Explain how the results show that the design solves the problem and meets the specifications and constraints.

⑦ Redesign the Solution

Respond to the questions in your Guide about how you would redesign your solution. Your redesign should be based on the knowledge and information that you gained during the activity.

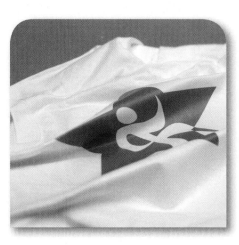

⑧ Communicate Your Achievements

In your Guide, describe the plan you will use to present your solution to your class. Show any handouts and/or Power-Point slides you will use.

UNIT 5

Energy, Power, and Transportation

Unit Outline

Energy and Power

SECTION 1 Work and Energy
SECTION 2 Resources for Energy
SECTION 3 Power and Power Systems

In this chapter, you will learn about the following renewable and nonrenewable sources of energy:

- coal and oil
- natural gas
- solar
- wind
- geothermal
- nuclear
- biomass

All industries, businesses, and homes use energy. Most often, we use energy in the form of electricity. Different forms of energy are converted into the electricity we use.

- Ancient plants and animals convert energy from the sun into stored chemical energy. A power plant converts this chemical energy into electrical energy.

- Solar cells convert the energy in sunlight into electrical energy.

- Windmills convert wind energy into mechanical energy that turns a turbine. The mechanical energy is then converted into electrical energy.

- Hydroelectric plants at the base of dams convert the potential energy of water into electrical energy.

- Nuclear power plants convert the energy contained in the nucleus of an atom into heat and then into electricity.

This electricity is what lights up homes, skyscrapers, and entire cities.

Work and Energy

Benchmarks for Learning

- Work is the product of force and distance.
- Energy is the capacity to do work.
- Energy cannot be created or destroyed.

Reading Strategy

Listing   Make a list of the different types of energy described in this section. For each entry on your list, write a quick definition of that term. You may also want to use a sketch with your definition.

1. Energy
- kinetic—energy that moves
- potential—stored energy

Vocabulary

work	potential energy	heat
energy	thermal energy	joule
kinetic energy		

Work

You probably think that doing two pages of math homework or writing a report is hard work. According to science, though, it isn't work at all. In science, the word *work* has a different—and very exact—meaning.

Work is the means by which energy is transferred from one object to another. Work is done whenever a force pushes or pulls on an object, causing the object to move. The amount of work done is equal to the distance the object moves times the amount of force applied, as the formula shows:

$$\text{Work} = \text{Force (F)} \times \text{Distance (D)}$$

In Figure 14.1, the work done is equal to the force of the girl kicking the ball times the total distance the ball moves while in contact with her foot.

Consider another situation. Two boys pull for an hour on ropes hitched to opposite ends of a cart. The boys are equally strong. Because neither is stronger, the cart does not move. How much work has been done? The answer is none. Even though the boys have exerted a great deal of force on the cart, it did not move.

Energy

How does work take place? Something or someone supplies the energy. **Energy** is the ability to do work. It is what supplies the force that is used to do work. When work is done on

Figure 14.1   Work is the product of force and distance.

Inferring   *As distance increases, does work increase or decrease?*

Work = Force x Distance

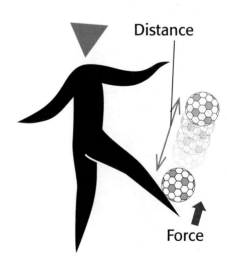

Distance

Force

an object, energy is transferred from the person or machine to the object. More energy is needed to do a large amount of work than a small amount of work. Much of today's technology changes energy from one form into another so that it can do useful work.

Types of Energy

Two kinds of energy are kinetic energy and potential energy. Energy that is moving is called **kinetic energy.** A falling stone has kinetic energy. So does a moving bicycle. The amount of kinetic energy an object has depends on its weight and its speed.

Energy that is stored and used later is called **potential energy.** Gravitational potential energy is energy that is stored in an object because of its position relative to the ground (Figure 14.2). The higher an object is above the ground, the greater its gravitational potential energy. Elastic potential energy is energy that is stored in an object that is stretched or compressed. A stretched rubber band and a compressed spring have elastic potential energy.

Conversion of Energy

Both kinetic energy and potential energy have many forms. Some of these forms are thermal, chemical, electrical, and nuclear. A toaster, for example, provides electrical energy. Most forms of energy can be converted to other forms. When you make toast, the toaster's electrical energy is converted to the thermal energy that heats the bread.

Figure 14.2 Potential energy is changed to kinetic energy as the box is dropped.

Interpreting *What type of energy does the person have as he is sitting on top of the pole?*

Comparing Potential and Kinetic Energy

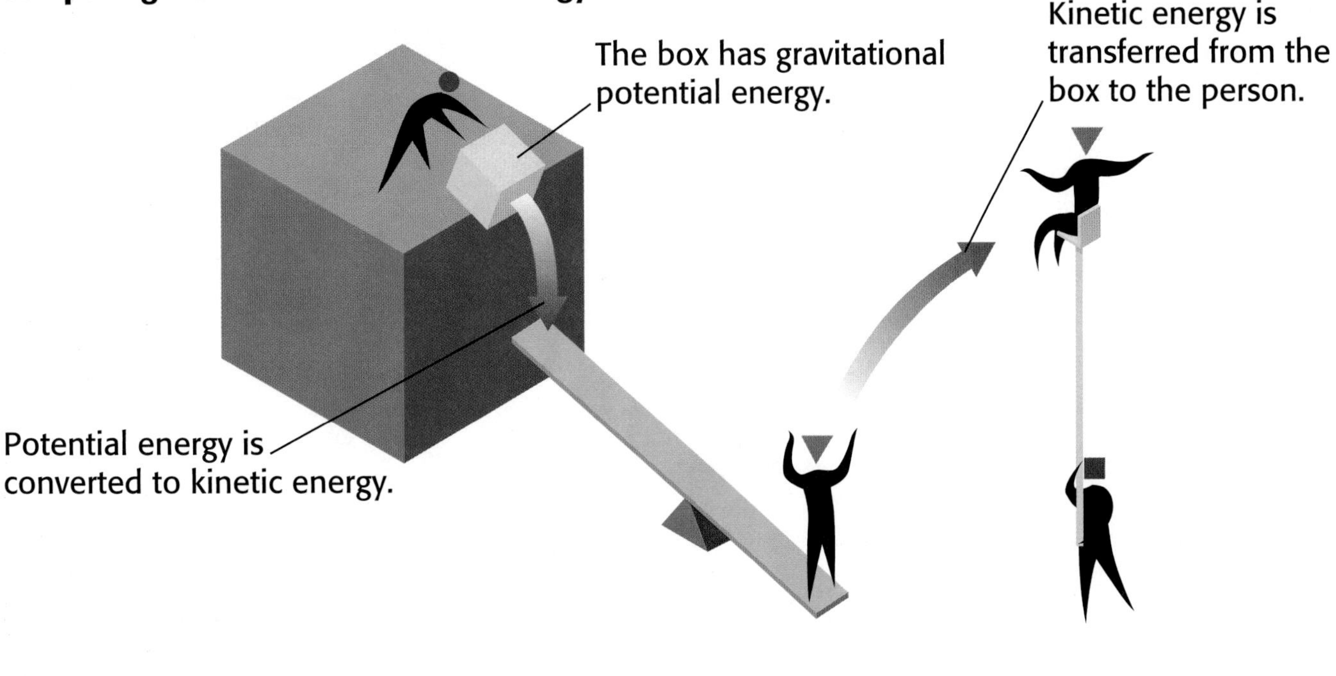

The box has gravitational potential energy.

Potential energy is converted to kinetic energy.

Kinetic energy is transferred from the box to the person.

Recall that most substances are made of small particles called atoms. **Thermal energy** (also called internal energy) is the energy of the atoms that make up a substance. When the thermal energy of an object increases, the atoms move faster and the object becomes warmer. A hot slice of toast, for example, has more thermal energy than a cold slice of bread.

Heat can also be thought of as a form of thermal energy. More exactly, **heat** is thermal energy that is transferred from something that is warmer to something that is cooler. With our example of making toast, a temperature difference exists between the wires inside the toaster and the slice of bread. Heat is transferred from the wires to the bread. Like work, heat does something to an object or substance and transfers energy.

Both heat and work are measured in joules. A **joule** (J) is the amount of energy required to move an object 1 meter using 1 newton (N) of force. Using the equation, Work = Force × Distance, we can determine that an object that is moved 10 meters using 20 N of force uses up 200 J of energy. Different types of work use up different amounts of energy (Figure 14.3).

Conservation of Energy

The law of conservation of energy states that energy cannot be created or destroyed, but that it can change from one form to another (Figure 14.4). When we stretch a rubber band, we change work into potential energy. We are doing work on the rubber band (a force acting through a distance), therefore its potential energy increases. When we drop a ball, its potential

Figure 14.3 The same amount of energy can be used to do different activities.

Calculating *If 315,000 joules are used up during a 6-minute run, how many joules are used up during each minute of running?*

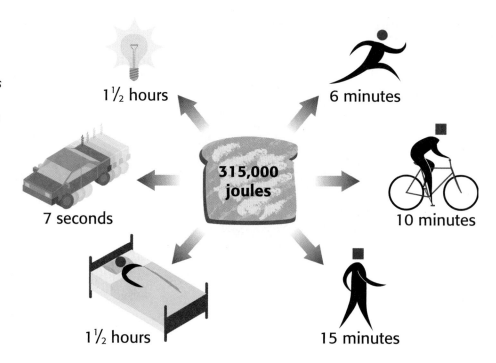

1½ hours

6 minutes

315,000 joules

7 seconds

10 minutes

1½ hours

15 minutes

Applying *As the glass cools down, where is the thermal energy transferred?*

energy decreases as it gets nearer the ground. However, its speed, hence its kinetic energy, increases. In both of these instances, there is a conversion from one energy form to another, but the total energy remains constant.

Energy is always conserved, but does it remain valuable to us in all its forms? Consider a pot of water boiling on the stove. When the gas was ignited, the chemical energy of the gas was converted into thermal energy in the flame. The flame heated the pot, which then transferred heat to the water. When the flame is turned off, the pot cools to room temperature. Did the energy disappear? No, the water's internal energy is transferred as heat into the kitchen's air and, eventually, to the air outside.

SECTION 1 Assessment

Recall and Comprehension

1. If you push against a wall until you are really tired, have you done any work? Explain your answer.
2. What is energy?
3. Explain the law of conservation of energy.
4. What is thermal energy? How is it related to heat?

Critical Thinking

1. **Calculating** A girl pushes on a bicycle with 40 newtons of force, moving it 30 meters. How much work has she done?
2. **Contrasting** Explain the difference between potential energy and kinetic energy.
3. **Applying** On a cold winter day, why does a thick coat make you feel warmer? What is the source of the heat?

QUICK ACTIVITY

Some materials have internal energy that can be released when mixed with another material. Over a large sink, add one teaspoon of baking powder to an empty 35-mm film cartridge. Now, add water until the container is 3/4 full and quickly snap the top in place. What happens? Why? Make sure you wear safety glasses and protective clothing.
For more related Design Activities, see pages 446–449.

Connecting to STEM

How We Use Energy

Our society uses a large amount of energy each day. We use energy to light our homes at night. We use it to heat our homes when it is cold and to cool them when it is hot. We use energy to get to and from school and work.

Any type of technology requires some form of energy. Large amounts of energy are needed to construct new buildings and to manufacture products. Energy is also needed to run equipment on farms, making them more productive. The energy we use comes from many different sources and is used in many different ways.

Comparing Graphs

The line graph below shows how energy use in the United States is increasing every year. The pie chart shows where our energy came from in 2008. The United States produces most of its natural gas, nuclear fuel, coal, and renewable energy supplies. To meet our energy needs, however, for every gallon of oil that is produced in this country, we import an additional 2 gallons of oil.

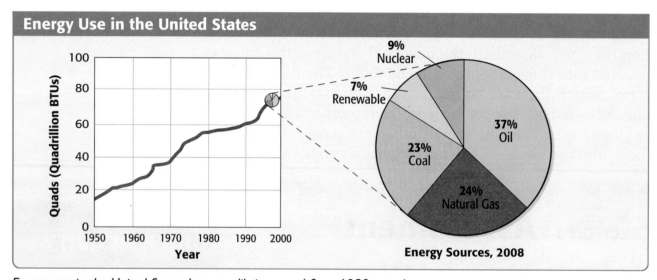

Energy Use in the United States

Energy use in the United States has steadily increased from 1950 to today.

Exactly How Much Energy?

The total amount of energy used in 2000 in the United States was 98 quads. A quad is one quadrillion (1,000,000,000,000,000) British thermal units (Btus). A Btu is the amount of energy needed to raise the temperature of 1 pound of water 1 degree Fahrenheit.

One quad, then, could heat up 1 quadrillion pounds of water by 1 degree. A quad equals the energy given off by burning about 20 gallons of gasoline every day for a year.

Critical Thinking

1. **Interpreting** If energy consumption continues to increase, as shown in the graph, approximately what will consumption be in 2010? In 2050?

2. **Inferring** The pie chart shows how the United States used different types of fuel in 2008. How does the pie chart relate to the line graph?

SECTION 2
Resources for Energy

Reading Strategy

Outlining As you read this section, make an outline of renewable and nonrenewable sources of energy. Include a brief definition of each term.

> I. Nonrenewable
>   Energy
>   A. Coal
>   1. formed from
>   compressed
>   plant remains
>   2.
>   B. Oil
>   1.

Benchmarks for Learning

- Energy resources may be renewable or nonrenewable.
- Many of our energy needs are met by burning nonrenewable fossil fuels.
- More renewable energy resources are being developed.

Vocabulary

fossil fuel	solar energy	geothermal energy
nuclear fission	photovoltaic cell	nuclear fusion
radioactive waste	turbine	biomass

Nonrenewable Energy Resources

Energy sources may be described as renewable or nonrenewable. Nonrenewable energy sources, such as oil, cannot be replaced. There is a fixed supply that will eventually be used up.

Most of the energy we use today comes from nonrenewable, or limited, energy sources. It takes millions of years for coal, oil, and natural gas to be created (Figure 14.5). As we use them up, more will not become available. At our present rate of use, it is estimated that we have only a hundred-year supply of oil and natural gas remaining.

Figure 14.5 Materials that accumulate on forest floors may eventually form coal or gas.

Clarifying *Why are coal and gas considered nonrenewable resources?*

Fossil Fuels

Oil, natural gas, and coal are all called fossil fuels. A **fossil fuel** is an energy-rich substance formed from the remains of plants and animals that lived millions of years ago (Figure 14.6). After the plants and animals died, they did not rot away completely because they were covered by fallen trees, leaves, and mud. Additional layers of soil material, along with movements of Earth's surface, further buried the plant and animal remains. The great pressure over millions of years plus heat changed them into oil, gas, or coal.

Fossil fuels are made primarily of hydrogen and carbon atoms joined together as molecules in a high-energy, chemical bond. During combustion, when fossil fuels are burned, the molecules break down into simpler molecules. As they break down, they release much of the energy contained in the high-energy bonds. This is the energy that we get from fossil fuels. When carbon burns, it produces carbon dioxide. When hydrogen burns, it produces water (hydrogen oxide).

The more energy that is released, the better the fuel. The bonds in gasoline, which is made from oil, release a great deal of energy when the gasoline burns. That is why gasoline is used so widely as a fuel.

Coal

Coal was one of the first fossil fuels to be used widely during the Industrial Revolution. Coal can be mined from the surface of the land by strip mining, or from deep below Earth's surface. In strip mining, very large but shallow holes are

Figure 14.6 Coal forms underground over millions of years. Lignite, bituminous, and anthracite are all forms of coal.

Interpreting *Which is formed more quickly—lignite or anthracite?*

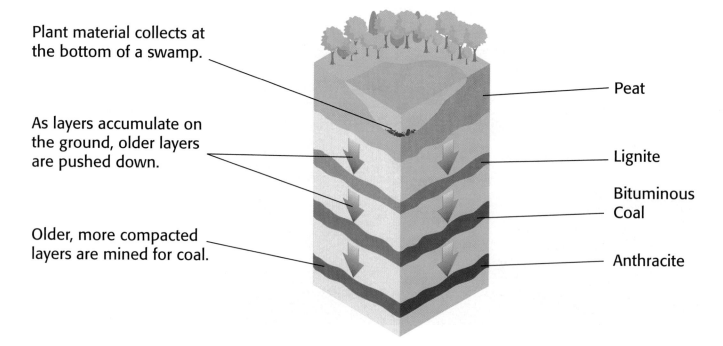

Plant material collects at the bottom of a swamp.

As layers accumulate on the ground, older layers are pushed down.

Older, more compacted layers are mined for coal.

Peat

Lignite

Bituminous Coal

Anthracite

dug into the ground and the coal is removed. The hole must be filled in and the land restored by planting trees and grass when mining is finished. To reach deep coal, miners have to tunnel into the ground, dig the coal out, and transport it back to the surface (Figure 14.7). Coal mining can be dangerous, but modern equipment and safety standards have helped make it safer.

Coal is the most abundant nonrenewable energy resource, but there are drawbacks to its use. Coal must be moved in large quantities by train, barge, or truck to its destination. Often there are traces of sulfur in the coal, which creates sulfur dioxide when burned. When this pollutant combines with water in the air, it forms sulfuric acid. This acid can form acid rain, which can cause serious damage to forests and aquatic life. Engineers and ecologists are working to solve these problems.

Oil

Often called "black gold," oil is the energy resource on which we depend most. Oil is used much more than coal because it is more easily taken from the ground, stored, and transported. It also has much more energy per pound than coal and can be refined into many other useful fuels. Some of these fuels are home heating oil, gasoline, and jet fuel.

Oil is in such demand that oil wells are drilled virtually anywhere oil is discovered. Wells are dug even in areas with harsh climates or hazardous conditions, such as the northern coast of Alaska and the deep ocean floor (Figure 14.8).

Natural Gas

Natural gas is used for home heating and cooking as well as in industry. Natural gas burns cleaner than other

Figure 14.7 Coal can be mined from Earth's surface or from deep underground.

Hypothesizing *What are some of the risks of underground coal mining?*

Figure 14.8 Oil is in such demand that it is drilled in all types of locations.

Summarizing *Why is oil used more commonly than coal?*

fossil fuels. It has a greater amount of hydrogen than carbon, so the products of combustion contain more harmless water vapor (burned hydrogen) than for other fossil fuels. Natural gas is transported by pipelines from its source in the ground to the places where it is used.

Natural gas is used as fuel in several industries. In the glass industry, huge gas-fired furnaces melt glass in large quantities. Gas is also used to make fertilizers, which has helped increase food production tremendously.

Nuclear Fission

Uranium is another important nonrenewable energy resource. Unlike oil, gas, and coal, uranium is not a fossil fuel. It is not burned to obtain energy. Uranium is a nuclear, or atomic, fuel.

Matter can be changed into energy by **nuclear fission,** the splitting of an atom's nucleus (Figure 14.9). Recall that an atom's nucleus is made up of protons and neutrons. In nuclear fission, an unstable atom, such as uranium-235, is hit with an external neutron. The atom splits and forms two smaller nuclei and extra neutrons, which give off a great deal of energy in the form of heat and light. The new neutrons hit more atoms around them. They make these atoms split, giving off more energy and more neutrons, creating a chain reaction.

Very large amounts of energy are produced during nuclear fission. In a nuclear power plant, the amount of energy released is controlled by controlling the number of neutrons. Without this type of control, an explosion could occur. In a nuclear power plant, the nuclear energy is converted into electricity.

Figure 14.9 Nuclear energy is released when a neutron hits a large atom, such as uranium.

Inferring *Why is this process of energy release referred to as fission, or splitting?*

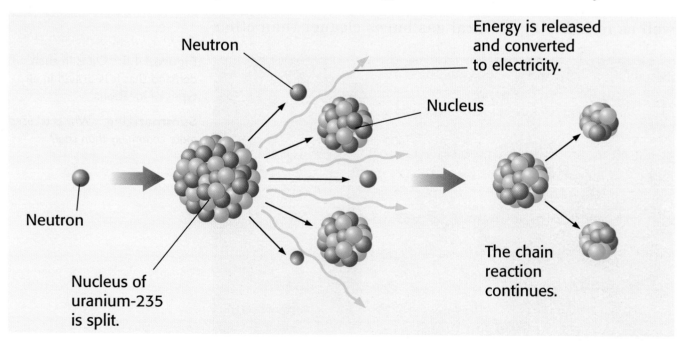

Neutron

Energy is released and converted to electricity.

Nucleus

Neutron

Nucleus of uranium-235 is split.

The chain reaction continues.

Figure 14.10 One of the challenges of using nuclear energy is determining how to store its wastes safely.

Applying *If your town were going to build a nuclear energy facility, would you support or oppose the decision? Why?*

A small amount of atomic fuel can produce a huge amount of energy. However, there are risks with nuclear power that are not present with fossil fuels. A nuclear power plant produces excess unusable material called **radioactive waste,** which contains unstable atoms that give off radiation. Radiation can cause burns and illness. At very high levels, it can even kill people. It also remains radioactive for thousands of years (Figure 14.10). Safety precautions must be taken to make sure that people are not exposed to too much radiation.

Renewable Energy Resources

Renewable energy resources are those that are so abundant or that replenish themselves so quickly that they will never run out. For example, solar energy is essentially unlimited because, although the sun will burn out in a few billion years, we could never deplete the amount of energy released by the sun in our lifetime.

It is a good idea to use renewable energy sources such as the sun. However, it can be difficult to find ways to harness these resources. Renewable energy resources have low energy values, so it takes much more energy to get the energy value needed. Research and development of renewable energy sources are ongoing.

Human and Animal Muscle Power

The first source of energy that people used was muscle power. Initially, human muscle power could be used only for small tasks. When people learned to work together, large jobs could be done using human power.

Later, people learned how to use animal muscle for the power they needed. They harnessed oxen, donkeys, and other animals to machines. These machines could plow the land, pump water, and do other jobs. In some parts of the world, human and animal power are still important sources of energy.

Solar Energy

We depend on **solar energy**—the energy from the sun—to support life on Earth through photosynthesis. The sun's energy is also purposefully used in many ways. Solar power plants, for example, use many special mirrors called parabolic reflectors to focus sunlight onto a single spot to generate very high temperatures. The heat causes water to boil and create steam, which can be used to generate electricity.

The sun's heat is also used directly for heating homes and other buildings. For example, some solar collectors are placed on the roof of a house to heat water for showers, washing, and cooking. Some buildings also use fans and pumps to distribute the collected heat throughout the building. Passive solar designs for houses use the sun for partial heating. Windows, walls, and doors are carefully placed so the sun can be used for heating during the winter.

A **photovoltaic cell** is a solar cell that can turn sunlight into electricity (Figure 14.11). These solar cells are used often in remote areas that are not served by power lines. They can also be used to charge batteries that run radio relay stations, radio telephones, and other electrical devices. Photovoltaic cells also are used on most satellites and space vehicles. About 10 percent of the sun's energy is converted into electricity in a photovoltaic cell.

Figure 14.11 A solar cell converts light energy into electrical energy.

Inferring *Why is solar energy used more commonly in some parts of the world than in others?*

Wind Energy

Wind has been used as an energy source for hundreds of years. Machines that harness the wind have been used to

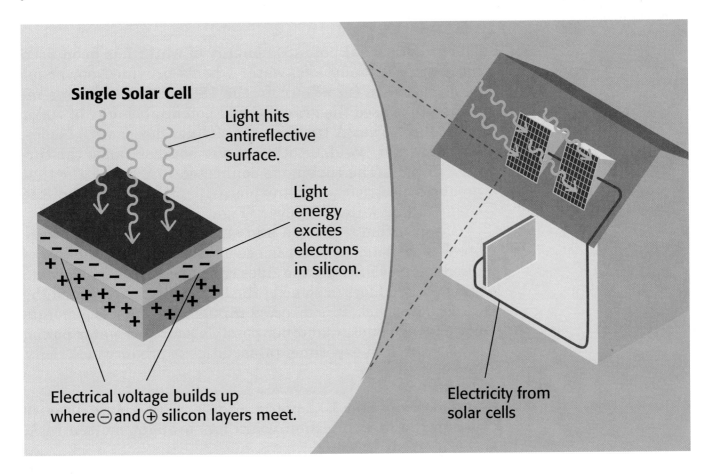

Single Solar Cell

Light hits antireflective surface.

Light energy excites electrons in silicon.

Electrical voltage builds up where ⊖ and ⊕ silicon layers meet.

Electricity from solar cells

Figure 14.12

Figure 14.13

Figure 14.12 A wind farm converts wind energy into electrical energy.

Clarifying *Is wind a nonrenewable or renewable source of energy?*

Figure 14.13 In a dam, the potential energy of water is converted to electrical energy.

Describing *What is the role of the turbine in the generation of electricity?*

pump water, grind grain, and move ships across the water. Today, wind is used to generate electricity (Figure 14.12). In some locations where it is windy, farmers lease their land for wind farms. The farmers earn income and can still grow crops on the land. However, wind-powered generators cannot make electricity when the wind isn't blowing or if it is blowing below a certain speed. The speed of the wind and the diameter of the blades determine the amount of energy produced. Due to this constraint, wind farms are useful only in places where the wind blows briskly most of the time.

Water Energy

The gravitational potential energy of water has been used for centuries. Long ago, water wheels provided power for grinding grain into flour. In the 1800s, a turbine was invented that used the gravitational potential energy of water to produce power. A **turbine** is a circular device with blades. Falling water, wind, or other power sources cause the turbine to turn. The turbine is connected to a generator that produces electricity. In many places, dams have been built to create falling water for power (Figure 14.13).

Gravity from the moon and sun causes tides, which are large movements of water in oceans and rivers. In the Bay of Fundy in Nova Scotia, the tides can cause the water level to rise nearly 60 feet in some of the bay's tributaries. When the water flows back, it is directed through turbines to generate power. Wave energy, another possible source of water power, is being used in experimental facilities to produce electricity.

Geothermal Energy

At Earth's center is a hot, molten core. Volcanic activity within the last 3 million years has brought molten rock, called magma, to within 3 miles of Earth's surface. **Geothermal energy** is thermal energy stored below Earth's surface (Figure 14.14). When magma is close to the surface, it can heat underground water to the boiling point, creating steam. It is possible to use this steam to power turbines, which in turn drive electric generators. The challenges to building geothermal plants are the relatively unstable land areas, in geological terms, on which they must be located and water that is very corrosive because of the dissolved minerals and salts. Also, the number of places with magma close to Earth's surface is limited.

Figure 14.14 The energy of hot steam from underground is called geothermal energy.

Identifying *How does the word* geothermal *describe this type of energy?*

Nuclear Fusion

Nuclear fusion is another atomic method of changing matter into energy. In **nuclear fusion,** the nuclei of two atoms are forced together to form a new nucleus, and a large amount of energy is released. However, a tremendous amount of energy is needed to force the nuclei together. Under the right conditions, more energy is given off than is used.

A successful fusion power plant has not yet been built, but fusion research continues. It is hoped that someday fusion will prove to be an efficient way to generate electricity. Fusion power plants will use two forms of hydrogen—deuterium and tritium—as fuel. Sources for these fuels are readily available. The waste products of fusion are not radioactive, and the process would not cause pollution. Many people hope that fusion plants will provide most of our electricity in the future.

Biomass

Biomass is accumulated vegetable and animal wastes. Biomass can serve as a major source of renewable energy. Three basic biomass processes are used to produce energy.

The first process is direct combustion of waste products. Burning wood is a biomass process. A cord of wood (a stack 4 feet by 4 feet by 8 feet) can provide the same amount of heat as about 200 gallons of oil. Wood is renewable, but forests must be managed carefully. Once trees are cut in an area, new trees should be planted to replace those cut down. In parts of the world where there is not much wood, dried animal wastes are gathered and burned to provide heat.

The second way to process biomass is gasification. Methane gas is produced as the biomass rots. It is collected and stored to be used as a fuel. Methane gas can be converted into electricity or used to power a combustion engine.

Figure 14.15 Gasohol, a type of car fuel, in part uses energy from corn.

Characterizing *Which biomass process is used to form gasohol?*

The third biomass process is fermentation, which uses microorganisms to turn biomass such as corn or other grain into alcohol and carbon dioxide gas. The alcohol can be stored and used as a fuel, or it can be added to other fuels to make them last longer. An example of this is gasohol, which is a mixture of gasoline and alcohol (Figure 14.15).

SECTION 2 Assessment

Recall and Comprehension

1. What is an energy resource?
2. Name three types of fossil fuels. How is each one used?
3. Identify four renewable energy resources. What makes each one renewable?
4. Describe nuclear fission and nuclear fusion.

Critical Thinking

1. **Analyzing** Why is oil used so often as an energy resource?
2. **Evaluating** Are photovoltaic cells a useful technology if only 10 percent of the sun's energy can be converted to electricity? Explain your reasoning.
3. **Comparing** What are the differences between renewable and nonrenewable energy resources?

QUICK ACTIVITY

As biomass decays, it can give off heat and flammable gas. Fill a heavy-duty lawn and leaf bag with freshly mowed grass clippings. Insert a long thermometer into the bag and tie the bag tightly closed. Let it sit for a few days, while periodically recording the internal temperature of the grass clippings. What happened? Can you explain? **For more related Design Activities, see pages 446–449.**

Technology in the Real World

Looking for Oil

Oil and gas form only where there is the right combination of rock types and structures. Scientists who search for oil look for this combination. In their search, they use sensitive tools such as magnetometers, gravimeters, and seismographs. In addition, photographs taken from satellites are used to find land formations that are likely to yield oil. Computers are used to enhance these photos to show different kinds of terrain in different colors.

Using Magnetometers and Gravimeters

Magnetometers are used to sense magnetic fields. Rocks that are near oil fields are generally not magnetic, so researchers look for low levels of magnetic energy. A gravimeter measures gravity. Rock formations that lie underground cause the gravimeter's readings to change. Geologists use this information as another tool in mapping underground rocks.

Using Seismographs

Seismographs measure vibrations in the Earth. Large vibrator trucks, as shown below, send out vibrations into the ground. The vibrations travel through the rock, bending where different layers of rock meet.

Scientists evaluate the time it takes the vibrations to travel through the rock and then bounce back to the surface. They also consider the strength of the vibrations. This information is used to make a map of the rock layers that lie deep underground.

Critical Thinking

1. **Summarizing** Why is the location of rocks important for finding oil?

2. **Describing** How are seismographs used to map layers of rock underground?

Vibrator ("elephant") trucks send seismic signals into the ground.

Motion sensors detect the waves as they bounce back to the surface.

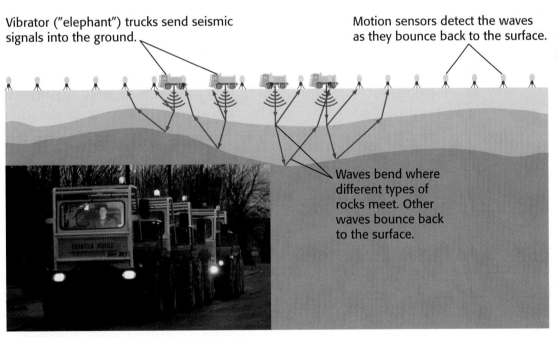

Waves bend where different types of rocks meet. Other waves bounce back to the surface.

Vibrator trucks send signals into the ground.

Power and Power Systems

Benchmarks for Learning

- Power is the rate at which work is done.
- Power systems drive other technological products and systems.
- Power systems convert energy from one form to another.

Reading Strategy

Listing Create a list to help you understand how a gasoline engine works. For each step in the process, create a bullet that describes what takes place.

How a Gasoline Engine Works
1. **Piston moves down.**
- Gas and air are pulled into the cylinder.

Vocabulary

power	gasoline engine	external combustion
internal combustion	diesel engine	electrical generator
engine	gas turbine engine	

Figure 14.16 Power is the rate at which work is done.

Inferring *As power increases, is energy used up more or less quickly?*

Power

In science, **power** is the rate at which work is done. More specifically, it is the amount of work done during a given period of time. Power measures how quickly work is done. This can be illustrated by the following equation:

$$\text{Power (P)} = \frac{\text{Work (W)}}{\text{Time (t)}}$$

The person jogging in Figure 14.16 has a certain amount of power. In Figure 14.3, we estimated that a person running for 6 minutes would use 315,000 joules. We can determine the amount of power needed for this run using the equation above.

$$P = \frac{315{,}000 \text{ J}}{(6 \text{ minutes})\,(60 \text{ seconds/minute})}$$

$$= \frac{315{,}000 \text{ J}}{360 \text{ seconds}}$$

$$= 875 \text{ watts (W)}$$

One watt of power is equal to 1 joule per second. If the person had a greater amount of power, less time would be needed to jog the same distance. Similarly, a car with a powerful engine can accelerate to a given speed faster than the same car with a less powerful engine. The same amount of work is necessary to move the car to that speed, but the power available determines how quickly the work will be done.

There are other units that measure power, such as horsepower. The term *horsepower* was originally derived from the amount of energy a horse needed to pull a load of coal. One unit of horsepower is equal to 746 watts.

The power to do work comes from energy. A machine that uses energy to do work is called a power system. An engine is a power system that converts various sources of energy into mechanical energy to do work (Figure 14.17).

Engines can do many different jobs. They supply the electricity to light homes. They power cars and planes. They carry rockets into space. There are many different kinds of engines, which use many different kinds of fuels. Two types of engines are internal combustion engines and external combustion engines.

Internal Combustion Engines

In an **internal combustion engine,** the combustion process occurs inside the engine. When the fuel is ignited, chemical energy is converted into thermal energy and then into mechanical energy. The two most common types of internal combustion engines are gasoline and diesel.

Gasoline Engines

The first successful internal combustion engine fueled by gasoline was invented in Germany by Nicolas Otto in 1876. It was smaller and more efficient than the steam engines of the time. In a **gasoline engine,** a spark ignites the gasoline, which is a highly flammable liquid derived from oil.

Cars use a gasoline internal combustion engine. In this type of engine, gasoline is burned to produce heat, and then this thermal energy is converted into mechanical energy. Because there are four steps in this process, it is called a four-stroke cycle.

- During **Step 1,** the intake stroke, the piston moves downward, creating a vacuum and pulling a mixture of gasoline and air into the cylinder through an open intake valve.

- During **Step 2,** the compression stroke, the piston moves up and compresses the air-fuel mixture. When the piston reaches the top, the spark plug fires to begin step 3.

- During **Step 3,** the power stroke, the firing spark plug ignites the air-fuel mixture. The heat from the rapid burning raises the temperature and pressure of the gas, and the high-pressure gas pushes on the piston. The gas expands, pushing the piston down.

- During **Step 4,** the exhaust stroke, the piston moves back up. The exhaust valve opens and the burned gases are forced out. Then, the cycle begins again.

The piston is connected to the crankshaft, which transforms the vertical piston motion to a rotary motion. The crankshaft is connected, via gearing and axles, to the car's wheels.

The gasoline engine has come into widespread use. It is used in nearly all cars and many trucks. Small gasoline engines are used in lawn mowers and go-carts. They are also used in portable generators, leaf blowers, snowblowers, and outboard engines in boats.

Diesel Engines

A **diesel engine** is similar to a gasoline engine, except that it burns fuel oil, which is heavier and oilier than gasoline (Figure 14.18). The fuel is injected into the cylinder at the end of the compression step. Because only air is compressed, the compression is much greater. This greater compression means that the temperature of the air is hotter than in a gasoline engine. As a result, the fuel is ignited as it is injected.

Diesel engines tend to be larger than most car engines. Although they are used in some cars, they are used mostly on heavy equipment such as tractors, earth movers, trucks, and buses, as well as on locomotives and ships. They are known for their low maintenance and long life, but they do not have the quick response of gasoline engines.

Figure 14.18 Diesel engines are used in most large trucks. Diesel engines are heavier than gasoline engines, but they are also more durable.

Contrasting *How is compression different in gasoline and diesel engines?*

For: Energy Bandit Activity
Visit: www.mytechedkit.com

Energy Bandits: Saving Our Energy Online

Energy sources can be nonrenewable or renewable. Most of the energy we use comes from nonrenewable energy sources. These sources cannot be replaced, so it is important for all of us to try to conserve them.

These Energy Bandits devour our power by gobbling up nonrenewable energy sources. Go online to find out how you can stop these thieves from taking away our power and energy.

Gas Turbine Engines

Gas turbine engines use pressurized natural gas, kerosene, or jet fuel as their energy source. Gas turbine engines are used on jet airplanes and in rockets.

Gas turbine engines are based on a principle that can be observed in a simple garden hose. Take a garden hose and turn on the water. As you hold the hose, there is very little force on your hand from the water leaving the hose. Now, put a nozzle on the end of the hose and turn on the water. You should feel much more force from the nozzle. This is the reactive force created by an increase in velocity across the nozzle.

Attaching a nozzle to a garden hose to increase the water's velocity is similar to what happens in a jet engine (Figure 14.19). A **jet engine** is simply a gas turbine engine that is attached to the wing of an airplane.

In step 1, air is compressed in the compressor, raising its pressure. In step 2, energy is added to the air by burning fuel in the combustion chamber, and the temperature and pressure of the air increase. In step 3, the high-temperature, high-pressure gas expands in the turbine, producing power. The turbine is used to drive the compressor, and the exhaust gases from the turbine go through a nozzle, increasing in

Figure 14.19 A jet engine propels the plane forward by expelling air at very high speeds.

Interpreting *What is the role of the compressor?*

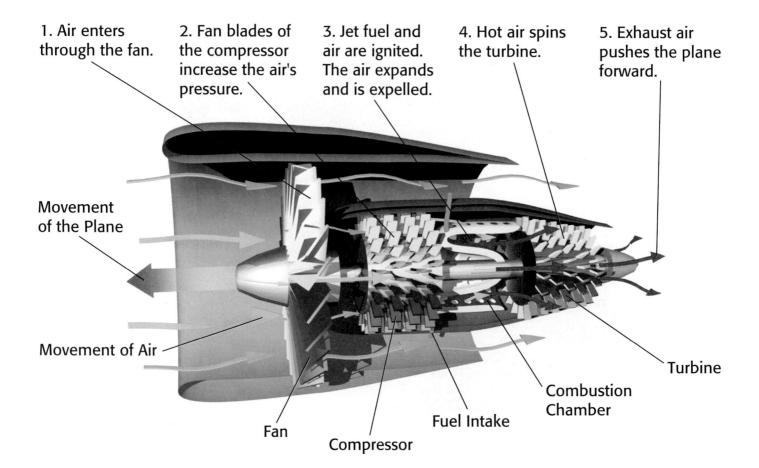

1. Air enters through the fan.

2. Fan blades of the compressor increase the air's pressure.

3. Jet fuel and air are ignited. The air expands and is expelled.

4. Hot air spins the turbine.

5. Exhaust air pushes the plane forward.

Movement of the Plane

Movement of Air

Fan

Compressor

Fuel Intake

Combustion Chamber

Turbine

velocity. This increase in velocity causes a force on the engine that propels the airplane forward.

Rocket engines work in the same way as jet engines except that they carry their own oxygen supply. This oxygen is used to burn the fuel, which is typically a form of liquid oxygen.

External Combustion

In **external combustion,** fuel is burned outside of an engine and provides heat to another liquid or gas. As with any type of combustion, fuel is ignited and chemical energy is converted into thermal energy. In external combustion, however, the thermal energy heats another liquid, which is converted into mechanical energy to produce the power.

Many power plants use steam as a source of power for external combustion (Figure 14.20). Fuel (gas, coal, or oil) is burned, and the heat causes water to boil under pressure in a tank, producing steam. The high-pressure, high-temperature steam leaves the tank and enters a turbine, causing the turbine to rotate. The spinning turbine is connected to a generator, producing electric power that is distributed along power lines. The steam leaves the turbine at a low pressure and

Figure 14.20 Power plants are used to convert different forms of energy into electricity.

Analyzing *Where in this power plant is water heated? Cooled?*

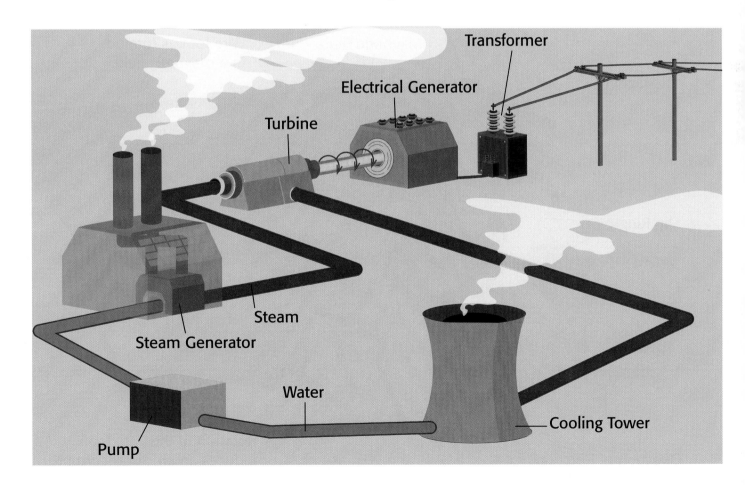

temperature and condenses back to liquid water. A pump is used to increase the pressure of the water so that it returns to the tank to be reheated.

Other energy sources besides combustion can be used to heat the water to produce steam. For instance, solar energy can be reflected with special mirrors, and nuclear energy can heat the water under pressure in a nuclear reactor.

Generating Electricity

Although steam is the most common source of mechanical energy for turning turbines, falling water from a dam is also used. In a dam, the energy from moving or falling water is used to turn underground turbines. An **electrical generator** converts the mechanical energy into electrical energy (Figure 14.21).

To understand how an electrical generator works, you first need to understand how electricity is produced. When a wire, called a conductor, moves through a magnetic field, it cuts the lines of magnetic flow. A force develops in the wire that pushes the charged particles to one end and induces an electrical current. Current can also be induced if the magnet moves around the conductor.

Figure 14.21 Electricity can be generated by rotating a wire in a magnetic field.

Applying *How can the strength of a generator be increased?*

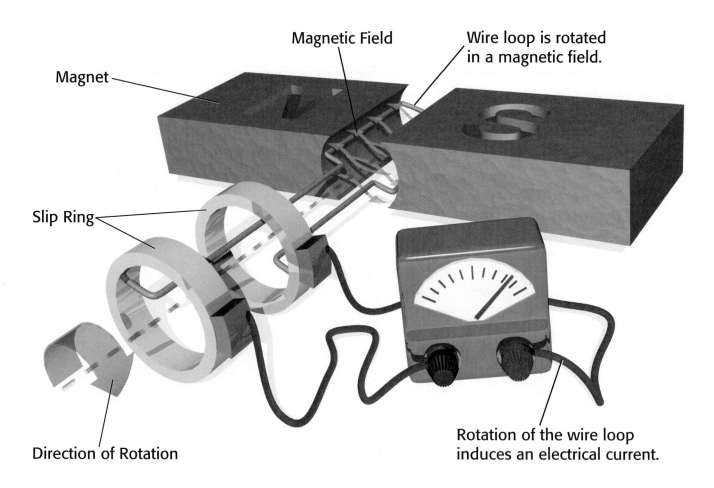

Magnetic Field

Wire loop is rotated in a magnetic field.

Magnet

Slip Ring

Direction of Rotation

Rotation of the wire loop induces an electrical current.

In an electrical generator, mechanical energy turns a turbine that spins a cylinder called a rotor, which has large magnets. The rotor spins within a stator, a stationary structure that contains coils of conducting wire. As the magnets spin within these wires, an electrical current is produced. Generators in power plants use hundreds to thousands of wires to generate high-voltage current.

Electrical Power Transmission

Electrical power is transmitted by wires that carry current to the places where it will be used (Figure 14.22). Motors and other electrical equipment are used to change the electricity into force and motion.

Wires carrying high-voltage electricity lose less energy to heat than do low-voltage lines. Therefore, high voltage is sent on wires that carry electricity over long distances. Transformers are devices that can increase or decrease the force of electrical current. They are used to change the high voltage to lower voltage near the place where the electricity will be used.

Power transmission lines from several power generating stations, or power companies, are often joined in a grid. The high-voltage wires from one power station are joined to the wires from other power stations through a series of switches. A grid is useful when demand is so high that a station or company cannot meet electricity needs. The station or company can then tap into electricity from other stations or companies to supply needed power.

Figure 14.22 Electricity is carried along high-voltage lines to reduce the loss of energy through heat.

Inferring *Why aren't high-voltage lines used directly in homes and offices?*

SECTION 3 **Assessment**

Recall and Comprehension

1. What is a gasoline engine? How does it produce power?
2. Describe the steps by which a four-stroke engine turns fuel energy into the mechanical energy that moves a car.
3. Explain how a rocket engine works.

Critical Thinking

1. **Contrasting** Explain the differences between gas turbine engines and power plants from a combustion viewpoint.
2. **Analyzing** Why is it necessary to send electric power at high voltage when it is sent over long distances?
3. **Comparing** How are gasoline and diesel engines similar?

QUICK ACTIVITY

Permanent magnet motors can easily be converted into electrical generators. Take a small permanent magnet motor and connect its terminals to a light-emitting diode (LED). Now give the motor a spin. What happened? What happens when you spin the motor faster? **For more related Design Activities, see pages 446–449.**

Chapter Summary

- Work is done only when a force moves an object. The amount of work equals the distance the object moves times the force.

- Energy is the ability to do work. It is the source of the force needed to do work. Kinetic energy is energy in motion. Potential energy is energy that is stored.

- Heat is thermal energy that is transferred from something at a higher temperature to something at a lower temperature. Thermal energy is created by the energy of the particles in a substance.

- Energy sources can be nonrenewable or renewable. Most of the energy used in the United States comes from nonrenewable energy sources, such as coal, oil, and natural gas. Renewable sources of energy include solar, wind, water, geothermal, and biomass.

- Power is the rate at which work is done. Machines that use energy sources to provide power are called power systems.

- An engine converts various forms of energy into mechanical energy to do work. Some different types of internal combustion engines are gasoline and diesel engines, gas turbines, jet engines, and rockets.

- Power plants use external combustion to convert thermal energy into mechanical energy, which generates electrical energy. Electrical power is transmitted over wires to where it will be used.

Building Vocabulary

Your teacher may give you a crossword puzzle. Complete the puzzle using the following words from this chapter. Exchange puzzles with a classmate to check each other's answers.

1. biomass
2. diesel engine
3. elastic potential energy
4. electrical generator
5. energy
6. external combustion
7. fossil fuel
8. gasoline engine
9. gasoline turbine engine
10. geothermal energy
11. gravitational potential energy
12. heat
13. internal combustion engine
14. jet engine
15. joule
16. kinetic energy
17. law of conservation of energy
18. nuclear fission
19. nuclear fusion
20. photovoltaic cell
21. potential energy
22. power
23. power system
24. radioactive waste
25. rocket engine
26. solar energy
27. thermal energy
28. turbine
29. work

 See your teacher for the Crosstech puzzle.

Reviewing Content

1. What is thermal energy?

2. How is energy stored in a fossil fuel?

3. What is power? How is it different from work?

4. Describe three different kinds of engines.

5. How do rocket and jet engines work?

Applying Your Knowledge

1. A boy pushes on a car with 100 newtons of force but does not move it. How much work has he done?

2. While playing pinball, a boy draws back the spring-loaded shooter to put a ball into play. Explain the energy changes that occur from the moment he pulls back the spring until the ball starts moving.

3. Is wood used for home heating in your area? Why or why not?

4. List all the energy resources used by your family, telling how they are used and whether each resource is renewable. Consider heating and cooling your home, running appliances and machines inside and outside your home, and your means of transportation.

5. Do research to find out how diesel fuel is different from gasoline and why it is used for diesel engines instead of gasoline.

Critical Thinking

1. **Analyzing** Will the energy generated by falling water in a hydroelectric plant be more, less, or the same as the gravitational energy stored in the water that falls (assuming there are no losses in the generator)? Explain your answer.

2. **Interpreting** Is solar energy really unlimited? Why do we call it an unlimited energy source?

3. **Inferring** Why are researchers trying so hard to find alternatives to using coal for energy?

4. **Comparing and Contrasting** How is nuclear fission different from nuclear fusion? Why is fission considered a limited energy source whereas fusion is considered an unlimited energy source?

5. **Describing** Describe how a power plant produces electricity.

 Connecting to STEM
science • engineering • technology • math

Nuclear Energy

Investigate how nuclear energy can be used to generate electrical power. Use digital resources—video, data, graphic images—to explain the process. Indicate areas of controversy and draw conclusions about the value of using nuclear power.

Design Activity 27

CATAPULTING BEANBAGS

Problem Situation

Catapults have been used as engines of war since Greek and Roman times. The catapults used then were quite large and could launch large stones, animals, and even kettles of burning material. When using a catapult, one of the challenges is knowing where the launched projectile will land. A catapult that can be adjusted to different distances would be ideal.

Materials

You will need:

- 10 sticks, each 60 cm long
- dowel rod
- drill
- foam board or thin wood
- foam cup
- glue gun
- rubber bands of various sizes
- saw
- small beanbag

Your Challenge

You and your teammates are to design and construct a catapult that will launch a beanbag.

> Go to your **Student Activity Guide, Design Activity 27.** Complete the activity in your Guide, and state the design challenge in your own words.

① Clarify the Design Specifications and Constraints

To solve the problem, your design must meet the following specifications and constraints:

- The catapult needs to be able to launch a beanbag to hit targets 1 foot in diameter at distances of 2 yds, 4 yds, and an intermediate distance (to be determined by your instructor).

> In your Guide, state the specifications and constraints. Add any others that your team or your teacher included.

② Research and Investigate

To better complete the design challenge, you need to first gather information to help you build a knowledge base.

> In your Guide, complete Knowledge and Skill Builder 1: Catapult Investigation.
>
> In your Guide, complete Knowledge and Skill Builder 2: Biomechanics of Catapults.
>
> In your Guide, complete Knowledge and Skill Builder 3: Hinges and Energy Sources.

③ Generate Alternative Designs

In your Guide, describe two possible solutions that your team has created for the problem. Your solutions should be based on the knowledge you have gathered so far.

④ Choose and Justify the Optimal Solution

Refer to your Guide. Explain why you selected the solution you did, and why it was the better choice.

⑤ Develop a Prototype

Construct your catapult. Include a drawing or a photograph of your final design in your Guide.

In any technological activity, you will use seven resources: people, capital, time, information, energy, materials, and tools and machines. In your Guide, indicate which resources were most important in this activity, and how you made trade-offs among them.

⑥ Test and Evaluate

How will you test and evaluate your design? In your Guide, describe the testing procedure you will use. Explain how the results show that the design solves the problem and meets the specifications and constraints.

⑦ Redesign the Solution

Respond to the questions in your Guide about how you would redesign your solution. Your redesign should be based on the knowledge and information that you gained during the activity.

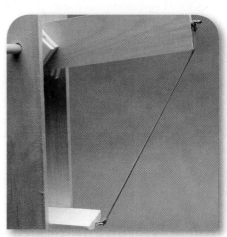

⑧ Communicate Your Achievements

In your Guide, describe the plan you will use to present your solution to your class. Show any handouts and/or PowerPoint slides you will use.

Design Activity 28
WINDMILL POWER GENERATION

Problem Situation

Windmills are becoming a popular source of electrical power. Imagine that you are working for a company that is interested in providing windmills as a power source for homeowners in remote locations. You have been asked to develop the initial design of a windmill that generates a great deal of electrical power. The voltage output from the generator will be an indication of its power production. The greater the voltage, the greater the power produced.

Materials

You will need:

- 3-volt DC motor
- box fan
- foam board
- gears or pulleys with rubber bands
- glue gun
- high-speed rotary tool
- multimeter
- plastic propeller blade
- wire with alligator clips
- wooden sticks, 60 cm long

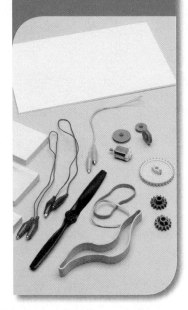

Your Challenge

You and your teammates are to design and construct a model windmill.

Go to your **Student Activity Guide, Design Activity 28.** Complete the activity in your Guide, and state the design challenge in your own words.

1 Clarify the Design Specifications and Constraints

To solve the problem, your design must meet the following specifications and constraints:
- The model windmill will generate the most electrical power possible.
- The model windmill will be tested using a box fan provided by the instructor.

In your Guide, state the specifications and constraints. Add any others that your team or your teacher included.

2 Research and Investigate

To better complete the design challenge, you need to first gather information to help you build a knowledge base.

In your Guide, complete Knowledge and Skill Builder 1: Windmills Around the World.

In your Guide, complete Knowledge and Skill Builder 2: Generator Testing.

In your Guide, complete Knowledge and Skill Builder 3: Gearing Up.

③ Generate Alternative Designs

In your Guide, describe two possible solutions that your team has created for the problem. Your solutions should be based on the knowledge you have gathered so far.

④ Choose and Justify the Optimal Solution

Refer to your Guide. Explain why you selected the solution you did, and why it was the better choice.

⑤ Develop a Prototype

Construct your model windmill. Include a drawing or a photograph of your final design in your Guide.

In any technological activity, you will use seven resources: people, capital, time, information, energy, materials, and tools and machines. In your Guide, indicate which resources were most important in this activity, and how you made trade-offs among them.

⑥ Test and Evaluate

How will you test and evaluate your design? In your Guide, describe the testing procedure you will use. Explain how the results show that the design solves the problem and meets the specifications and constraints.

⑦ Redesign the Solution

Respond to the questions in your Guide about how you would redesign your solution. Your redesign should be based on the knowledge and information that you gained during the activity.

⑧ Communicate Your Achievements

In your Guide, describe the plan you will use to present your solution to your class. Show any handouts and/or PowerPoint slides you will use.

Transportation

In this chapter, you will learn about the following different types of transportation:

- trains
- cars
- subways
- ships
- hydrofoils
- submersibles
- blimps
- planes
- space travel

Jules Verne's novel *Around the World in Eighty Days* is set in the 1800s. In his novel, Verne describes a journey around the world using the various means of transportation of the day—hot air balloon, train, steamship, and even an elephant. Today, the story of transportation is just as varied, but much faster:

- The military fighter aircraft, the F-22 Raptor, can fly at two times the speed of sound, Mach 2.0, or 1317 miles per hour.

- Astronauts can orbit the Earth in a Space Shuttle in only 90 minutes.

Faster transportation has brought us closer to people throughout the world, in turn making the world a much smaller place:

- We can enjoy fresh foods from other countries and buy goods made in other countries.

- We can sell our food and products to other countries.

- We can visit places as tourists throughout the world and learn about other cultures.

Resources for Transportation

Benchmarks for Learning

- Transportation systems are made up of subsystems.
- Transportation systems provide the means of moving people and goods from place to place.
- Government regulations influence the design and operation of transportation systems.

Reading Strategy

Listing As you read about different resources for transportation systems, list as many resources as you can. Use the headings in this section to begin your list.

Resources for Transportation
- A. People
  - 1. Motormen
  - 2. Pilots

Vocabulary

vehicle   transmission

Transportation Systems

Transportation systems have become indispensable to our modern lives. We depend on cars, buses, and trains to bring us to work and school. We similarly depend on trucks, trains, airplanes, and ships to bring us the fuel, food, clothing, and other goods we need. Over time, the transportation systems on which we depend have evolved in complexity and interdependency.

Many of today's transportation systems are based on the use of a **vehicle,** a device that carries people and objects from one place to another. Vehicles include automobiles, trucks, trains, ships, and airplanes, which move on roadways, railways, waterways, and airways.

We can think about transportation systems at several levels of detail. On a broad level, consider the highway transportation system. It consists of many subsystems, such as roadways and vehicles. Roadways include streets, roads, highways, and expressways. Vehicles used in these systems include cars, trucks, buses, motorcycles, and bicycles.

We can also focus on the more individual transportation systems, such as within a vehicle. In a car, for example, the structural system (the frame and body) contains the power system (the engine and powertrain) as well as the steering, suspension, brake, and other systems. These systems interconnect and must work together for safe and effective operation.

At both the broad and individual level, transportation systems often rely on computer technology to ensure the systems are running properly and on schedule.

Resources for Transportation Systems

Transportation systems and their subsystems provide a means of moving people and goods from one place to another. These include such systems as conveyor belts, escalators, and elevators, as well as cars, trucks, trains, ships, and planes. All transportation systems draw on the seven technological resources in many different ways.

People

Transportation systems are designed, built, operated, and used by people. Drivers operate trains, and pilots fly jets. Workers sell the tickets, make the schedules, clean and maintain the vehicles, and buy supplies.

People in government help regulate transportation systems to make sure they meet safety and environmental standards. They work in federal, state, and local agencies that regulate our country's transportation systems. People in these agencies recommend, develop, and enforce laws and policies to maintain the safety of each transportation system. They also investigate the causes of accidents so that they can find ways to prevent them in the future.

People are needed to keep transportation safe and reliable. Most vehicles must meet government safety standards before being sold or used. Designers and engineers are needed to test and improve new designs for safety (Figure 15.1). Features such as air bags and improved bumpers provide for safer automobiles. However, they also

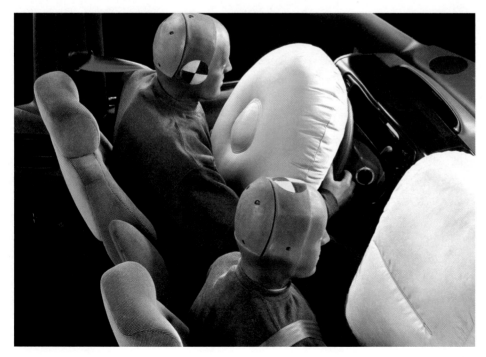

Figure 15.1 Many different tests are used to design an air bag.

Describing *Why is there a trade-off between cost and safety in most transportation systems?*

increase manufacturing costs and, therefore, the capital needed to include them. Designers, lawmakers, and public-interest groups are constantly debating the trade-offs between costs and improved safety.

Capital

A large amount of capital is needed for transportation systems. Building ports, roadways, and vehicles is costly. Maintenance also is costly, but it is especially important in systems that carry people. Money is used to repair damaged roads, or expand highways by adding lanes to lessen the impact of large amounts of traffic. Although people and companies pay for the vehicles they operate or manufacture, transportation systems also receive capital from the government. Roadways and ports are built and maintained using money collected through taxes or tolls.

Time

In a transportation system, travel time depends on distance and technology. Time can be the few seconds it takes to move parts from one station to the next on an assembly line. It can be the number of days it takes a truck to cross the continent, or the weeks it takes a ship to cross the ocean.

Mass transit systems move thousands of people on a regular basis (Figure 15.2). Time is important to the smooth operation of these systems. Vehicles or other types of transportation often move on a set schedule, ensuring that people get to their destination on time.

Figure 15.2 Moving sidewalks help people move around an airport more quickly.

Extending *What facilities other than airports might use moving sidewalks?*

Figure 15.3 Air-traffic controllers juggle a great deal of information to make sure that planes take off and land safely.

Making Judgments *Would it be wise to have air traffic control run by a computer rather than a person? Why or why not?*

Information

Many kinds of information are used to design and operate transportation systems (Figure 15.3). People who drive cars, fly planes, or sail ships need information to guide them. Information about location, course or route, speed, and vehicle operation is important for a safe and rapid trip. Many tools are used to provide such information to both drivers and those who monitor transportation routes. These include road signs, radar, two-way radios, satellite systems, and on-board computers.

Energy

Vehicles today generally use stored energy. Cars, for example, have engines that use chemical energy in the form of gasoline, which is converted to mechanical energy for motion. Oil, in the form of gasoline and diesel fuel, is one of the most efficient ways of supplying energy. It burns easily and produces a large amount of energy. Because it burns easily, however, oil must be used carefully. It can also cause extensive damage to the environment when it spills in transit. Airplanes, trucks, and ships all use oil-based fuels.

Some vehicles use electricity for power. In these vehicles, electric motors change electricity into rotary motion. The shaft of the motor is connected through gears or belts to the wheels of the vehicle. These connecting gears or belts are called the **transmission,** or drive (sometimes called the drive train). As the motor turns, the wheels turn and the vehicle moves. Subways and other mass transit systems use electric motors that get their electricity from overhead wires or an extra (third) rail.

Figure 15.4 Composite materials are used in many types of vehicles, such as this racecar.

Characterizing *What are the advantages of using composites over other materials?*

Electric motors have certain advantages. They are easy to control because they are small, and they do not pollute the area around the vehicle. They are not yet used widely, however, because they are relatively expensive and the batteries have to be recharged often. Hybrid vehicles, which use a combination of electrical and gasoline energy, are becoming more affordable and popular. Engines that convert hydrogen into electricity are another option being explored.

Materials

The materials used for vehicles, roadways, and ports depend on the transportation system. Airplanes must be made of strong, lightweight materials, such as aluminum and titanium. New composite materials—fibers mixed with epoxies—are also being used because they are stronger and lighter than metals (Figure 15.4).

Cars and trucks are primarily made of metal, usually steel. Other lightweight materials such as aluminum, plastics, fiberglass, and composites are used to reduce weight, thus requiring less fuel to move them.

Tools and Machines

A wide variety of tools and machines is used to create and maintain transportation systems. Highways are built and repaired using heavy equipment. The guidance systems in aircraft are checked using computerized electronics. Bridges are checked for strength using embedded strain gauges, and drainage systems are cleaned using special equipment.

SECTION 1 Assessment

Recall and Comprehension

1. Name four different types of subsystems found in a car.
2. Why is capital needed in transportation systems, and from what sources do the funds come?
3. How have materials improved so that stronger but lighter airplanes and cars can be built?
4. What type of energy is most commonly used for transportation?

Critical Thinking

1. **Extending** Traffic signs are an important source of information for drivers. Make a list of five highway signs you commonly see. Include a sketch of each one, and tell what information each one provides.
2. **Applying** In what ways have modern transportation systems affected the way your family lives?

QUICK ACTIVITY

The use of the global positioning system (GPS) has greatly increased the safety and efficiency of many transportation systems. Use the Internet to research how GPS works. If possible, use a moving-map GPS to investigate the area around your school or neighborhood. What problems, if any, did you encounter?
For more related Design Activities, see pages 476–479.

Land and Water Transportation

Reading Strategy

Outlining   As you read, make an outline of the basic types of land and water transportation. Use the headings in this section to get started.

I. Land
   Transportation
   A. Railroads
   B. Motor vehicles

Benchmarks for Learning

- Various transportation methods are used, depending on the environment in which they operate—in land, water, air, or space.
- Substances have characteristic qualities, such as density.

Vocabulary

maglev train   air-cushion vehicle   submarine
hydrofoil   submersible   buoyancy

Classifying Transportation Systems

The various transportation systems are alike in many ways. In most, vehicles carry people or goods from one location to another. Transportation systems are also different— especially in the routes that are used to move people and cargo. One way to classify systems is by the environment in which they move—land, water, air, or space.

Land Transportation

The earliest forms of transportation took place on land. People walked from place to place. Animals were used to drag heavy loads on sleds. Later, the wheel made it possible to pull heavier loads more quickly, and people and goods were transported by wagon or coach. Some cultures today still use animal and human muscle power, such as in bicycles and carts, for most of their transportation needs.

The Development of Railroads

It is difficult to overestimate the importance of railroads as a transportation system (Figure 15.5). From the 1800s until the early 1900s, trains provided the only means, other than animal-drawn wagons, of moving people and products long distances on land.

The first railroad to use steam engines was opened in England in 1830. The engine could pull the train at speeds up to 30 miles per hour. Railroad systems quickly came into widespread use to transport people and goods, or cargo. In 1869, the first transcontinental railroad connected the eastern and western parts of the United States.

Figure 15.5   The first transcontinental railroad contributed to the development of the American West.

Inferring   *What effect did the invention of the car have on the importance of trains?*

After the Civil War, several inventions made railroads much safer. One was the air brake, which allowed all cars on the train to be stopped at the same time. This invention prevented traincars from hitting each other during an emergency stop. In 1893, a law was passed that required the use of air brakes on all trains. After World War II, U.S. railroads quickly changed over from steam engines to diesel engines. Railroad companies preferred the better mileage, cleaner burning fuel, and lower maintenance of diesel engines.

Railroads Today

Today, high-speed rails provide an attractive alternative to air travel. High-speed trains in Europe, China, and Japan routinely operate at nearly 200 miles per hour, providing much faster and safer transportation than provided by cars. The fastest high-speed rail travel was 469 mph by the TGV (train à grande vitesse) in 2010. The U.S. federal government and some states are upgrading tracks and infrastructure to develop high-speed rail corridors. The forces created by high-speed trains require that the tracks and the bed supporting the tracks be redesigned to withstand the higher forces and that roadway crossings be eliminated, so trains do not need to slow down.

Maglev trains float above the tracks using magnetic levitation (Figure 15.6). They can travel even faster than high-speed trains—up to 300 miles per hour. These trains are lifted, propelled, and guided by magnetic forces, so there is no friction between the railcar and the track. Maglev trains are still in development, particularly in Germany, Japan, and the United States. Because of their cost, governmental support is required to develop the maglev trains.

Figure 15.6 Maglev (magnetic levitation) trains are propelled by large magnetic forces on the track.

Drawing Conclusions *If maglev trains are expensive to develop and build, why is this technology used?*

Motor Vehicles

The development of motor vehicles in the twentieth century dramatically changed the way people live. Today, 500 million cars are in use globally. The automotive industry is the largest one in the United States. Many people who work in a city use cars to commute to and from the city's suburbs and within the city each day. Some people travel more than four hours a day. Car pools, in which several people ride together and take turns driving, save fuel and wear and tear on cars and help reduce pollution and congestion.

Trucks are another important part of the transportation industry. They deliver parts to manufacturers and products to stores. They use refrigerated trailers to transport fresh foods, and tankers to carry gasoline and diesel fuel to service stations.

Mass Transit

Many cities have mass transit systems designed to transport thousands of people at a time. These systems improve travel for people who commute to and from work (Figure 15.7). They also help control traffic congestion. Today, most city subways, trolleys, and elevated trains are powered by electricity using a third rail. Plans are being developed to use maglev trains for fast travel to airports in urban areas. This would not only help lessen traffic but also streamline arrivals and departures from the airports.

Figure 15.7 Mass transit systems are important for getting people to and from work.

Summarizing *What are the advantages of mass transit over driving? The disadvantages?*

Water Transportation

People have always traveled on water. Large cities were often located on natural harbors because water made travel and trade easier. Throughout history, different forms of energy have been used to move vehicles along rivers and across oceans. At first, people used muscle power to move small boats with poles, paddles, and oars. They soon learned how to use energy from the wind. The early Egyptians, Phoenicians, and Romans had ships with dozens of rowers as well as sails. With sails, people could travel farther, and their transportation system increased, which created new trade routes, and people to trade with. Much later, with the development of the steam engine, ships came into use.

Ships

The *Clermont*, which carried people and cargo between New York City and Albany, New York, was the first successful steam-powered boat. Its engine pushed a paddle wheel in the water, which moved the boat forward.

Today, ships are used to carry most intercontinental freight (Figure 15.8). Ships can carry large, heavy cargoes more cheaply than planes can. Tankers carry crude oil, and freighters carry everything from automobiles to bananas. Because ships can carry such tremendous loads, extra care must be used to avoid spilling any of the load into the sea, where it might damage the environment. Countries work together to develop international laws to regulate ships.

Figure 15.8 Ships are used to carry much of the world's freight.

Summarizing *What factors need to be considered when deciding how to send a load of cargo?*

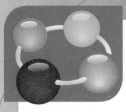

Connecting to STEM

What Makes a Boat Float?

Early boats were made of wood. Because wood is lighter than water, it floats. If you were to weigh a cubic inch of wood, it would weigh less than a cubic inch of water. **Density** tells us how much mass is in a given volume by the following formula:

$$\text{Density} = \frac{\text{Mass}}{\text{Volume}}$$

An object that is less dense than water will float on water, and an object that is more dense than water will sink.

Making Steel a Little Lighter

Today, boats are made of many different materials, including steel and fiberglass. Many of these materials are more dense than water. A solid piece of most of these materials would sink in

water, but boats made from them float. Why? Because boats use the density of air to float.

Consider a boat made of steel. Steel is about 8 times more dense than water. Because it is denser, a solid cube of steel placed in water sinks.

As shown in the illustration, the weight is greater than the buoyant force. Air is about 850 times less dense than water. This difference in densities is what helps keep boats afloat. By adding air to a steel boat, the overall density of the boat is decreased. The steel and the air together are lighter than the water. As a result, the boat floats.

Adding Cargo

The density of a boat is found by adding the mass of the steel and the mass of the air and then dividing by the boat's volume. If cargo is added, the mass of the ship is increased, but not by enough to make the boat more dense than water. International regulations indicate how much cargo a ship can carry safely. These regulations are based on how much of the ship is allowed to be underwater.

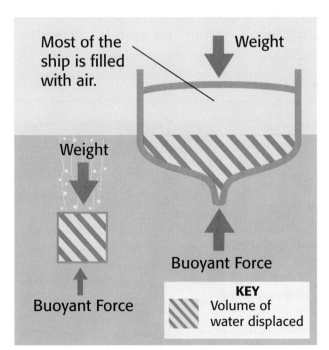

A steel cube sinks when placed in water. A steel ship with the same weight floats.

Critical Thinking

1. **Inferring** Based on the principles of density, why is it critical that boats do not fill with water while at sea?

2. **Hypothesizing** Very salty water is more dense than fresh water. Based on what you know about density, how might swimming be different in saltwater than in freshwater?

Hydrofoils and Air-Cushion Vehicles

A boat with a fairly flat bottom rises up in the water as it goes faster. This is the principle used in a hydrofoil (Figure 15.9). A **hydrofoil** is a flattened fin that is attached to the bottom of a boat to lift it above the water. Once a boat with hydrofoils is moving fast enough, it rises above the surface of the water, where there is little resistance. An **air-cushion vehicle** (ACV) is a boat with large fans that push air underneath to lift it on a cushion of air above the water. This type of boat travels fast and does not roll with the sea.

Figure 15.9 Because of its shape, a hydrofoil skims along the surface of the water.

Comparing *How are a hydrofoil and a maglev train similar?*

Submersibles and Submarines

Most ships travel over the water's surface. Some, however, can also operate below the surface. A **submersible** is a small vehicle that can stay underwater for short periods of time. Submersibles are usually tethered to a surface ship or platform. They are used for deep-ocean exploration and research. A **submarine** is a large type of submersible that can operate above or below the surface of the water. Submarines are used primarily by a government as part of its navy.

Submersibles and submarines can operate either on or below the water's surface. They do this by changing their buoyancy. **Buoyancy** is the tendency of an object to float when it is submerged in a fluid. Special tanks fill with water, causing the ships to sink. Then, the water is removed and the tanks fill with air, causing the ships to rise. These ships can float at any depth by changing the amount of water in their tanks.

SECTION 2 Assessment

Recall and Comprehension

1. Which two new types of trains allow for faster travel by rail?
2. Why are trucks important to the transportation system?
3. Before the steam engine was invented, what forms of energy did boats use?
4. How does a hydrofoil work?

Critical Thinking

1. **Drawing Conclusions** Why were steam engines replaced by diesel and electric engines in trains? What advantages did diesel and electric engines provide?
2. **Describing** Describe how submarines and submersibles are able to sink and rise.

QUICK ACTIVITY

Go online and test the Ultralight virtual activity. Examine what is happening to the flaps on the wing and the tail when you have the ultralight turn, when you have it increase in altitude, or decrease in altitude. The digital simulation should correctly indicate how the aircraft would behave. Did it?
For more related Design Activities, see pages 476–479.

For: Flight Activity
Visit: www.mytechedkit.com

Ultralights: Flight Control Online

Advances in construction and materials technology have enabled designers to create single-seat aircraft. These allow people to experience the feeling of flight from a bird's-eye perspective. Go online to learn more about one such aircraft—the ultralight.

Air and Space Transportation

Benchmark for Learning

- Transportation services and methods have led to a population that is on the move.

Reading Strategy

Listing Make a list of all the different types of air transportation that are available today. Next to each term, make a sketch or write a quick definition.

Vocabulary

lift   weight   thrust
blimp   drag

Air Transportation

People have always dreamed of flying. In Greek mythology, Icarus flies too near the sun and the wax holding his wings melts. The Italian painter and scientist Leonardo da Vinci conceived of helicopters and flying machines in the fifteenth century (Figure 15.10). Today, air transportation includes vehicles as different as hot-air balloons and jets.

Lighter-Than-Air Vehicles

Recall that buoyancy is the tendency of an object to float when submerged in a fluid. The fluid can be water or air. Objects that float in air have **lift,** an upward force equal to the weight of the air displaced by the object.

For an object to float in air, it must weigh less than the air that it has displaced, just as an object that floats in water must weigh less than the water displaced. This occurs when a lightweight container is filled with a gas (hot air, hydrogen, or helium) that is lighter than air. The two substances together weigh less than the air that they displace. Lighter-than-air (LTA) vehicles use passive lift. They float in the air because of their volume and weight.

Hot-Air Balloons

Air travel began in 1783 when two Frenchmen built a hot-air balloon. The Montgolfier brothers sent a sheep, a duck, and a chicken on an 8-minute flight over France. They powered the balloon—heating its air to make it less dense—by burning a mixture of straw and manure. Today, hot-air balloons are powered by propane gas.

Figure 15.10 In 1483, Leonardo da Vinci sketched out an idea for a helicopter.

Hypothesizing *Why do you think a working helicopter was not built until the twentieth century?*

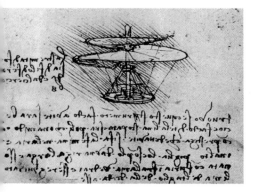

Blimps

Another type of LTA vehicle, a **blimp,** is filled with helium and is propelled forward by an engine. Helium is heavier than hydrogen but much lighter than air, and it does not burn. Blimps are used for advertising, for some kinds of cargo lifting, and as platforms for cameras. Often, we see them used at sporting events such as football games.

Active-Lift Aircraft

Active-lift vehicles create lift by their movement through the air. They must have power to fly. The first powered flight was made by Orville Wright on December 17, 1903. Orville and his brother Wilbur had been experimenting with gliders, which are unpowered planes. To build their airplane, they added to a glider a 12-horsepower engine driving two propellers.

Early airplanes were made of wood and cloth. They had two or three sets of wings to increase lift. Heavier and stronger materials were used when engines became more powerful. The number of sets of wings was reduced to one to reduce air friction. Air passenger service started in the United States in 1914.

Flight Technology Improves

World War II brought many advances in airplane design and manufacture. Airplanes were mass-produced, and thus they became cheaper and more consistently designed. Airframe design and electronics were improved. An important advance was the development of the jet engine. Jet engines were used on military planes right after World War II, but they were not used on passenger planes until 1952.

The early passenger jet airplanes had structural problems that caused crashes. The jet airplanes of today have an excellent safety record for many reasons. Modern jet engines require less maintenance than internal combustion engines. Flight crews are trained extensively. Modern air traffic control systems keep airplanes from flying too close to each other in the sky and direct them to a safe landing. These government-run control systems are an important part of the air transportation system.

Faster Than the Speed of Sound

The first person to fly faster than the speed of sound was Chuck Yeager. In 1947, he flew an X-1 experimental jet to more than the speed of sound (about 700 miles per hour).

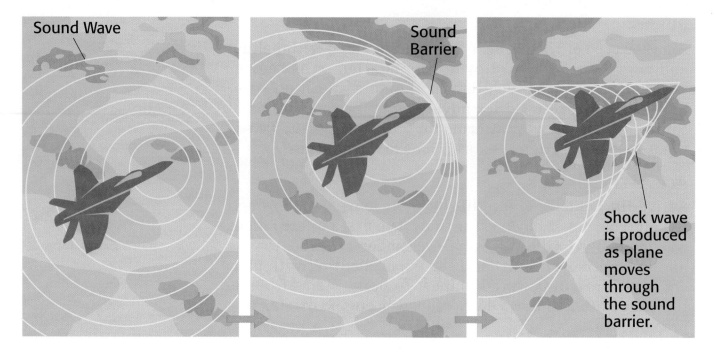

Sound Wave

Sound Barrier

Shock wave is produced as plane moves through the sound barrier.

Figure 15.11 As a plane approaches the speed of sound, sound waves pile up and form a sound barrier.

Describing *What is a sound barrier? How is it related to individual sound waves?*

When a plane approaches the speed of sound, the air it pushes ahead of it forms a shock wave called the sound barrier (Figure 15.11). The shock wave makes the airplane hard to handle. Once it exceeds the speed of sound, the airplane is easier to control. A supersonic transport (SST) flies at twice the speed of sound.

Air Travel Today

Many different kinds of airplanes are used today (Figure 15.12). Small private planes carry two to six people over short distances. Airliners that can land and take off on short runways handle commuter traffic. Jumbo jets carry hundreds of people at a time over long distances. Military aircraft carry large cargoes and refuel in flight. They can travel anywhere without having to land to refuel. Small jets that travel at twice the speed of sound protect our borders.

Modern jet airplanes have increased world travel dramatically. Railroads had allowed travel across countries and continents, but travel on ships to countries across oceans was slow. The jet plane carries people quickly to any place in the world for vacation or business. Air travel takes many different forms today. Each represents a technological solution to the constraints of moving against gravity and wind.

- **Helicopter**—This aircraft can lift off vertically from the ground and land in remote and inaccessible places. They are used for short distance transportation.

- **Airliner**—Jet engines provide the necessary power to lift tons of cargo or passengers.

Figure 15.12   Figure 15.13

- **Light aircraft**—These aircraft have short wings that produce good lift and low amounts of drag. Propellers are used to power the aircraft.

- **Glider**—A glider has no source of power and instead uses wind currents and temperature changes to move through the air.

Space Transportation

The world entered the Space Age in 1957 when the Soviet Union launched *Sputnik,* a satellite that circled Earth once every 90 minutes. Yuri Gagarin of the Soviet Union made the first manned flight in 1961. Two American astronauts followed him later that year.

During the 1960s, both the United States and the Soviet Union made more space flights. In July 1969, Americans Neil Armstrong and Edwin Aldrin became the first people to set foot on the moon. The United States placed its first space station into orbit in 1973.

How Space Travel Works

Because space is airless, space vehicles must carry the oxygen that is needed to burn fuel. So far, all space vehicles have been launched using rocket engines (Figure 15.13). The engines use both liquid and solid fuels.

To attain orbit, a rocket must reach a speed of more than 17,000 miles per hour. An additional thrust is needed to accelerate the vehicle to 25,000 miles per hour. This minimum speed is required to prevent the rocket from being pulled back to Earth because of gravity.

The Space Shuttle is the first reusable space vehicle. It is used as a space truck, carrying objects and people back and

Figure 15.12 Every day, jets are used to transport people and cargo over great distances.

Hypothesizing *How do you think air travel has affected where people live?*

Figure 15.13 Rockets and other space vehicles must carry their own oxygen to burn fuel.

Extending *What new sources of energy might be explored for space travel?*

LIVING GREEN
Earth-Friendly Travel

Over the last century, the technological advances in transportation have allowed humans the increased ability to travel across land, air, and water masses with great ease. As a result, travel for pleasure (tourism) has also grown dramatically. Because all travel by vehicle, whether car, train, plane, or bus, requires the combustion of fuel, all travel has an effect on the environment. There are many ways that vacationers can take the environment into consideration when planning travel, and this new industry is most often referred to as ecotourism, geotourism, or sustainable travel.

MAKE A CHANGE

Whether you're travelling across the state or across the ocean, there are many ways you can reduce the impact your travel has on the environment. If travelling by airplane, try to find direct flights. Take off and landing requires the most amount of fuel in a flight, so one that has stopovers and connections will burn more fuel, and create more carbon dioxide gases in the environment. If your trip doesn't require a flight, think about using mass transportation. Buses and trains carry many more people than a car, so the impact on the environment is less. If you have to travel by car, make sure it's in good working condition. Dirty air filters and tires with low air pressure both force the car to use more gas. Be sure there are no oil, antifreeze, or other fluid leaks, avoiding putting toxic materials directly into soils and water supplies.

TRY THIS

Choose your dream vacation destination and use the Internet to plan the most eco-friendly trip you can. Start with your transportation, and then expand your plans to choosing hotels that practice environmentally friendly energy use and water conservation. You can even choose a working vacation with environmental groups that take on challenges such as cleaning up beaches, protecting endangered wildlife, and monitoring coral reefs.

forth from Earth to low Earth orbit. The Space Shuttle uses three engines and two solid-fuel rocket boosters to reach escape velocity.

Challenges of Space Travel

Space travel poses many unique problems. In space, there is no air. Space travelers must carry their atmosphere with them. Once away from Earth's gravity, space travelers also

Technology in the Real World

How Airplanes Fly

When an airplane is in flight, four forces act on it: weight, drag, lift, and thrust. **Weight,** which is caused by gravity, is the force pulling the plane toward the Earth. Acting alone, it would pull the plane down to the ground. **Drag** is the wind resistance. Drag tends to hold the plane back as it moves forward.

Lift and thrust move a plane up and forward. **Lift** is the upward force that must be created to make the airplane fly. Lift is created in part by the shape of the airplane and its wings. **Thrust** is the forward force produced by the engines that move the airplane. An increase in any of these variables, shown below, causes the plane to move in that direction.

What's in the Shape of a Wing?

Airplane engines move airplanes forward at speeds from 100 to more than 1,000 miles per hour. The wings are shaped so that air rushing over them produces a higher pressure on the bottom side than on the top.

Bernoulli's principle describes this effect. It states that as air flows over a surface, the pressure of the air on that surface decreases as air speed increases. The curve at the top of the wing makes the air move faster over the top of the wing than under the bottom. Lower pressure is created above the wing. Higher pressure underneath the wing results in an upward force on the wing, lifting the plane.

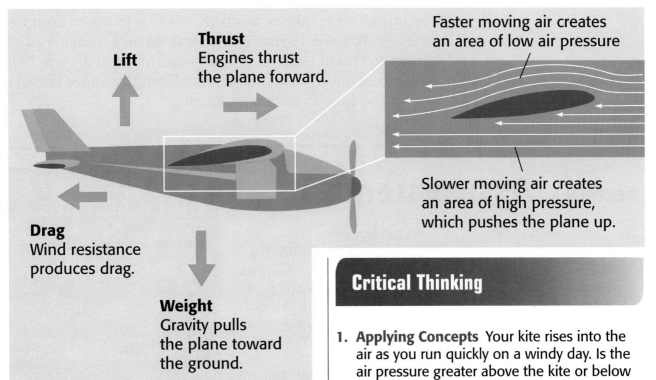

Thrust Engines thrust the plane forward.

Lift

Drag Wind resistance produces drag.

Weight Gravity pulls the plane toward the ground.

Faster moving air creates an area of low air pressure

Slower moving air creates an area of high pressure, which pushes the plane up.

The four forces acting on a plane in flight are weight, drag, lift, and thrust.

Critical Thinking

1. **Applying Concepts** Your kite rises into the air as you run quickly on a windy day. Is the air pressure greater above the kite or below it? Explain your answer.

2. **Relating Cause and Effect** You are riding in a car on a highway when a large truck speeds past. Explain why your car is forced toward the truck.

Figure 15.14 The lack of gravity is one of the biggest challenges of traveling in space.

Applying *Which aspects of your morning routine would be difficult without gravity?*

experience weightlessness. This can be fun and exciting, but it also presents problems, such as how to take a shower or drink a liquid (Figure 15.14).

Because of the great distances to be traveled, space exploration often requires months or even years to complete a journey. It took *Voyager*, an exploratory space vehicle, 12 years to travel from Earth to Neptune. Research is being conducted to determine the effects of lengthy space travel on humans and how to deal with them.

SECTION 3 **Assessment**

Recall and Comprehension

1. Name two kinds of lighter-than-air vehicles, and explain what keeps them afloat.
2. How did airplanes change during and after World War II?
3. What were the two major events in space travel in the 1960s?
4. What are some of the challenges of space travel?

Critical Thinking

1. **Contrasting** Explain the difference in the way that passive-lift and active-lift vehicles travel in the air.
2. **Clarifying** Why do rocket engines carry oxygen as well as fuel?
3. **Comparing** In an airplane, how is weight similar to drag? How is lift similar to thrust?

QUICK ACTIVITY

Many banks use a vacuum system to send small objects from one location to another. Cylindrical cartridges carrying the objects are pulled through a pipe by air pressure. Using a shop vacuum and a hose, devise a similar system to send your own message through the hose. Check with your teacher before using the device, and remember to always wear your safety glasses.
For more related Design Activities, see pages 476–479.

Intermodal and Nonvehicle Transportation

Reading Strategy

Listing As you read this section, list the different types of transportation systems that are described. Include brief definitions to help you remember the material.

Vocabulary

intermodal transportation system

pipeline

conveyor belt

Benchmarks for Learning

- Intermodalism is the use of different methods of transportation in an interconnected system of moving goods and people.

- Transporting people and goods involves a combination of individuals and vehicles.

- Processes such as loading, moving, unloading, and evaluating are necessary for a transportation system to operate efficiently.

Intermodal Transportation Systems

Many different methods of transportation are used to move people and goods. Sometimes, several modes of transportation are combined to ship goods across the country and around the world. When different systems are used in combination, it is called an intermodal system.

In an **intermodal transportation system,** freight is hauled by a combination of trucks, trains, and ships. Freight is loaded into a special container at a factory and does not have to be reloaded for the remainder of the journey. The container is a semitrailer that attaches to a tractor truck to travel by road or is loaded on a flatcar to travel by railroad. The semitrailer may be transported to a seaport, where it is loaded on a container ship to travel to another port. From there, it may be loaded on a train or moved by truck to its destination. The freight itself stays within the container and does not have to be loaded and unloaded.

With an intermodal system, there is less damage and loss than in regular shipping. An intermodal system is so reliable that it is often used to supply parts for "just-in-time" manufacturing. Stores can order goods, such as jackets or shirts, from a factory in China. The goods are produced, packaged, and labeled with prices before being shipped. The container might move from train, to truck, to ship, and back to train and truck before it reaches the store.

Nonvehicle Transportation Systems

Not all transportation takes place on roadways, railways, waterways, and airways. Some types of transportation take place within factories, warehouses, stores, skyscrapers, and airports. Others take place in oil fields and coal mines. Nonvehicle transportation systems that move materials and people include pipelines and conveyor belts.

Pipelines

A **pipeline** is a system that is used to transport crude oil or natural gas. It extends from the field where these resources are pumped to a refinery or to trucks or ships for further transport.

The Trans-Alaska pipeline moves crude oil from northern Alaska to tankers at Valdez, a port on Alaska's southern coast (Figure 15.15). The pipeline is heated and insulated. This feature keeps the oil liquid enough so that pumps can move it even in cold weather. Great care must be taken when designing and maintaining pipelines so that they do not leak any oil or gas.

Conveyor Belts

On most assembly lines, individual parts are moved from workstation to workstation by a **conveyor belt.** Most cars are assembled using this system. Large metal parts are cut, drilled, and machined at stations along conveyor belts. Finished products also may be moved on conveyor belts for

Figure 15.15 The Trans-Alaska pipeline transports oil across hundreds of miles.

Extending *What other methods are used to transport large amounts of oil?*

packing and shipping. Small objects can be moved along a guide path, with their motion driven by small vibrations. Coal and electronic parts are often moved this way.

People Movers

Some transportation systems move people over short distances. These people movers include escalators, elevators, and personal rapid transit systems (PRTs).

Freight elevators had been used for some time before 1852, when Elisha Otis invented a device to prevent them from accidentally falling. Otis's invention and other advances helped make high-rise buildings possible. Without elevators to transport people many stories high, skyscrapers would be impractical. Imagine walking all those flights of stairs! Today, most modern cities have a central cluster of skyscrapers, due in part to the technology of elevators.

PRTs move people horizontally rather than vertically. They generally use small cars similar to the cars of a train. They move along tracks from one part of an airport to another, or from one part of a city to another. Most PRTs are controlled automatically.

All types of people movers exist to help people get from place to place more easily or quickly. People can move a little more quickly on these devices than they can by simply walking or climbing stairs. People movers are often used where people are likely to be carrying packages or luggage, such as in department stores or airports. This is a good example of how technology has made our lives easier.

SECTION 4 Assessment

Recall and Comprehension

1. Describe how intermodal transportation works and what its advantages are.
2. What part do pipelines play in a transportation system?
3. How might small objects be moved by a conveyor belt?
4. Name three types of people movers.

Critical Thinking

1. **Extending** Describe how both intermodal and conveyer methods of transportation are used in the manufacture of a car.
2. **Comparing** Compare conveyor belts and moving sidewalks.

QUICK ACTIVITY

Work in groups to design an intermodal transportation system for moving oranges from rural Florida to a supermarket in San Francisco. Describe every type of transportation that will be needed, such as trucks, trains, or air cargo, and then decide on the best sequence for moving the cargo. You may want to research the relative cost and speed of different types of transportation. **For more related Design Activities, see pages 476–479.**

CHAPTER

15 Review and Assessment

Chapter Summary

- Most transportation systems use vehicles to carry people or cargo from one place to another. All transportation systems use the seven technological resources, but each one uses them in different ways.

- American railroads opened the West to settlement. Railroads today transport freight throughout the country as well as people, primarily in mass transit systems in urban areas.

- Cars have become an essential means of transportation for many Americans. Trucks are important for transporting freight.

- Ships transport freight internationally. Submarines are used by the military, and submersibles are used for exploration and research.

- The first powered flight in 1903 led the way toward commercial air transportation. Passenger air service began in 1914, and jet engines were first used in passenger planes in 1952. Jet engines made air travel safer and faster.

- The Space Age began in 1957. Today, the Space Shuttle is used as a space truck, carrying objects and people back and forth from Earth to low Earth orbit. Space transportation systems must solve problems of great distance, weightlessness, and airlessness.

- Intermodal transportation systems combine different forms of transportation. Freight is packed into containers that are moved by trucks, trains, and ships.

- Other modes of transportation include pipelines and conveyor belts to transport goods, as well as elevators, escalators, and moving sidewalks to transport people.

Building Vocabulary

Your teacher may give you a crossword puzzle. Complete the puzzle using the following words from this chapter. Exchange puzzles with a classmate to check each other's answers.

1. **air-cushion vehicle**
2. **blimp**
3. **conveyor belt**
4. **hydrofoil**
5. **intermodal transportation system**
6. **lift**
7. **Mach 1**
8. **maglev**
9. **pipeline**
10. **submarine**
11. **submersible**
12. **transmission**
13. **vehicle**

See your teacher for the Crosstech puzzle.

Reviewing Content

1. What is a vehicle in a transportation system? Give three examples other than a car.

2. How does the government regulate transportation systems, and why?

3. What vehicles are used to explore and travel underwater?

4. Explain the importance of traffic controllers to the air transportation system.

5. Name two kinds of transportation systems that do not use vehicles. Describe how they work.

Applying Your Knowledge

1. Make a map of the various roads you travel to get to school, and explain how your route is a small part of a highway transportation system.

2. Research the importance of railroads in opening the American West to settlement. Describe their role in moving people, goods, and mail.

3. Access the Internet to research the current state of the orbiting space station. Include a sketch or a photo that you have downloaded.

4. Investigate the Trans-Alaska pipeline. Find out what environmental concerns were raised in constructing and operating the pipeline and how they were addressed.

5. Describe the possible modes of transportation that a letter mailed to a pen pal in a foreign country might take. Assume that you deposit the letter in a mailbox on a local street, and include transportation within the post office as well as between locations.

Critical Thinking

1. **Hypothesizing** How have advances in transportation systems made countries interdependent?

2. **Evaluating** Is a car a necessity or a luxury in your family? Why?

3. **Inferring** What kind of human-powered transportation is widely used for sport or leisure travel in this country but is used as basic transportation in other countries? Explain your answer.

4. **Extending** The government is responsible for establishing safety regulations for transportation systems. Who else has responsibility for developing and maintaining safety standards? Explain your answer.

5. **Analyzing** Why is electricity a better power source for mass transit than for personal transportation?

 Connecting to STEM
science · technology · engineering · math

"Sea" of Air

Similar to living at the bottom of the ocean, we live at the bottom of an ocean of air. Do research to find out the weight of air compared with the weight of water. How different are they? How deep is the ocean of air in which we live?

Design Activity 29
MAGLEV

Problem Situation

Magnetically levitated (maglev) trains travel at speeds of up to 300 miles per hour. They achieve this speed in part because they do not touch the rails. They float suspended above the rails by the repulsive force between magnets with the same poles. The trains are pulled by magnets as well. One goal for a transportation company is to use maglev trains to move as many people and goods as quickly as possible for the least cost.

Your Challenge

You and your teammates are to design and construct a maglev vehicle that will transport as many pennies (the load) as quickly as possible using the fewest possible magnets (cost).

> Go to your **Student Activity Guide, Design Activity 29.** Complete the activity in your Guide, and state the design challenge in your own words.

① Clarify the Design Specifications and Constraints

To solve the problem, your design must meet the following specifications and constraints:
- The maglev vehicle must be less than 12" long.
- The maglev vehicle must be 2.5" wide.
- The maglev vehicle must transport as many pennies as quickly as possible using the fewest possible magnets.

> In your Guide, state the specifications and constraints. Add any others that your team or your teacher included.

② Research and Investigate

To better complete the design challenge, you need to first gather information to help you build a knowledge base.

> In your Guide, complete Knowledge and Skill Builder 1: Maglev Trains.

> In your Guide, complete Knowledge and Skill Builder 2: Magnetic Properties.

Materials

You will need:
- foam board
- glue gun
- pennies
- piece of plywood
- rectangular ceramic magnets with center hole
- ruler
- small plastic or paper cup
- string
- thin plastic sheeting
- wooden dowels
- wooden strips, 1–2" wide

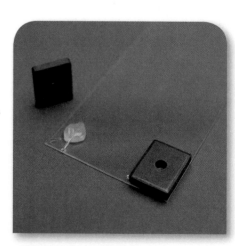

In your Guide, complete Knowledge and Skill Builder 3: Testing Out Your Track.

③ Generate Alternative Designs

In your Guide, describe two possible solutions that your team has created for the problem. You will want to consider where the sails are located, how many magnets you plan on using, their location, as well as where and how the load will be placed on the vehicle.

④ Choose and Justify the Optimal Solution

Refer to your Guide. Explain why you selected the solution you did, and why it was the better choice.

⑤ Develop a Prototype

Construct your maglev vehicle. Include a photograph of the vehicle or a drawing of your final design showing the location of the magnets, sail, and load in your Guide.

In any technological activity, you will use seven resources: people, capital, time, information, energy, materials, and tools and machines. In your Guide, indicate which resources were most important in this activity, and how you made trade-offs among them.

⑥ Test and Evaluate

How will you test and evaluate your design? In your Guide, describe the testing procedure you will use. Explain how the results show that the design solves the problem and meets the specifications and constraints.

⑦ Redesign the Solution

Respond to the questions in your Guide about how you would redesign your solution. The redesign should be based on the knowledge and information that you gained during the activity.

⑧ Communicate Your Achievements

In your Guide, describe the plan you will use to present your solution to your class. Show any handouts and/or PowerPoint slides you will use.

Design Activity 30
REDESIGNING A MOUSETRAP CAR

Problem Situation

Many products exist that can be improved. Indeed, technological innovation is often the result of improving on the works of others. Automotive engineers have carefully improved such areas as engine design and aerodynamics to achieve better performance and gas mileage. In this problem situation, the power source for a small toy car is a mousetrap. It is a simple idea with lots of room for improvement.

Materials

You will need:

- epoxy glue
- mousetrap
- plastic straws
- rubber bands
- small toy car
- string
- thin plywood, CDs, or container covers for making wheels
- wire coat hanger or wire rod

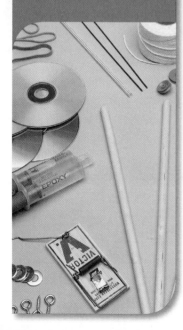

Your Challenge

You and your teammates are to redesign a mousetrap car so that it will travel the greatest distance possible.

> Go to your **Student Activity Guide, Design Activity 30.** Complete the activity in your Guide, and state the design challenge in your own words.

① Clarify the Design Specifications and Constraints

To solve the problem, your design must meet the following specifications and constraints:

- The mousetrap car must travel the greatest distance possible.
- The distance is to be measured in a straight line from start to finish. (If the car goes in a circle and ends up where it started, the distance traveled is zero feet.)
- The mousetrap must have a release mechanism to hold the arm in place before starting.
- The lever arm on the mousetrap can be no longer than 10".

> In your Guide, state the specifications and constraints. Add any others that your team or your teacher included.

② Research and Investigate

To better complete the design challenge, you need to first gather information to help you build a knowledge base.

In your Guide, complete the following Knowledge and Skill Builders:

1: Force and Distance
2: Friction
3: Big Wheels

③ Generate Alternative Designs

In your Guide, describe two possible solutions that your team has created for the problem. You will want to consider the size of the wheels, the length of the lever arm, additional weights, additional length, and frictional effects.

④ Choose and Justify the Optimal Solution

Refer to your Guide. Explain why you selected the solution you did, and why it was the better choice.

⑤ Develop a Prototype

Construct your redesigned mousetrap car. Include a scale drawing of your final design that shows the size and location of the wheels, arm, mousetrap, and body. You may also include a photograph of the car.

In your Guide, indicate which technological resources were most important in this activity, and how you made trade-offs among them.

⑥ Test and Evaluate

How will you test and evaluate your design? In your Guide, describe the testing procedure you will use. Explain how the results show that the design solves the problem and meets the specifications and constraints.

⑦ Redesign the Solution

Respond to the questions in your Guide about how you would redesign your solution. Your redesign should be based on the knowledge and information that you gained during the activity.

⑧ Communicate Your Achievements

In your Guide, describe the plan you will use to present your solution to the class. Show any handouts and/or PowerPoint slides you will use.

UNIT 6

Biotechnical and Chemical Technology

Unit Outline

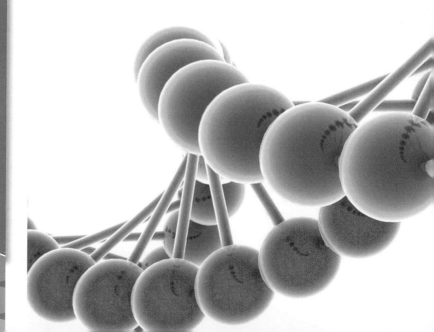

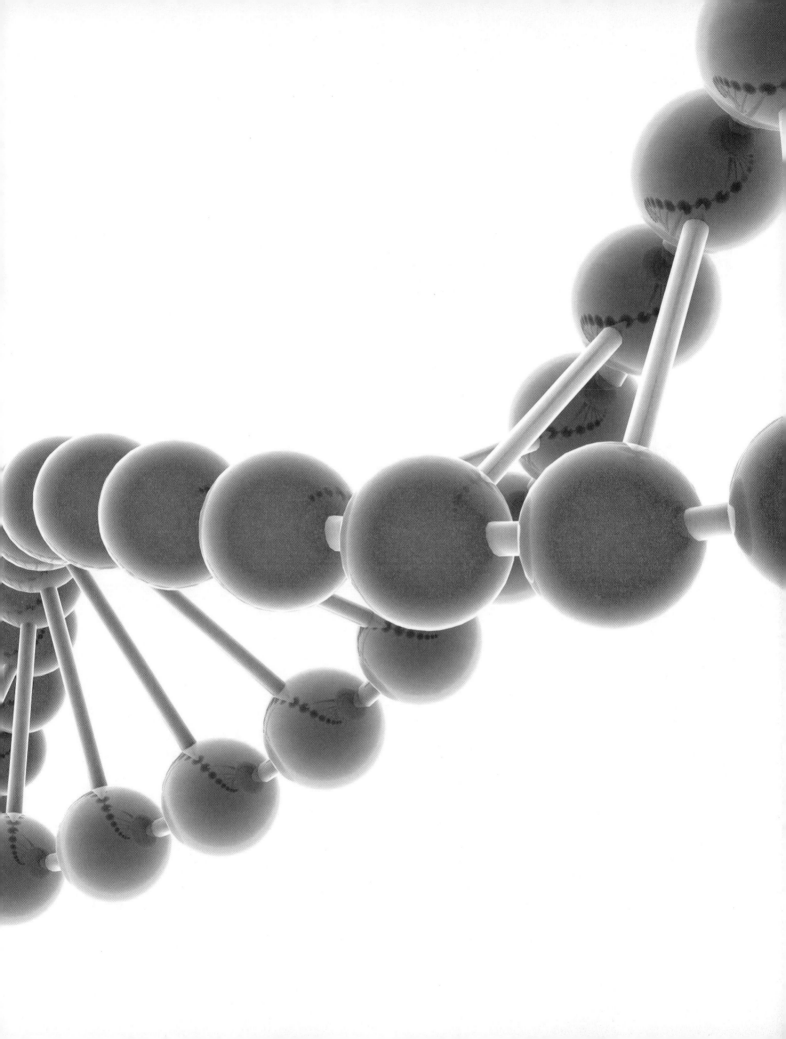

Biotechnical Systems

In this chapter, you will learn about the following different technologies used in biology, medicine, and agriculture:

- genetic engineering
- bioprocessing
- bionics
- CAT scans and ultrasound
- hydroponics
- food production

Biotechnology applies technological processes to living things—from sheep to garden peas to the bacteria that are used to clean up oil spills. Using biotechnology, the specific features of an animal, a plant, or a microbe can be transferred to another organism, or even used to create entirely new organisms. Today, biotechnology has many different applications, such as

- enhancing and preserving foods

- creating varieties of fruits and vegetables that are more nutritious and flavorful

- making medicines for diseases such as cancer and diabetes

- removing waste materials

- recovering oil from deep underground

   Biotechnology may slowly change our world in a number of different areas. As research continues, our values will determine how biotechnology is used.

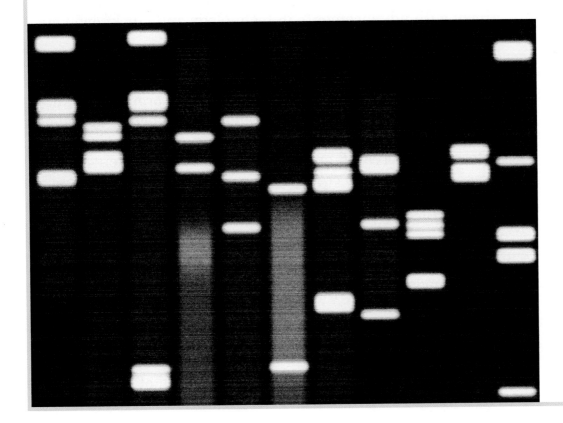

Biotechnology

Benchmarks for Learning

- Biotechnology applies the principles of biology to commercial products and processes.
- The instructions for specifying every organism's characteristics are carried in its genes.
- In genetic engineering, the structure of genes is modified to produce new genetic makeups.

Reading Strategy

Mapping   Make a map showing different ways in which biotechnology is used. Describe each technique using a sketch or a quick definition.

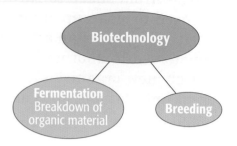

Vocabulary

biotechnology	hybrid	DNA
fermentation	bioprocessing	genetic engineering
crossbreeding	chromosome	genome

What Is Biotechnology?

The word *biotechnology* combines the words *biology* (the study of living things) and *technology*. **Biotechnology** is the use of technology to make living things into products or new forms of life. It is used to create new medicines or to improve plants or animals. Biotechnology is not new, but it has grown with advancements in technology and our knowledge of biology.

In recent years, tremendous advances have occurred in medicine, agriculture, and food production. Scientists have gained a better understanding of how living organisms function. In addition, biotechnical developments such as artificial hearts and other body parts, cloning, and genetic engineering of foods have created opportunities for improvement in our lives. They have also created ethical and moral challenges.

First Uses of Biotechnology

Biotechnology has existed for at least 8,000 years, since people first learned that yeast could be used to make bread. People made bread using the process of fermentation.

Fermentation

Foods such as bread, yogurt, and kimchee (a type of pickled cabbage) are made through a biological process called fermentation. **Fermentation** is the breakdown of organic material by microorganisms such as the bacteria in yeast.

Bread is made by using yeast to cause fermentation. Yeast is a living organism that feeds on sugar and starch. It breaks down the sugar and starch and produces alcohol and carbon dioxide gas. These products create air pockets and cause the bread to rise, while also giving bread its flavor. Yogurt is made by adding a small amount of cultured bacteria to heated milk, and then allowing the ingredients to ferment. The fermenting process causes the milk to thicken and turn slightly sour. Cheese, beer, and wine have been made throughout history with a similar fermentation process.

Breeding Techniques

Farmers also have used biotechnology for centuries, in a process called selective breeding. Early farmers found that they could produce better crops if, at planting time, they selected seeds from only the best plants. Using this simple technique, valuable traits of the best plants were passed on in the seed. Farmers used the same principle for animals. Over generations, they bred animals with certain traits, such as speed, agility, or intelligence (Figure 16.1).

Crossbreeding is another form of biotechnology that has been used for a long time. **Crossbreeding** combines the traits of one plant or animal with those of another. For example, one kind of wheat may have large seed heads but weak stems. Another kind may have strong stems but small seed heads. By crossing the plants, a new kind of wheat with strong stems and large seed heads might be produced. Crossbred plants or animals are called **hybrids.** They have the traits of both parents.

Figure 16.1 Many generations of selective breeding have produced dogs that have very specific traits.

Extending *What traits might a breeder look for in dogs?*

Biotechnology Today

Modern biotechnology blends biological knowledge with engineering techniques. Today, biotechnology allows people to create crops that are resistant to pests and weeds. It has led to new strains of food that provide better nutrition. It has also led to new medicines and medical technologies. Important techniques used in biotechnology include bioprocessing, and genetic engineering, and products include antibiotics and vaccines.

Bioprocessing

Bioprocessing is a modern variation of the ancient process of fermentation. **Bioprocessing** uses living organisms to process materials, breaking them down and changing them into other materials (Figure 16.2). Garbage, for example, is broken down through bioprocessing. Microorganisms feed on the plant and animal wastes in garbage, giving off a gas called methane. This gas is similar to the gas we use for heating and cooking. In this way, bioprocessing can turn garbage into a useful source of energy.

Through bioprocessing, living cells also can produce useful materials. One new bioprocess uses microorganisms to recover oil from old oil wells. When oil wells are first drilled, underground gases push oil to the surface. Once the gases are used up, there is no mechanical way of bringing the oil to the surface. With modern techniques, microorganisms are pumped into the oil-carrying rock, along with sugar and water, to create new gases that force the remaining oil to the surface.

Bioprocessing also plays a part in sewage treatment. Treatment plants use microorganisms to break down the solid matter in sewage. The microorganisms help purify the water so that it can be recycled.

Genetic Engineering

Genetic engineering allows scientists to give plants or animals new traits, or features. They do this by working with the individual cells that make up every organism. Inside each living cell are many chromosomes. A **chromosome** is a rod-shaped body made up of long, threadlike molecules of genetic material. This genetic material is called **DNA** (deoxyribonucleic acid) (Figure 16.3). A DNA molecule has two connected spiral strands. Both strands are twisted together like a spiral staircase, forming a double helix. DNA strands are made up of many genes. These genes combine

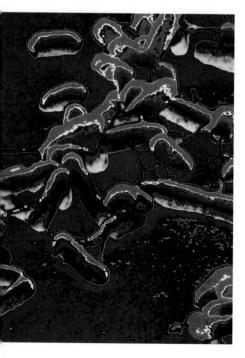

Figure 16.2 These bacteria, called *Pseudomonas putida*, are oil-eating bacteria that are often used to clean up oil spills.

Summarizing *How can bioprocessing be beneficial to the environment?*

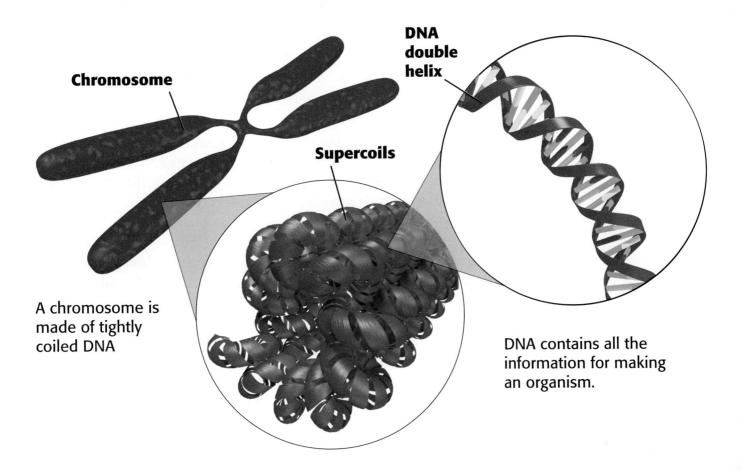

Chromosome

DNA double helix

Supercoils

A chromosome is made of tightly coiled DNA

DNA contains all the information for making an organism.

in different ways to give organisms their traits. They determine whether a life-form is a plant or an animal. They determine what it will look like and how it will function.

In **genetic engineering,** a gene is taken from one cell and spliced (inserted) into another cell. The second cell then carries the traits of the spliced gene. This technique is also called gene splicing. Gene splicing is used to produce certain medications, such as insulin and interferon. Someday, gene splicing may be used to cure rather than simply treat certain diseases.

The Human Genome Project

One important research effort of biotechnology is called the Human Genome Project. A **genome** is an organism's complete set of DNA. The goal of the Human Genome Project was to sequence, or find the order of, the entire human genome. Now that it is complete, we know the exact order of the 3 billion chemical building blocks that make up our DNA. The Human Genome Project helps explain exactly what makes up genes and how they work.

A set of DNA maps is being developed by the Human Genome Project. These maps show the small genetic variations that can occur within a person's DNA sequence.

Figure 16.3 Every cell in the body contains DNA, which is coiled tightly into chromosomes.

Interpreting *Which is bigger—a chromosome or a gene?*

More than 30 genes related to diseases and conditions such as breast cancer, muscle disease, deafness, and blindness have been identified.

Antibiotics and Vaccines

As a result of genetic engineering, many medicines have been developed to treat or prevent disease. Antibiotics are medicines that kill bacteria within the body. They are used to treat many diseases, such as strep throat and Lyme disease.

Vaccines help our bodies fight disease and infection (Figure 16.4). As a result of modern vaccines developed using biotechnology, plagues and epidemics are rare in most of the world. By the 1970s, smallpox had been eliminated through mass vaccination. Polio, which once threatened populations with paralysis and death, is now prevented with vaccines. Scientists are using biotechnology to develop vaccines for new diseases, such as AIDS. They also are developing better ways to manufacture and deliver vaccines.

Today, the cure rate for many major diseases is on the rise. In the 1960s, fewer than 10 percent of children with cancer survived. Today, more than 50 percent survive. Due to biotechnologies, people often have a choice of therapies for treating serious diseases such as cancer, diabetes, heart disease, and multiple sclerosis. Scientists have also had success in preventing aging in rats, resulting in old rats whose hearts and lungs are as strong as those of young rats.

Figure 16.4 Many vaccines are produced using genetic engineering techniques.

Summarizing *Why are vaccines important?*

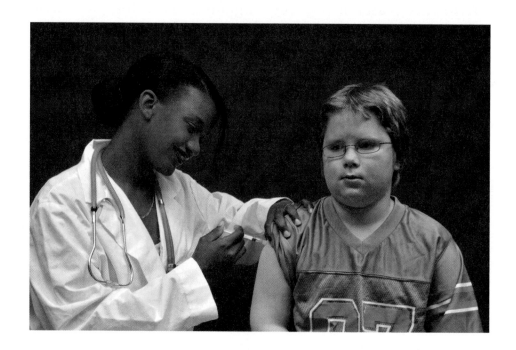

Connecting to STEM

Gene Splicing

Gene splicing is becoming one of our most important tools for treating disease. For certain diseases, such as cystic fibrosis, healthy genes can actually be inserted into the body to replace defective genes. Once inside the body, the healthy genes make proteins and other substances that the body could not otherwise make.

Using Interferon to Treat Disease

Gene splicing is also used to manufacture medicines. The medicine interferon is used to treat many types of diseases, such as cancer, hepatitis, and multiple sclerosis.

Interferon is a protein made naturally by your body to boost your immune system. Manufactured interferon is used to strengthen the body's natural defenses. Cancer occurs when cells in the body multiply rapidly and out of control. Interferon can kill cancer cells without hurting healthy cells. In patients with multiple sclerosis, interferon prevents the disease and its symptoms from getting any worse.

Getting Bacteria to Do the Work

Using gene splicing, scientists can make large amounts of interferon fairly inexpensively. *E. coli* is a type of bacteria that lives in the body. It is used in the manufacture of interferon. The interferon gene is inserted ("spliced") into the bacteria's DNA. Then, the bacteria are allowed to reproduce, or replicate.

E. coli bacteria divide every 20 minutes. After a short time, billions of cells have produced large amounts of interferon. The substance is collected and prepared for injection.

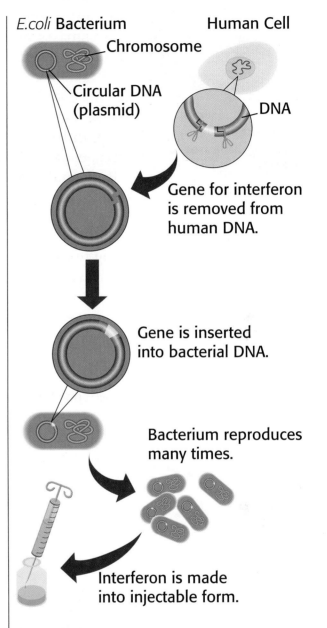

E.coli Bacterium Human Cell

Chromosome

Circular DNA (plasmid)

DNA

Gene for interferon is removed from human DNA.

Gene is inserted into bacterial DNA.

Bacterium reproduces many times.

Interferon is made into injectable form.

Critical Thinking

1. **Summarizing** In your own words, describe the process of splicing a gene.

2. **Inferring** Why does the rapid life cycle of bacteria help to produce large amounts of interferon?

Resources for Biotechnology

Biotechnical systems use each of the seven technological resources—people, capital, time, information, energy, materials, and tools and machines. Of these, people, information, and materials are the most important.

Scientists study living things to learn more about them and how they might be used. Scientists are educated in fields such as biology and chemistry. Engineers design the tools that are needed to turn research ideas into large-scale production. Technicians run the equipment and help scientists and engineers in their work. Nontechnical workers such as managers, salespeople, lawyers, and financial experts are also needed.

Information is the most important of all resources for biotechnology (Figure 16.5). Information leads to new ideas and discoveries. Workers must work both to keep up with these developments and to generate new and creative ideas. They must bring to their work a good knowledge of science and math.

Many of the materials used in biotechnology are living cells, or materials that come from them. For example, a special type of substance called a restriction enzyme acts like a pair of scissors. This material is used to cut a single gene from a piece of DNA. Another substance called a ligase is used to stick different pieces of DNA together. Other biotechnical tools include laboratory equipment, computers, and specialized electronic tools. All these tools are used to do research and to monitor biotechnical processes.

Figure 16.5 The Human Genome Project provides information for many areas of biotechnology.

Summarizing *Why is new information important in biotechnology?*

SECTION 1 Assessment

Recall and Comprehension

1. What are two biotechnology processes from the past that are still used today?
2. What was the goal of the Human Genome Project?
3. Name three applications of bioprocessing.
4. Give an example of each of the seven technological resources as used in biotechnical systems.

Critical Thinking

1. **Taking a Position** Some people think that genetic engineering should be stopped because plants and animals should remain "natural." Other people support it because it may improve the traits of these organisms. What do you think?
2. **Contrasting** Explain how the tools used in biotechnology differ from the tools used in other technologies.

QUICK ACTIVITY

Observe how bread mold reacts to materials such as antibacterial ointment, saltwater, vinegar, soap, and alcohol. Cut the crust off a piece of fresh white bread, then flatten the bread with a rolling pin. Use a butter knife to score 1" squares on the bread. Use a cotton swab to apply a different material to each square. Place in a sealed container in a warm, dark area. What happened after 1 week? **For more related Design Activities, see pages 506–509.**

For: DNA Activity
Visit: www.mytechedkit.com

Basic Biotechnology: DNA Discovery Online

Biotechnology has a major impact on the field of medicine. Technologists are using the principles of biology to help improve much of the equipment and processes used by scientists and researchers. However, it all starts with understanding the basic building blocks of living things—the cell. Go online and discover for yourself the inner workings of cells.

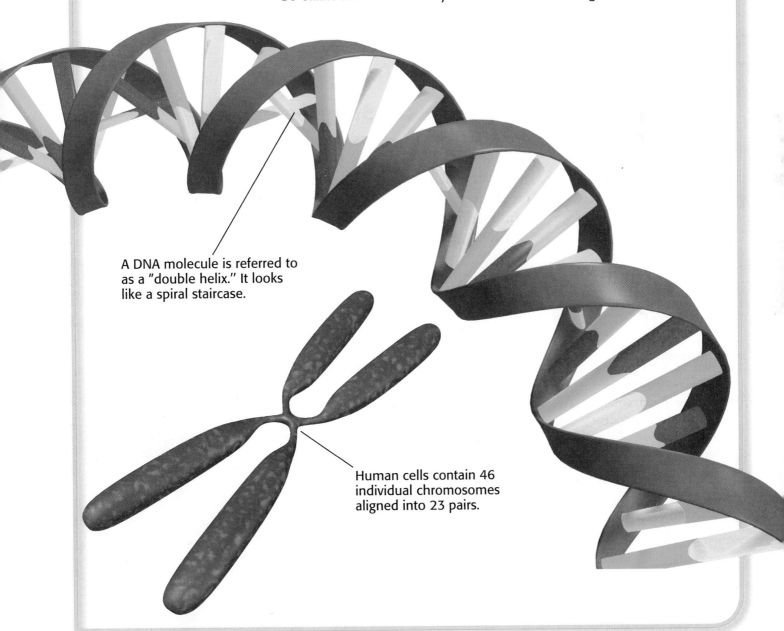

A DNA molecule is referred to as a "double helix." It looks like a spiral staircase.

Human cells contain 46 individual chromosomes aligned into 23 pairs.

Medical Technology

Benchmarks for Learning

- Advances in medical technologies are used to improve healthcare.
- Technological advances have made possible the creation of new devices to repair or replace certain body parts.

Reading Strategy

Listing   As you read the section, make a list of the different types of medical technologies available today. Next to each entry, write down a description of how it is used.

Medical Technologies
  A. Tools for surgery
     1. Lasers—used instead of knives and scalpels
     2. Heart and lung machines—

Vocabulary

bionics   CAT scan   ultrasound

New Tools for Surgery

Medical technology has revolutionized the tools used to treat disease and injury. New technologies are helping doctors improve medical care (Figure 16.6). Surgery can be performed on almost every part of the body. Diseased body parts can be repaired or replaced.

Instead of using scalpels (small, sharp knives) in operations, doctors are using lasers more often. A laser beam can be used to remove tissue and to repair broken blood vessels. Lasers can also be used to unclog arteries that are blocked by fatty deposits. A few years ago, a person might have had

Figure 16.6   Medical technology has provided doctors with extremely precise machines. Here, a doctor examines the eye of a diabetic patient.

Predicting   *How do you think medical care will change during your lifetime?*

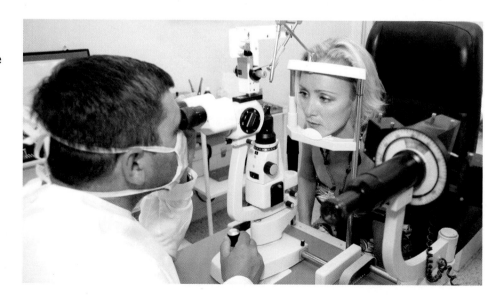

to undergo a leg amputation because of poor blood circulation. Today, circulation can often be restored through laser surgery.

Electronic tools also have greatly improved medical care. Heart and lung machines keep people alive during surgery on these organs. The heart and brain are monitored by electronic devices during and after surgery. Doctors often use fiber-optic tools and cameras to see inside the body without cutting it open. Battery-operated pacemakers are placed under the skin to control an irregular heartbeat. Computers help to diagnose diseases.

Natural Transplants

Many body parts can be replaced by transplanting natural parts. Each day in the United States, an average of one heart, 20 kidneys, and 65 corneas (the clear covering of the eye) are transplanted. Organs are donated, usually by families when a family member has been killed in an accident.

Much of the human body can be recycled. Organs that can be transplanted include the heart, liver, kidneys, intestines, and lungs. Tissues that can be transplanted include bone marrow and corneas as well as skin, which is usually used to replace the skin of people who have been badly burned. Blood from one person can be used to replace blood lost by another. Even the lining of the brain can be reused.

For a long time, transplants were tricky and largely unsuccessful. Because immune systems react against foreign tissue, transplanted organs and tissues were usually rejected. Today, with the help of donor networks, doctors can find a donor who has tissues that better match the patient's. New drugs also have been developed to suppress the immune reaction so that more transplants succeed.

Bionics

Some body parts can be made artificially. The technology of making functional body parts is called **bionics.** Hands, legs, bones, joints, teeth, and hearts can all be made artificially. Materials that are used to make body parts include plastic and alloys of titanium and cobalt. Some bionic parts work in much the same way as those they were made to replace.

In some types of artificial limbs, computer chips are embedded in the limb and connected via circuitry to the brain. This technology allows people with artificial limbs to mimic natural motion. Signals from muscles are amplified and serve as signals to the computer chips. The chips can then direct more complicated actions.

Using Technology to View the Body

The development of the X-ray allowed doctors to look inside the human body for the first time without having to operate. The X-ray helped doctors identify broken bones and find signs of disease. However, X-ray radiation in large amounts can be harmful to the body. Also, X-ray images are not always clear enough to be useful. Medical technologies today provide a clearer window into the body than ever before.

CAT Scan

A more advanced type of X-ray is called computerized axial tomography (CAT). A **CAT scan** uses a rotating X-ray device that takes almost three hundred pictures as it moves around the patient. The image provides a cross section of the body, which is displayed on a computer screen (Figure 16.7). A CAT scan is much clearer and more complete than an ordinary X-ray image. It also uses lower radiation levels.

NMR

Nuclear magnetic resonance (NMR) is another body imaging system. NMR uses powerful magnets to make atoms in the body line up. High-frequency radio waves make these

Figure 16.7 A CAT scan assembles multiple X-rays into a single image.

Comparing and Contrasting
How are X-rays and CAT scans similar? How are they different?

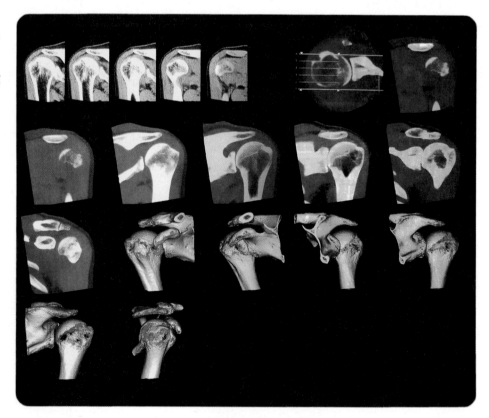

Technology in the Real World

Replacement Hearts

Creative medical technology has allowed doctors to replace hands, arms, or even entire legs to provide new mobility and independence with human-made prosthetic devices. But it is a bit trickier to replace organs that lie inside the body. Every year, people with various organ failures wait for human donors to provide healthy replacements of a kidney, liver, or heart. The list of donors, however, is much shorter than the need for transplants. Even after an organ has been transplanted successfully, patients must take dozens of medications to ward off their body's response to the new organ, and it is still sometimes rejected.

Instead of Muscle, Titanium

A new direction in medical technology is to replace the diseased organ with a completely human-designed organ. A medical technology company, Abiocor, has done just that with its new replacement heart.

In Abiocor's heart, form is entirely functional. This technological masterpiece is a two-pound mass of clear plastic and titanium.

Robert Tools was the first person to receive a fully implanted artificial heart.

A replacement heart's design mimics the valves and arteries of a real heart.

Unlike previous technologies, the Abiocor heart is powered by a wireless system that requires no direct opening to the outside body. This design drastically reduces the chance of infection. The Abiocor heart also responds to the body like a natural heart: Its pump speeds up and slows down according to the body's demands.

Making a Real Difference

Every year, more than 300,000 people in the United States die from congestive heart failure. Although the Abiocor heart is a new, experimental technology, it holds the promise of changing the lives of thousands of people who suffer from heart disease. As medical technology grows, the list of human-made organs may grow as well.

Critical Thinking

1. **Hypothesizing** How might the possibility of human-made organs change people's outlook on growing older and staying healthy?

2. **Extending** What are the main functions carried out by the heart? Do research to learn why replacing this organ is such a complex task.

Figure 16.8 New ultrasound techniques allow doctors to learn more and more about how a baby grows.

Clarifying *If ultrasound uses sound waves, why can't we hear it?*

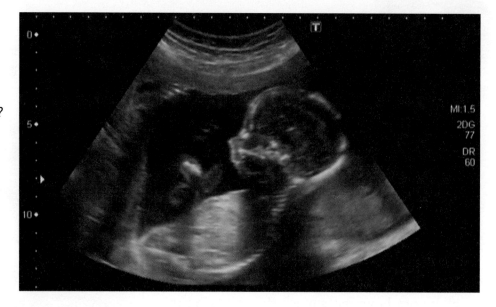

atoms vibrate. The length of time it takes for the atoms to stop moving determines the kind of image displayed on the computer screen. The image tells a great deal about body tissues.

Ultrasound

Ultrasound uses sound waves that are too high for humans to hear. These sound waves are passed through body tissues. The echo that is received as the sound wave bounces off body tissue is changed into electrical signals. These signals create an image on the computer screen (Figure 16.8). Ultrasound is commonly used to examine a fetus inside the womb.

SECTION 2 Assessment

Recall and Comprehension

1. Name three tools that have improved medical care, and explain what each one does.
2. Explain how tissues and organs can be recycled, giving specific examples.
3. Describe how ultrasound is able to create pictures of soft tissues.

Critical Thinking

1. **Hypothesizing** Is organ donation likely to become more or less common in the future? Provide a specific example in your answer.
2. **Comparing and Contrasting** Compare how a CAT scanner and an NMR imaging machine work.

QUICK ACTIVITY

In cryogenics, extremely cold liquids are used to freeze the bodies of people who have recently died of a disease, in the hope that they might be brought back to life once a cure has been found. Imagine that a cave man was discovered trapped in a glacier and brought back to life today. Write a creative story about what challenges the cave man would have to face in modern times.
For more related Design Activities, see pages 506–509.

Agriculture and Food Production

Reading Strategy

Listing What important changes have taken place in farming? As you read this section, make your own list of the new techniques used in farming.

Changes in Farming
1. Engine power substituted for animal power
2. Fewer, smaller farms

Vocabulary

pesticide	combine	irrigation
herbicide	aquaculture	hydroponics
fertilizer	cloning	

Benchmarks for Learning

- Technological advances in agriculture directly affect the amount of resources required to produce food.
- Refrigeration, freezing, dehydration, and irradiation provide for long-term storage of food.

Changes in Farming

During the twentieth century, farming went through important changes. In the early 1900s, during the era of mechanization, farms substituted engine power for animal power. Farmers were able to cultivate more land and produce a greater amount of food. Farming became more productive. As a result, when the mechanization era began around 1900, there were 6 million farms in the United States. Fifty years later, there were 600,000 fewer farms. Farmers could farm more land with the new machines that were available, but not all farmers could pay for the latest machinery. Smaller farms could not compete and began to disappear.

Chemicals and the Green Revolution

The next important period in agriculture, called the chemical era, began about 1950. Farmers began to use chemicals on their land. **Pesticides** were used to kill insects that attacked crops. **Herbicides** destroyed weeds. **Fertilizers** manufactured from chemicals were added to the soil to improve plant growth. Because these chemicals increased food production tremendously, the term "Green Revolution" was coined. As a result of chemicals and improved machinery, the techniques used by millions of farmers around the world changed dramatically.

During the Green Revolution, crossbreeding created improved strains of wheat, rice, and corn. New hybrid strains yielded more grain per acre than the older strains. Scientific research continued to improve important varieties of plants grown for food. Today, farmers raise five times as much corn per acre as they did just 50 years ago. Also, dwarf varieties of fruit trees made caring for orchards and harvesting fruit easier and less expensive.

Improving Agriculture with Biotechnology

The era of biotechnology is the third and current era in farming. Crossbreeding and other genetic techniques have had an effect on animal production, and on the productivity of farms. Dairy farms, for example, use fewer cows to produce more milk. In 1900, there were 20 million cows in the United States. Today, there are only 11 million cows, but they produce twice as much milk. This change is due to the use of bovine growth hormone, which can increase a cow's milk production by up to 40 percent. Detractors worry that the hormone may have detrimental effects on the humans consuming the milk.

Other areas of farming have also been transformed by technology. Modern plows that use steel blades to break up the soil are pulled by tractors. Planting is carried out by machines. Harvesting, which is the process of gathering crops when they are ready, is done using a combine. A **combine** is a machine that cuts the crop and separates grain from the straw or hay.

In the United States, about 70 percent of working people were farmers in 1820. Today, less than 3 percent are farmers. These farmers provide food not only for the U.S. population but also for export to countries around the world. With this

Figure 16.9 The number of farms has declined over the last several decades.

Interpreting *When did the number of farms drop more—in 1975 or in 1985?*

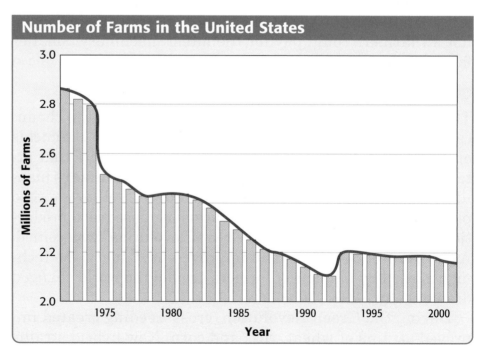

Number of Farms in the United States

decrease in the number of farmers, there has also been a steady decline in the number of farms (Figure 16.9). Today, there are just over 2 million farms in the United States. The average size of a farm is gradually increasing. These large farms are expensive to start up and maintain. To be profitable, they must use the latest technology and equipment.

Modern Agriculture and Food Production

Today, technology is central to how we grow, harvest, process, and store food. Machinery is continually being improved to make farming easier and more productive (Figure 16.10). Foods that were once harvested from the wild are grown in controlled conditions. This allows items, such as fruits and vegetables, to be picked at just the right time. Food is kept fresh for long periods using special packaging. Very little about farming is the same as it was 100 years ago.

Meat and Fish Production

More and more of the fish and shellfish we eat are raised on farms. **Aquaculture** is the farming of aquatic organisms, such as salmon, oysters, and plants. A model environment must be created so the fish grow well, the water remains clean, and food is supplied appropriately. Often, the productivity of fish farms is higher than it is in nature.

Many of the crops produced on farms provide feed for animals. Animals that are raised for food, such as cattle, fish, and poultry, are fed grain. Cattle are typically range-fed for nearly a year and then brought to feedlots, where they consume feed grain to gain weight. Poultry and fish both are given feed grain. The efficiency of the conversion process from grain to meat varies with the type of animal. For cattle in feedlots (not on the range), the conversion is about

Figure 16.10 Modern farms use sophisticated machinery to increase production.

Identifying *What kinds of tasks on this dairy farm might be controlled by computers?*

Food Processing and Preserving

Process	How It Is Done	Examples
Cooking	Most modern cooking appliances use electric or gas heat. Microwave ovens cook by sending radio waves through the food. Particles of water in the food vibrate, heating the food.	All types of foods
Canning	Food is steamed or cooked for a few minutes to kill bacteria. Food is then put into steel cans that have a tin coating.	Fruits and vegetables, beans, soups
Pasteurization	Milk is heated to 160°F for 20 seconds, then quickly cooled to about 38°F.	Milk that is to be used within days
Ultra Heat Treatment (UHT)	Milk is heated to about 175°F and then to 300°F.	Milk that will be stored for several months
Refrigeration	Temperatures slightly above freezing are used to slow down the growth of bacteria. Refrigeration is used for short-term storage of food.	Milk, fruits and vegetables, meats

Figure 16.11 Many different techniques are used to process and preserve food before we buy it in a supermarket.

Interpreting *If you wanted to take milk on a camping trip, what type of milk would you buy?*

7 pounds of grain per pound of additional weight. For pigs, the ratio is less than 4 to 1, for chickens the ratio is 2 to 1, and for fish the ratio is less than 2 to 1, the most efficient conversion of feed grain. The intense gathering of animals in feedlots can create problems with disease and pollution.

Food Processing and Preserving

Eating provides people with essential nutrients such as carbohydrates, fats, minerals, protein, vitamins, and water. Most fresh food eventually spoils. Bacteria, mold, and insects feed on it, making it unhealthy to eat. Today, foods are processed to keep them safe to eat, tasty, and easy to prepare. Foods are also preserved so they can be stored for a long time. Common methods of food processing and preserving are shown in Figure 16.11.

Cloning and Tissue Culture

Cloning is the creation of an exact genetic duplicate of another organism. Genetic engineering has enabled people to clone organisms, such as bacteria. This allows scientists to perform research experiments on many copies of the same organism. Scientists have also cloned mice, which has proved to be useful in medical research. In 1996, Scottish scientists cloned the first large mammal, a sheep named Dolly. Since then, other complex mammals, such as monkeys, have been cloned. There is great promise in using cloned animals in animal breeding because of the high-quality animals that can be produced. An overriding concern, however, is that the

Process	How It Is Done	Examples
Freezing	Uncooked food is steamed first to destroy all bacteria. Most foods are frozen by dipping the packaged food into a tank of freezing salt water.	Vegetables, fish, poultry, canned juices, red meat
Drying	Water is removed, causing the concentrations of salt and sugar to increase. The higher salt and sugar concentrations kill bacteria living in the food.	Powered milk, soups, potatoes, orange juice, all types of dried fruits
Using Additives	Food additives are chemicals that are added to food to improve it or to keep it from spoiling. They include preservatives, flavorings, and dyes.	Lunch meats, fast foods, all types of prepared foods
Irradiation	X-ray and gamma-ray radiation kills microorganisms in food. Foods can be irradiated after they are packaged.	All types of packaged foods

technology may someday be used to clone people. This issue presents tremendous ethical problems for society.

Another method of reproduction used to produce new and better plants is called tissue culture. A piece of leaf tissue is grown in a test tube. The millions of new cells are exposed to many different conditions, such as heat or salty soil. The cells that survive are used to grow new plants. The new plants have the same traits that let the parent cell survive the harsh conditions.

Figure 16.12 These crops will survive because they were genetically altered to withstand freezing temperatures.

Hypothesizing *What other traits might be added to crop plants?*

Genetically Altered Crops

The use of genetically altered crops is one of the most important characteristics of modern farming. Gene splicing is used to transform plants, animals, and bacteria. For example, a normal type of bacteria causes crystals of ice to form on the leaves of crops during frosts, killing the plants. A gene-spliced bacteria prevents this from happening (Figure 16.12). This new kind of bacteria may save crops during cold spells.

There are concerns, however, about the effects of genetically altered crops. Some people think that the insects will mutate eventually, creating a super insect that is resistant to the corn's insecticide as well as to insecticides applied by farmers. In the case of corn, there was concern that its pollen contains trace amounts of insecticide that are harmful to migrating Monarch butterflies. To attempt to counteract this problem, farmers also planted traditional corn, with no insecticides, in adjacent fields, though there was no way for the butterflies to determine which were safe.

LIVING GREEN
Organic Alternatives

Since the 1940s, small groups of farmers in the United States have practiced organic farming as an alternative to conventional farming. Organic farming refers to the practice of growing crops without the use of pesticides or chemical fertilizers or using genetically altered seeds. Organic raising of livestock (including milk and egg production) does not use antibiotics or hormones, ensures that animals have access to the outdoors and that they are fed from products that are 100 percent organic. This type of farming gained popularity in the 1980s, and in 1990 the Organic Foods Production Act became law, requiring organic farmers to meet national standards of production. In 2008, organic food sales were over $18 billion, up from $3.6 billion in 1997.

MAKE A CHANGE
Organic foods currently make up about 3 percent of total food production in the United States, and on average cost 10-30 percent more than conventional foods due to the cost of producing foods that meet organic standards. People that choose to eat organic foods typically believe that these types of foods are healthier, due to the lack of chemical residues in the food, that it is better for the environment, and that it is more humane to animals. The health benefits of eating organic have not been studied for long enough to show proof; however, the benefits of organic farming on the environment include less pollution due to runoff, enriched soil, increased biodiversity, and healthier animal populations.

TRY THIS
Are organic foods important in your community? What portion of the foods available in your local market is organic? Does your family eat any organic foods? The organic foods debate is growing as fast as the organic farming industry. How do you feel about this type of farming versus conventional methods? Take the time to inform yourself, understand any misconceptions, and then organize a debate in your class.

Irrigation and Hydroponics
Water is critical for plant growth. Too much or too little can decrease crop production. In places where water is in short supply, irrigation is used often. **Irrigation** is the artificial application of water to help plant growth. Water for irrigation comes from rivers, streams, lakes, and reservoirs.

Figure 16.13 In hydroponics, plants are grown in water rather than soil.

Summarizing *What are the advantages of growing food hydroponically?*

Center-pivot irrigation is used to supply water to large land areas. Water is sprayed through pipes that rest on wheels and travel around a center point. Systems can be as large as a quarter of a mile long. Where water is extremely scarce, drip irrigation is used. Each plant receives its own trickle of water through plastic tubing at root level, so no water is wasted. This method is effective only for small farms.

Hydroponics involves growing plants in water, sand, or gravel rather than in soil (Figure 16.13). The word *hydroponics* comes from Greek words that mean water and farming. Nutrients are supplied in the water to the plants' roots. Hydroponic farming is done in greenhouses where weeds and pests can be eliminated. Air temperature and humidity can be kept at the best levels for growth. Tomatoes and lettuce are examples of plants grown by hydroponics.

SECTION 3 Assessment

Recall and Comprehension

1. Explain how technology has advanced agriculture. Give examples.
2. What are two methods of irrigation?
3. Describe three ways food is processed.

Critical Thinking

1. **Relating Cause and Effect** Explain several reasons why farmers today can produce larger amounts of crops on less land.
2. **Making Judgments** Do you think irradiation of food is a helpful or harmful technological advance? Explain your answer.

QUICK ACTIVITY

Hydroponic gardening can be done using inexpensive materials, such as PVC pipe, seeds, peat moss starter pods, a circulator pump, a hose, grow lights, and hydroponic fertilizer. Using the Internet, research the topic of hydroponic gardening. Use the information you have learned to set up a small hydroponic garden in your classroom. **For more related Design Activities, see pages 506–509.**

16 Review and Assessment

Chapter Summary

- Biotechnical systems are technologies that relate to living things. Biotechnical systems include biotechnology, medical technology, agriculture, and food production.

- Biotechnology includes genetic engineering and bioprocessing. In genetic engineering, a gene is taken from one cell and spliced into another cell. Bioprocessing uses living organisms to process materials.

- Modern medicine has wiped out diseases such as smallpox using vaccines. Progress is being made in finding cures for cancer, heart disease, and other diseases.

- Electronic tools such as the laser have made surgery safer and better. Body parts may be transplanted from one person to another. Artificial parts can be manufactured to take the place of missing body parts.

- Computers have made it possible to see inside the human body using CAT scanning, nuclear magnetic resonance (NMR), and ultrasound to help diagnose diseases.

- During the twentieth century, mechanical, chemical, and biotechnology innovations increased agricultural change. Today, fewer farmers produce far more food than they could have years ago.

- Most foods must be processed either industrially or at home. Food is processed to be ready to eat or to be ready for preparation. Processed foods sometimes contain food additives.

- Food is sometimes preserved by using heat to kill bacteria, as in pasteurization of milk. Cold temperatures also preserve food by slowing down or stopping bacterial growth.

Building Vocabulary

Your teacher may give you a crossword puzzle. Complete the puzzle using the following words from this chapter. Exchange puzzles with a classmate to check each other's answers.

1. aquaculture
2. bionics
3. bioprocessing
4. biotechnology
5. CAT scan
6. chromosome
7. cloning
8. crossbreeding
9. DNA
10. fermentation
11. fertilizer
12. genetic engineering
13. genome
14. herbicide
15. heredity
16. hybrid
17. hydroponics
18. irrigation
19. pesticide
20. ultrasound

See your teacher for the Crosstech puzzle.

Reviewing Content

1. What three technologies make up the field of biotechnology?

2. Give two possible applications of genetic engineering.

3. What were the three eras of agriculture in the twentieth century?

4. Identify four methods used to preserve food.

5. Explain how technology can provide substitutes for body parts.

Applying Your Knowledge

1. Research DNA and make a sketch of its double-helix structure.

2. Look in your local grocery store for fresh products with labels identifying the following agricultural methods. Then, make a list of the products grown using each method.
   (a) hydroponics
   (b) aquaculture (labeled "farm raised")

3. Research the names of food additives and their possible side effects. Then, look at food labels in your home to find out what products use these additives.

4. Many athletes, particularly football players, have arthroscopic knee surgery at some time in their professional lives. Investigate this type of surgery and explain how it is done.

5. Find an anatomical drawing showing the organs of the human body. Make a copy or make your own drawing. Mark the body parts that can be replaced by artificial parts.

Critical Thinking

1. **Analyzing** Why are agriculture, food production, and medical technology referred to as biologically related technologies?

2. **Interpreting** How can agricultural productivity be rising in the United States even though the number of farms is decreasing?

3. **Making Judgments** Considering both the advantages and disadvantages of chemical preservatives in processed foods, do you think the use of preservatives should be allowed?

4. **Drawing Conclusions** What are some of the impacts of modern medical technology on society?

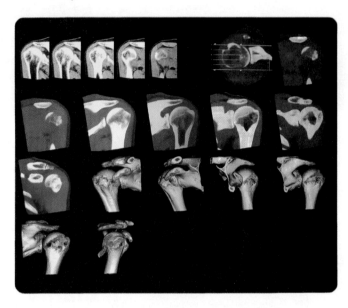

5. **Posing Questions** What questions need to be asked to explore the safety and ethical issues surrounding genetic engineering?

Connecting to STEM
science · technology · engineering · math

"Old-Style" Breeding

Although genetic engineering is the focus of much of biotechnology, simple breeding techniques are still in use today. Using the library and the Internet, research how dogs or cats are bred. What is meant by a pure-bred? How do breeders select the male and female for a particular litter? What traits characterize a particular breed? Prepare a report to share with your class.

Design Activity 31

DRYING BY DESIGN

Problem Situation

A local elementary school group is planning a weekend hike. During the hike, each group member will have to carry his or her own supplies, including food and clothing. The school is concerned about the heavy load each person will have to carry. Unable to think of ways to substantially reduce the weight of the group's backpacks, teachers at the school have turned to you for advice.

Materials

You will need:

- 120W bulb (floodlight or heat lamp)
- apples cut into 1/16", 1/8", and 1/4" slices
- fan
- graph paper or computer spreadsheet
- scale
- slicing tool, such as a knife
- wire mesh

Your Challenge

You and your team members are to design, construct, and test a dehydrator that dries fruit in as short a time as possible while maintaining the quality of the food.

> Go to your **Student Activity Guide, Design Activity 31.** Complete this activity in your Guide, and state the design challenge in your own words.

① Clarify the Design Specifications and Constraints

To solve the problem, your design must meet the following specifications and constraints:

- Dehydrate 250 g of apple slices to 20 percent or less of their initial weight.
- You may use only approved materials and tools. In addition, the drying surface your team uses must have an area no greater than 144 sq. in.

> In your Guide, state the design specifications and constraints. Add any others that your team or your teacher included.

② Research and Investigate

To better complete the design challenge, you need to first gather information to help you build a knowledge base.

> In your Guide, complete Knowledge and Skill Builder 1: Investigating Food-Drying Techniques.

> In your Guide, complete Knowledge and Skill Builder 2: Graphing Food Drying.

In your Guide, complete Knowledge and Skill Builder 3: Factor Analysis.

3 Generate Alternative Designs

In your Guide, describe two possible solutions your team has created for the problem.

4 Choose and Justify the Optimal Solution

Refer to your Guide. Explain why you selected the solution you did, and why it was the better choice.

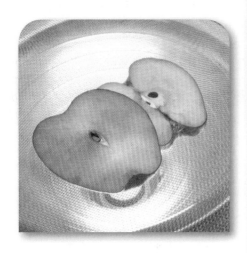

5 Develop a Prototype

Construct a working model of your dehydrator. Put a photograph or sketch of your final design in your Guide.

In any technological activity, you will use seven resources: people, capital, time, information, energy, materials, and tools and machines. In your Guide, indicate which resources were most important in this activity, and how you made trade-offs among them.

6 Test and Evaluate

How will you test and evaluate your design? In your Guide, describe the testing procedure you will use. Explain how the results show that the design solves the problem and meets the specifications and constraints.

7 Redesign the Solution

Respond to the questions in your Guide about how you would redesign your solution. Your redesign should be based on the knowledge and information that you gained during the activity.

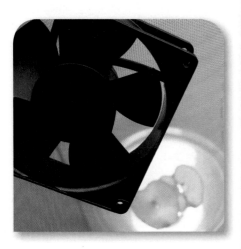

8 Communicate Your Achievements

In your Guide, describe the plan you will use to present your solution to your class. Show what handouts and/or Power-Point slides you will use.

Design Activity 32

MANUFACTURING CASEIN GLUE

Problem Situation

Bioprocessing uses living organisms to make or modify a product, improve a plant or animal, or develop microorganisms for a specific use. Casein glue is an adhesive that was used by people before the development of modern synthetic glues. The glue is manufactured from casein, a protein found naturally in milk. In this process, casein is isolated and removed from milk by allowing the milk to curdle. Curdling is caused by lactic acid, which is produced by bacteria that are normally found in milk. The curd is the isolated casein.

Your Challenge

You and your team members are to develop a formula for strong casein glue as well as a testing procedure to determine its strength.

> Go to your **Student Activity Guide, Design Activity 32.** Complete this activity in your Guide, and state the design challenge in your own words.

1 Clarify the Design Specifications and Constraints

To solve the problem, your design must meet the following specifications and constraints:
• You will create and test three different formulations to determine the optimal glue formulation.
• The testing procedure will use 1"-wide wooden slats as the glued material.

> In your Guide, state the design specifications and constraints. Add any others that your team or your teacher included.

2 Research and Investigate

To better complete the design challenge, you need to first gather information to help you build a knowledge base.

> In your Guide, complete Knowledge and Skill Builder 1: Preliminary Casein Glue Recipe.

> In your Guide, complete Knowledge and Skill Builder 2: Tensile and Shear Testing.

Materials

You will need:
• baking soda
• bucket
• cheesecloth
• cold water
• hot water
• nonfat dry milk
• sand or weights
• twine
• vinegar
• wooden slats, 1" wide, 6" long

In your Guide, complete Knowledge and Skill Builder 3: Factor Analysis.

3 Generate Alternative Designs

In your Guide, describe two possible solutions that your team has created for the problem.

4 Choose and Justify the Optimal Solution

Refer to your Guide. Explain why you selected the solution you did, and why it was the better choice.

5 Develop a Prototype

Make your optimum glue and test it. Indicate the recipe that you used in your Guide.

In any technological activity, you will use seven resources: people, capital, time, information, energy, materials, and tools and machines. In your Guide, indicate which resources were most important in this activity, and how you made trade-offs among them.

6 Test and Evaluate

How will you test your glue formulation? In your Guide, describe the testing procedure you will use. Explain how the results show that the design solves the problem and meets the specifications and constraints.

7 Redesign the Solution

Respond to the questions in your Guide about how you would redesign your solution. Your redesign should be based on the knowledge and information that you gained during the activity.

8 Communicate Your Achievements

In your Guide, describe the plan you will use to present your solution to your class. Show what handouts and/or Power-Point slides you will use.

Chemical Technology

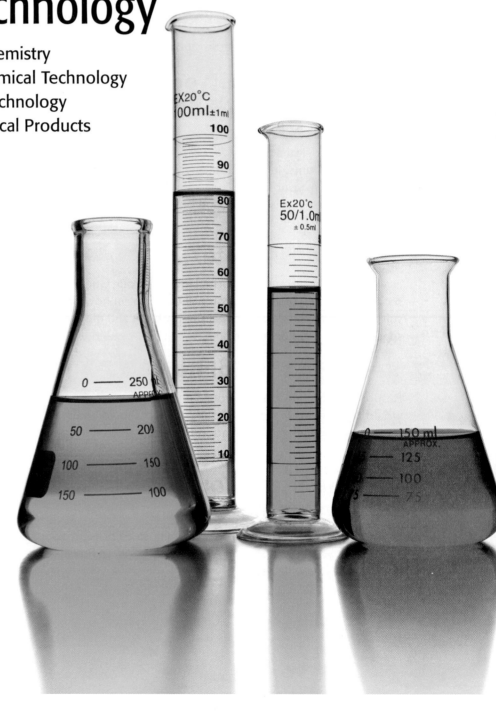

In this chapter, you will learn about the following different chemical products:

- personal care products
- food preservatives
- fossil and synthetic fuels
- construction materials
- photographic film
- paints and sealants

Many of the things we use every day are created and produced using chemical technology. The water we drink, the food we eat, and the clothes we wear are all modified or made using chemicals. Medicines and personal-care products, such as toothpaste, are also made using chemicals. Many industrial processes also rely on chemical technology, such as

- processing iron ore into steel
- making rubber products from natural rubber
- deriving paper from wood pulp
- making glass and ceramic tiles
- preparing foods and soft drinks
- dyeing textiles

Engineers and scientists study and apply chemical processes in order to improve products and lower their manufacturing costs. This type of development helps raise our standard of living and quality of life.

SECTION 1
Understanding Chemistry

Benchmarks for Learning

- Everything we see, smell, touch, and taste is composed of chemicals.
- All substances are composed of atoms. Often, atoms bond together into molecules.
- Molecules created by nature are natural products; those created by human-designed processes are synthetic products.

Reading Strategy

Mapping Make a map to show the relationships among different chemical units, such as atom, molecule, and compound.

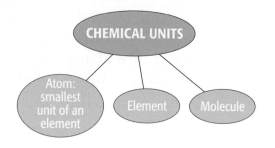

Vocabulary

element	compound	natural product
periodic table	molecule	synthetic product
chemical		

What Is Chemistry?

Chemistry is the study of the composition and properties of matter. Using chemistry, scientists explore methods of combining various substances to create new substances. Chemists look at the interactions between different substances to find processes for creating new products. The clothes you wear, for example, may be made of human-made fabric or dyed using chemical processes. The hair spray or styling gel you use on your hair was made using chemistry. Many foods you eat have been preserved and made more flavorful using chemicals. Everything we see, smell, touch, and taste is made of chemicals.

Chemical Elements and the Periodic Table

Recall from Chapter 11 that all matter is made up of atoms. An atom is the smallest unit of any substance that still retains the properties of that substance. An **element** is a pure substance made up of only one type of atom (Figure 17.1).

The **periodic table** lists all the known elements in a way that reflects the special properties of each (Figure 17.2). Each element has its own chemical symbol. For example, look at the information shown for H, the symbol for the element hydrogen. The atomic number, which is 1 for hydrogen, indicates how many protons are contained in that element. Take a look at other symbols in the table. The symbol for the element oxygen is O. The symbol for gold is Au, which comes from the Latin word for gold, *aurum*. Not all abbreviations

Figure 17.1 Neon, which is used to light up this sign, is a chemical element.

Contrasting *How do elements and compounds differ?*

for the elements in the periodic table are derived from their English spelling.

Chemical Compounds

A **chemical** is a substance that contains any of the elements found in the periodic table. Atoms of these elements can bond together to form chemical compounds. A **compound** is a substance that is formed by the combination of two or more elements. Water, for example, is a compound formed from the elements hydrogen and oxygen. The symbol for a chemical compound indicates which elements make up that compound. The symbol for water, which is H_2O, indicates that it is made of hydrogen and oxygen.

Molecules and Molecular Models

Just as the smallest unit of an element is an atom, the smallest unit of a chemical compound is called a **molecule.** Molecules are made of two or more atoms that act as a unit. A molecule of water contains two atoms of hydrogen and one atom of oxygen (Figure 17.3).

The molecular structure of a substance can be thought of as the skeleton of the molecule. It is made up of arrangements of atoms that are unique to that substance.

Figure 17.2 The periodic table shows information for every known element.

Interpreting *Are there more metal or nonmetal elements?*

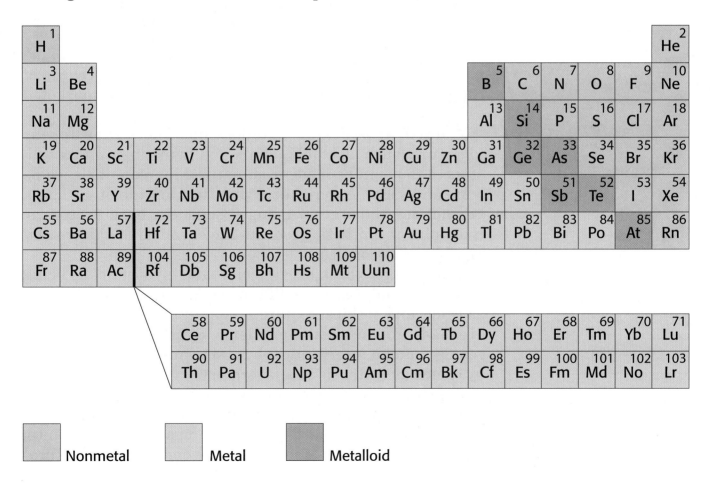

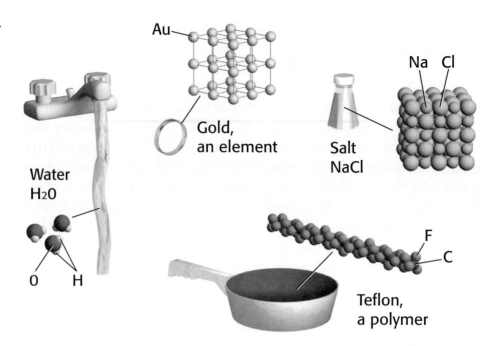

Figure 17.3 These molecular models show which elements make up common compounds.

Comparing *Which of the materials shown is not a compound?*

Water
H₂O

Au

Gold, an element

Na   Cl

Salt
NaCl

F
C

Teflon, a polymer

O   H

These unique groupings provide clues about how the molecule will react with other molecules. Chemists and chemical engineers use models and pictures to depict molecules.

Natural and Synthetic Chemicals

Chemicals can be extracted from a natural plant or animal source, or they can be made (synthesized) in a laboratory or chemical factory. There, chemical elements are combined to form compounds with the desired properties. Because of these different sources, chemicals are sometimes marketed as either natural or synthetic products.

A **natural product** includes substances that appear in nature (such as edible plants and fruits) as well as substances that can be made from natural ingredients (such as wine and soap). A **synthetic product,** on the other hand, is made in a laboratory using specific chemical processes. Examples of synthetic products are chemical fertilizers and most prescription drugs.

Many people think of "artificial" and "synthetic" as being equivalent, but there is a difference. Artificial products, such as cakes and cookies, require human intervention, but they are made from natural ingredients. Synthetic products also require human intervention, but they must also be produced by chemical synthesis in a laboratory.

Synthetic chemistry uses sophisticated knowledge of chemistry to design new compounds. Using computer modeling, chemists can predict molecular properties. For instance, synthetic oils such as Mobil 1™ use a combination of natural

Figure 17.4 Natural products, such as these vitamins, are made of chemicals that have been derived from plants.

Making Judgments *Why might people be willing to pay more for natural products than for synthetic products?*

and synthetic lubricants to create a very slippery oil that reduces friction between moving parts.

More Similar Than You Might Think

The public often considers natural chemicals to be better than synthetic chemicals. However, a synthetically produced chemical is made up of precisely the same molecules as those found in that same chemical produced in nature. In fact, the goal of chemists is often to figure out how to produce synthetic products that contain the same molecules as in the natural product. Vitamin C (ascorbic acid) is an example of a chemical that can be either extracted from plants or manufactured in a chemical plant. The largest demand for ascorbic acid is in the food industry. However, about one third of the total supply of vitamin C is used in vitamin preparations (Figure 17.4).

SECTION 1 Assessment

Recall and Comprehension

1. What is a chemical?
2. Using the periodic table, give the symbol and atomic weight of the following elements: copper, tin, lead, and uranium.
3. In your own words, define the following terms: atom, element, molecule, compound.
4. How are elements and compounds formed?

Critical Thinking

1. **Analyzing** Why is the chemical structure of molecules important to chemists?
2. **Comparing and Contrasting** Explain the ways in which natural and synthetic chemicals are both different yet similar.

QUICK ACTIVITY

You've heard that vitamin C can help fight colds and other illnesses. Find information to help determine the best form of vitamin C for you. Make a chart comparing the various forms of vitamin C (timed-release, ester-C, with rose hips, all natural, etc.). Compare prices and the amount of vitamin C in each product. What is the price per mg for each?
For more related Design Activities, see pages 536–539.

For: Acids and Bases Activity
Visit: www.mytechedkit.com

Chemical Technology: Acids and Bases Online

Everything we see, smell, touch, and taste is made up of chemicals. Many of our daily routines are really lessons in chemistry. Go online to learn more about the chemistry of common household products.

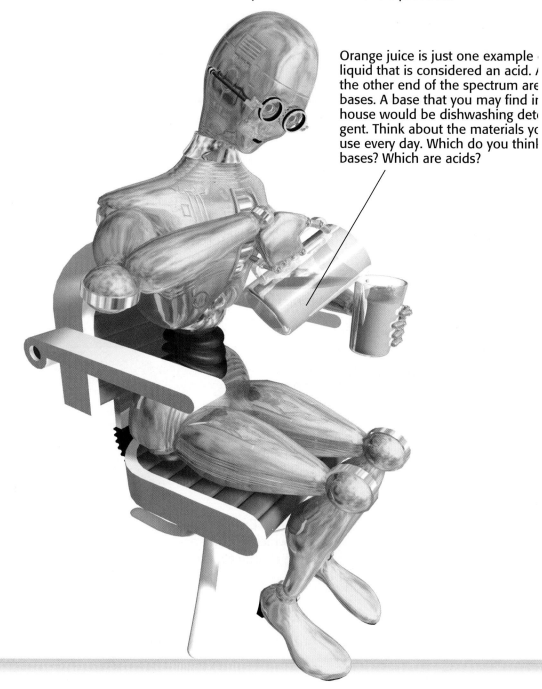

Orange juice is just one example liquid that is considered an acid. the other end of the spectrum are bases. A base that you may find in house would be dishwashing det gent. Think about the materials you use every day. Which do you think bases? Which are acids?

Resources for Chemical Technology

Reading Strategy

Listing   Make a list of as many different chemical resources as you can think of. Start your list by writing down each of the green headings in this section.

Resources for Chemical Technology
- People—chemical industry workers, industrial chemists, chemical engineers
- Capital—processing plants

Benchmarks for Learning

- Catalysts accelerate chemical reactions.
- Chemical reactions may release or consume energy.

Vocabulary

endothermic reaction	synthesize	spectrometer
exothermic reaction	catalyst	gas chromatograph

Using Chemical Resources

Chemical technology systems produce more than 70,000 commercial products—from paints to plastics to prescription drugs. They do this by using the seven technological resources.

People

Nearly 1 million workers are employed by chemical industries in the United States. These people work in 12,000 manufacturing facilities throughout the 50 states. Industrial chemists also work in many areas outside the chemical industry (Figure 17.5).

Capital

Industrial processing plants are huge industrial complexes costing many millions of dollars (Figure 17.6). One of the biggest expenses is the cost to control pollution and reduce hazardous wastes.

A sizable amount of money is also spent on basic research. New products and improved processes drive the industry. The chemical industry spends more on research and development than any other industry in the United States.

Time

Time is important in chemistry. Each chemical reaction has its own specific rate of reaction. This rate—how quickly or

Figure 17.5 This pie graph shows the percentages of people working in different chemical industries.

Interpreting *Do foods and cleaning products make up a large or small part of the chemical industry?*

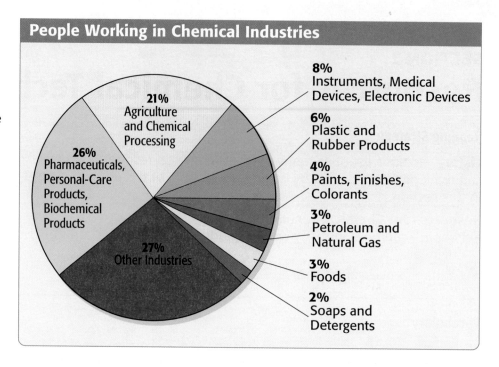

People Working in Chemical Industries

- 21% Agriculture and Chemical Processing
- 26% Pharmaceuticals, Personal-Care Products, Biochemical Products
- 27% Other Industries
- 8% Instruments, Medical Devices, Electronic Devices
- 6% Plastic and Rubber Products
- 4% Paints, Finishes, Colorants
- 3% Petroleum and Natural Gas
- 3% Foods
- 2% Soaps and Detergents

Figure 17.6 The cost of a large, highly automated chemical plant often runs into many millions of dollars.

Extending *In what other ways is capital used in chemical technology?*

slowly the reaction takes place—is dependent on the temperature and concentration of the materials. Different chemical reactions take varying amounts of time to occur.

Time is also important in production. Time is usually the determining factor for success in the marketplace. Once a market has been identified, a product is developed as quickly as possible to meet the need and, in turn, make a profit for the company. Some products are ready for sale within a few months of creation. Other products, such as medicines, may take many years before they are approved for sale because extensive testing is required to see if they are safe.

Information

Information must be gathered about consumer markets for new products and about the sources and costs of various chemicals. Information is also needed to determine methods of developing products and procedures for monitoring and testing the quality of finished products, determining their effectiveness, and quantifying any harmful reactions to humans or the environment.

A great deal of scientific and technical information is needed to understand how chemicals are produced and processed into intermediate chemicals and into specialty products. Much of this information is gathered through basic research. For example, research is done to figure out ways to make natural products artificially. In that way, the availability of these chemicals is ensured and the price of these important chemicals can be kept to a minimum.

Information is also required about the best and safest ways to handle, use, and dispose of chemicals without harming humans and the environment. This information is important not only for chemical workers who may come in contact with harmful chemicals but also for the average person. For example, anyone who changes the oil in a car should know the proper way to dispose of used motor oil, which is considered a toxic material. Motor oil should *not* be dumped into a drain or buried in the yard. It should be returned to a service station that collects used motor oil for recycling.

Some chemical industry employees need to perform analyses of chemical samples. In doing so, they draw on technical information to prepare the samples and use special techniques to separate mixtures into their component chemicals.

Energy

Some chemical reactions require energy, whereas others give off energy (Figure 17.7). A reaction that absorbs energy in the form of heat is called an **endothermic reaction.** An example of an endothermic reaction is the decomposition of water: Energy is needed to decompose, or break down, a molecule of water into its elements, hydrogen and oxygen. Most chemical reactions are endothermic.

A reaction that releases energy in the form of heat is called an **exothermic reaction.** An example of an exothermic reaction is the formation of water from hydrogen and oxygen gases.

This reverse reaction releases the same amount of energy that is absorbed in decomposing the water molecule. It is important for chemists and engineers to take into account the

Figure 17.7 Especially after exercise, the human body releases heat through breathing.

Inferring *Is this an example of an exothermic process or an endothermic process?*

energy needed in an industrial process because equipment must be designed to handle those requirements. For example, if a reaction is exothermic, conditions need to be right to prevent the reactor and pipes from blowing up.

Materials

Chemicals are produced from many different kinds of raw materials (Figure 17.8). These materials include petroleum and natural gas, metals, minerals, vegetable oils, and animal fats. The chemical industry uses these materials to manufacture more than 70,000 different products. Most products are made in huge quantities, totaling tens of billions of pounds every year. Some products, however, such as those used in specialized medicines, may be produced in quantities of only a few pounds each year.

Chemists **synthesize,** or create, molecules for a particular material by combining two or more substances in a chemical reaction. Once the molecules are formed, they are usually mixed with additives or fillers and then processed into other forms that are useable by consumers. Molecules can be transformed into other substances through chemistry.

Certain chemical materials called **catalysts** speed up a chemical reaction. One example of a catalyst is a chemical called ptyalin, which is found in human saliva. Ptyalin speeds up the conversion of starch, which is contained in many of the foods you eat, such as pasta and cereals. Starch is converted to sugar rapidly in the presence of ptyalin. Without this catalyst, it would take weeks for your body to digest certain foods.

Figure 17.8 Copper is a raw material that is used in many chemical processes.

Applying *What products are made with copper? List as many as you can.*

Tools and Machines

A great deal of specialized equipment is used to process chemicals (Figure 17.9). This equipment is used to synthesize, analyze, purify, and manufacture products. Reactors are used to contain chemical reactions. Separators and filtering devices are used to purify products.

Other types of equipment include instruments for analyzing, or evaluating, chemical samples. A **spectrometer** measures the light energy absorbed or transmitted through a sample. Spectrometers are often used to measure the amount of pigment or dye in a liquid sample. A **gas chromatograph** tests the purity of a substance by heating a product past its boiling point to create gases. The gases are then passed through a tube filled with tiny ceramic or polymer beads. Individual components of the mixture are separated, and a signal indicates the purity of the sample.

Safety equipment, such as exhaust hoods and shields, is used for all types of chemical processing. People working with chemicals wear safety goggles and special clothing for protection.

In the classroom, you should also be careful when using chemicals. It is important to develop good safety practices. For instance, do not inhale chemicals—they might be dangerous. Some chemicals undergo a chemical reaction when they are heated, so what you smell could be a different substance. You should always wear goggles and a safety apron. Also, dispose of chemicals in appropriate containers, not down the sink (unless directed to do so).

Figure 17.9 A gas chromatograph is one of the tools used to process chemicals.

Summarizing *What information does a gas chromatograph provide?*

SECTION 2 **Assessment**

Recall and Comprehension

1. Explain why different chemical reactions take varying amounts of time to occur.
2. List the kinds of information that are important in chemical technology.
3. What are catalysts, and why are they important to chemists?

Critical Thinking

1. **Drawing Conclusions** Why does the chemical industry spend more on research and development than any other industry?
2. **Contrasting** What is the difference between endothermic and exothermic reactions?

QUICK ACTIVITY

Red cabbage juice turns acidic solutions red and basic solutions green. Fill three different glasses with white vinegar, ammonia, and water. Pour a small amount of red cabbage juice into each glass. What do you observe? What can you conclude? Test other liquids—lemon juice, orange juice, clear soda, liquid soap—and record your results. **For more related Design Activities, see pages 536–539.**

Connecting to STEM

Reducing Pollution with Catalytic Converters

Automobiles burn gasoline, which is a type of hydrocarbon fuel. A hydrocarbon fuel is made up of hydrogen and carbon atoms. In an ideal situation, all these atoms would combine with the oxygen in the air to form water (H_2O) and carbon dioxide (CO_2). However, the combustion process is not ideal, and some harmful emissions are also produced in car engines. These emissions include carbon monoxide (CO) and nitrous oxides (NO and NO_2).

forms O_2, the oxygen molecule found in air. The nitrogen atom bonds with another nitrogen atom and forms N_2, which is part of the air we breathe.

Removing Carbon Monoxide

The second stage of a catalytic converter converts carbon monoxide into carbon dioxide and water. Hydrocarbons are burned to form water and carbon dioxide. Because oxygen is required for this reaction, an oxygen sensor monitors the amount of oxygen in the exhaust. If oxygen is not sufficient for the catalytic converter to work properly, more air is admitted into the engine.

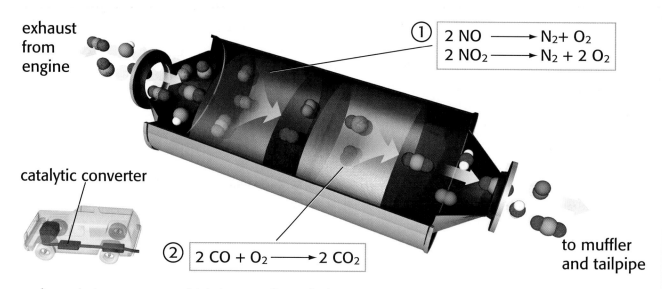

exhaust from engine

catalytic converter

①
$$2\ NO \longrightarrow N_2 + O_2$$
$$2\ NO_2 \longrightarrow N_2 + 2\ O_2$$

②
$$2\ CO + O_2 \longrightarrow 2\ CO_2$$

to muffler and tailpipe

A catalytic converter, which is part of a car's exhaust system, is designed to remove these pollutants. A catalytic converter is a ceramic structure shaped like a honeycomb and coated with the metal catalysts platinum, rhodium, and palladium. The honeycomb structure creates a lot of surface area for the exhaust to contact.

Removing Nitrous Oxides

The first stage of a catalytic converter converts NO_2 emissions. The catalyst holds onto the nitrogen molecule, freeing the oxygen atom. The oxygen atom bonds with another oxygen atom and

Critical Thinking

1. **Summarizing** How does a catalytic converter work? What goes into the converter? What is removed?

2. **Inferring** What would happen if cars did not have catalytic converters?

Using Chemical Technology

Reading Strategy

Outlining As you read this section, create an outline of the ways we use chemicals today. List examples given in the text, and then add any others of your own.

> I. Uses for Chemical Technology
> A. Making Polymers
>   1. Plastics
>   2. Rubber
>   3. Synthetic fibers and clothing

Vocabulary

cracking   pharmaceutical   surfactant

Benchmarks for Learning

- Chemical reactions occur all around us—for example, in cooking, cosmetics, and cars.
- Chemical technologies are used to modify or alter chemical substances.
- Chemical technology has improved many aspects of our lives, but it also involves risks.

What Is Chemical Technology?

One characteristic of the chemical industry is that it provides chemicals to be used by other industries in manufacturing products. The chemical industry provides resources to virtually all sectors of the economy.

The numerous products supplied by different areas of chemical technology impact our lives every day. Chemical reactions are all around us. Cooking produces a chemical change in food, as when you bake a cake. When our bodies absorb oxygen from the air, another chemical reaction is taking place.

Advances in Chemical Technology

The chemical industry in the United States advanced tremendously during the last century. It moved from an industry based on coal and organic chemicals to one based mainly on natural gas and oil.

Accompanying this change was a better understanding of the role of catalysts in chemical reactions. This led to the development of synthetic products such as fertilizers (Figure 17.10). New plant strains were created that were highly responsive to fertilizers, providing higher crop yields, not just bigger plants.

Polymers and Synthetics

The study of polymers, which are large molecules, has become an important branch of chemistry. Just after

Figure 17.10 Most fertilizers today are made from chemicals. Fertilizers are used to increase the productivity of farms.

Extending *What are some possible disadvantages of using chemical fertilizers?*

Figure 17.11 Plastics, such as these Legos, are one example of a product made using chemical technology.

Describing *In a typical day, what types of plastic products do you use?*

World War II, research on polymers led to the development of plastics (Figure 17.11). At the same time, the development of synthetic rubber became important because the source of natural rubber in the Far East was cut off during World War II. The rubber industry was transformed from one that depended on a supply of natural raw material to one that draws on the chemical processing of oil for its primary material inputs.

The knowledge gained during World War II also led to transformations in the clothes we wear. Synthetic fibers—nylon, acrylics, and polyesters—were developed and now are commonly used in the textile industry.

Chemically Based Fuels

The expanding automobile industry created a demand for more and improved gasoline and oil. The petrochemical industry again rose to the challenge and developed high-quality liquid fuels. In the process, the industry developed the ability to build large-scale production facilities that were more efficient than earlier, smaller-scale plants. Improved design information allowed this expansion of scale. New processes were defined that led to the development of polymers and synthetics. Today, the petrochemical industry makes up just one of the many different chemical industries (Figure 17.12).

Category	Examples of Products
Agriculture and Food	Pest and weed control, fertilizers, animal feed, food flavorings, sweeteners, preservatives, new foods, food packaging
Intermediate and Specialty Chemicals	Acrylic acid (used to produce superabsorbent polymers), Biocides (used to slow the growth of bacteria, molds, fungi, algae), Amines (used to manufacture pharmaceuticals, cosmetics, plastics, and dyes), Catalysts (used to speed up other chemical reactions)
Electronics	Semiconductors, superconductors, integrated circuits, photocopier cartridges, coated wiring
Energy and Fuel	Fossil fuels (natural gas, oil, coal), synthetic fuels, solar cells
Environmental Technologies	Pollution control methods, hazardous waste disposal, water and sewage treatment, recycling processes
Health and Medicine	Vaccines, vitamins, medicines
Housing and Construction	Insulation, roofing materials, concrete, vinyl siding, floor tiles and coverings, countertops
Industrial Coatings and Colorants	Primers, paints, stains, finishes, solvents, dyes, caulking, ceramic glazes, resins, glues
Paper and Printing	Pulp production, papermaking, pens, pencils, inks, packaging
Polymers, Materials, and Textiles	Plastics, fiberglass products, composites, nylon, denim, permanent-press and spandex fabrics
Personal-Care Products	Mouthwash, cosmetics, hair-care products, detergents

Figure 17.12 Chemical technologies are used in many different ways.

Extending *Do you use any of the products listed in this table?*

LIVING GREEN
A Clean Chemical Fuel?

Imagine a fuel for your car that burns so clean, the only exhaust is water. You probably think it sounds like science fiction, but hydrogen fuel cells are real, and they are already in use across the country. A hydrogen fuel cell uses hydrogen as its fuel and oxygen as its oxidizing agent. In this chemical reaction, when the hydrogen reacts with oxygen, the hydrogen is reduced and combined with the oxygen, creating H_2O (water) as its end product. In the fuel cell, as the hydrogen is used it must be replaced. The hydrogen fuel cell sounds like an environmentally friendly renewable energy source, but there are concerns. It takes an enormous amount of energy to separate the hydrogen from its source (natural gas, water, biomass) and convert it into useful electricity. Some scientists argue using this amount of energy to produce one cell is wasteful.

MAKE A CHANGE

Hydrogen fuel cells are still in the research phase to ensure their safety and reliability in passenger cars. A huge hurdle to using fuel cells in passenger vehicles is the cost of the cells. Currently, cells cost about $3,000 per kilowatt, and vehicle systems need about 10 kW to operate. There is still much more work to be done, and many companies predict that within the next 10–20 years, hydrogen fuel cell cars may be an affordable reality. That puts current students in technology and engineering in a good position to work on a controversial possibility in renewable energy!

TRY THIS

The U.S. Department of Energy hosts the National Junior Solar Sprint/Hydrogen Fuel Cells competition for teams of middle school students. The object is to build solar or fuel cell cars with the guidance of a teacher or mentor and compete in either design or race categories. If you don't think a competition is right for you, try a hydrogen fuel cell project for this year's science fair. Kits are available online for about $70.

Cracking is a chemical process in which a complex chemical compound is broken (cracked) into simpler compounds. This change is accomplished using high temperatures and pressure (thermal cracking) or by using a catalyst (catalytic cracking). These processes allow chemists and chemical engineers to create processes to convert crude oil, which is very thick, into gasoline and oils used in automobiles and homes.

Pharmaceuticals

The pharmaceutical industry is another important chemical industry. **Pharmaceutical** products include medicines, vaccines, and drugs used for healthcare. Again, World War II provided the impetus to link pharmaceutical companies with universities that were advancing the state of the art in biomedical research. For instance, penicillin was discovered in 1928, but its production was done on a small scale. Chemical engineers linked with microbiologists to solve the problems posed by large-scale production of penicillin, creating the first major achievement in biochemical engineering.

Because the development of new medicines had become a national priority, the National Institutes of Health (NIH) and the National Science Foundation (NSF) provided research funding that spurred the industry's growth. The advances in pharmaceuticals resulted in new antibiotics, anti-inflammatories, anti-ulcer drugs, antidepressants, and antihistamines.

Surfactants

In ancient times, clothes were cleaned by beating them on rocks in streams. Certain soaps, such as the saponin found in the plant soapwort, were used to facilitate the cleaning process. Soap is a **surfactant,** or *surface-active agent*, and is normally made from plant and animal fats. Soaps are naturally biodegradable (i.e., they break down into environmentally safe chemicals) and inexpensive, and they are processed from renewable resources. However, during World War II, these fats were in short supply and were very expensive. During this time, the first synthetic detergent was made. Detergents are synthetic versions of soaps derived from petrochemicals.

Surfactants have a unique chemical structure (Figure 17.13). Part of the molecule loves to combine with water and hates to combine with fats. The other part hates to combine with water but loves to combine with fats. When these chemicals are in the laundry, part of the molecule bonds with the dirt and part of the molecule bonds with the water so that the dirt is washed away with the water. If air is trapped in water that contains a surfactant, bubbles may

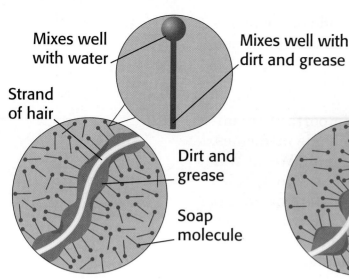

Mixes well with water

Mixes well with dirt and grease

Strand of hair

Dirt and grease

Soap molecule

Soap molecules in the shampoo loosen the grease and dirt on your hair.

Soap molecules break the dirt into tiny pieces.

Water carries away the dirt, which is surrounded by soap molecules.

Figure 17.13 Soap has a special structure that allows it to surround and attach to dirt and grease.

Interpreting *Which part of the soap molecule attaches to dirt?*

be formed. The amount of foam produced by a soap is not necessarily a measure of its cleaning power, although many consumers believe this to be true.

Using Chemicals Today

Chemical technology has improved many aspects of our everyday lives. Medicines and vaccines, for example, are allowing people to live longer, healthier lives (Figure 17.14). However, chemicals are used in many different ways—some of which people may not be aware of. To be wise consumers of technology, we need to understand how chemicals are used and what risks they pose.

Commonly Used Chemicals

Many of the chemicals that may pose health risks are used as food preservatives. Sulfites are used as a preservative in many fruits, vegetables, seafoods, and wines. BHA is used in dry foods, such as cereals. Nitrites are used to preserve red meats, poultry, and fish.

The levels of chemicals in a food that can be considered acceptable, or safe for human consumption, are regulated by the Food and Drug Administration (FDA). The FDA determines what levels are appropriate based on the results of many laboratory tests. Often, the effects of a chemical are difficult to assess. BHA, for example, has been shown to cause cancer in laboratory mice when they were given high levels of the chemical. However, in these same tests, BHA also reduced the effects of other carcinogens.

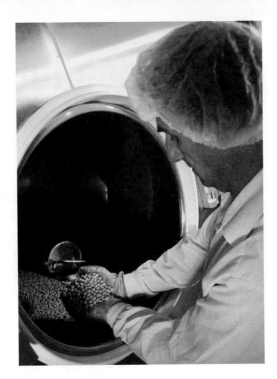

Figure 17.14 Prescription drugs play a key role in the medical industry.

Extending *How might our lives be different if we did not have medicines?*

Balancing the Risks with the Benefits

Although the public perceives chemical additives as a major cause of cancer, only 1 to 3 percent of cancer cases can be attributed to human-made chemicals. Additional testing and monitoring are needed to determine the exact effects of chemicals on our long-term health.

SECTION 3 Assessment

Recall and Comprehension

1. Describe the chemical process of cracking and how it is used to produce gasoline.
2. Name three synthetic products developed during and shortly after World War II.
3. Explain what a surfactant does and where it might be used.

Critical Thinking

1. **Extending** The various types of chemical technology are listed and described on pages 523–528. What photographs would you pick to illustrate each of these areas? Explain your choices.
2. **Calculating** If you were to dilute 10 milliliters of a chemical solution to measure one part per million (ppm), how many liters of water would you need to do this?

QUICK ACTIVITY

Do research on a chemical preservative, such as BHA, nitrites, or sulfite. Determine what levels have been established as safe for use in foods, then find out the levels of this preservative in several foods that you eat on a regular basis. Is the level safe according to guidelines? How do you feel knowing that you are "eating" this chemical? Explain your answer.
For more related Design Activities, see pages 536–539.

People in Technology

Wallace Hume Carothers

Wallace Carothers

In 1934, only a few years after he had come to DuPont, Carothers and his team of researchers produced the first long, flexible strands of nylon.

Today, the clothes we wear are made using all types of chemicals. Gore-Tex is used in raingear and boots, lycra and Spandex are used in athletic wear, and neoprene gel is used as a cushion in everything from wristbands to bicycle seats.

One of the first fabrics to be produced chemically was nylon, which was first made into women's stockings. Nylon was developed and produced due in large part to the efforts of a chemist named Wallace Carothers.

From Harvard to DuPont

Carothers was born in Iowa, where his father was a college professor. Carothers was always interested in science and chemistry. In high school, he built a chemical laboratory in his bedroom. He went on to earn his Master's and Ph.D. degrees, and then became an instructor at Harvard.

Nylon stockings were shaped on wooden forms.

In the 1920s, while Carothers was still at Harvard, he was approached by the chemical company DuPont. DuPont wanted to develop a synthetic fiber that could replace silk, which had come into short supply because of the United States' trade problems with Japan. DuPont hired Carothers to pursue basic research on polymers—long, chainlike molecules.

Developing and Making Nylon

In 1934, only a few years after he had come to DuPont, Carothers and his team of researchers produced the first long, flexible strands of nylon. This material was both strong and flexible—features that natural silk also possessed.

Once the formula for nylon had been determined, DuPont set about to produce millions of tons of nylon fiber at a relatively low cost. In large quantities, nylon had to be made in a hot, steel-lined vessel called an autoclave.

Nylon Hits the Shelves

The first nylon stockings, known simply as "nylons," were sold on October 27, 1939. They sold out within hours.

Due to his persistent efforts, Carothers brought to the world not just an incredible new material but also the knowledge to develop many more chemical products.

Critical Thinking

1. **Analyzing** What characteristics made Carothers successful as a researcher?

2. **Describing** How did historical events affect the production and use of nylon?

Developing Chemical Products

Product Development

You may know that furniture and buildings are products that are designed, but chemical products are also carefully designed by scientists and engineers. Designing and developing a chemical product is a complicated, time-consuming procedure. There are several steps in the product development cycle that must be completed before a product can be sold on the open market.

Market Analysis

As with most other technological designs, human needs and desires drive the development of a chemical product. A team of people, from business managers to environmentalists, works together. They determine the need for a new product, its possible impact, and whether it will be profitable.

Product Design and Synthesis

Once the market need is established, researchers design the new product. The appropriate starting materials are identified, and chemical reactions are performed to synthesize the desired product.

The product must have certain properties—it must meet specific requirements or design criteria for the particular application. For example, a dish detergent might have to have a certain viscosity (thickness), color, and appearance (Figure 17.15). These specifications are set directly by the product manufacturer and indirectly by the consumer.

Figure 17.15 The color and texture of dishwashing liquid are the result of careful testing and market research.

Inferring *Why is it important for companies to get people's opinions about a product?*

Figure 17.16 Scientists perform many tasks in a chemical laboratory.

Summarizing *What is the chemical process of purification?*

For example, consumers probably would not want to buy a cloudy, gray, two-layer dish detergent, so manufacturers set tight specifications to ensure product desirability.

Characterization and Purification

A team of scientists, engineers, and technicians will characterize a product to determine its physical properties (Figure 17.16). This work is done through a series of tests to find out exactly which molecules are in the product.

Once the exact properties of a product are known, scientists **purify** the product so that the substance is made of only one kind of matter. A product can be purified by filtering, distilling (evaporating and then condensing it), or subliming (heating a solid material into a gas and then condensing it back into a solid).

Formulation and Testing

Once the key ingredients have been identified, synthesized, characterized, and purified, they are formulated with other chemicals to create a product with the desired properties. A **formulation** is like a recipe, a systematic method of creating a result. The ingredients and their amounts are varied until the desired result is obtained. Formulators (people who do formulation) use computerized methods to create the desired product using a minimum number of trial formulations.

Sample formulations are thoroughly tested before the new product is manufactured. Some of the tests carried out on a formulation try to answer some of the following questions:

• Does the product work equally well in cold or hot temperatures?

• How does the product react with water?

- How does the product react chemically with the container in which it will be stored?
- How does the product look, feel, and smell, or what is its viscosity?

Scaling Up and Processing

Once the desired properties have been achieved, the formulation and its ingredients can be scaled up, or manufactured in large quantities. A pilot batch is usually made as a further test before sending the formulation to the manufacturing plant. A pilot batch is larger than the quantity produced for research, but many times smaller than the actual production run. The pilot batch is tested to make sure the formulation works for larger quantities.

Chemical formulations can be manufactured in a batch process or in a continuous process. In a **batch process,** all materials are included to complete one operation. A batch process is similar to making a giant pot of spaghetti sauce. All the ingredients are added and cooked in the pot (the chemical reactor) until the sauce (formulation) is done. At the end of the reaction, you have one giant pot of spaghetti sauce (the product).

In a **continuous process,** materials are constantly added to the reactor, where they are mixed and heated according to the formulation, continuously producing the desired product. A continuous process is usually more cost-effective than a batch process.

SECTION 4 Assessment

Recall and Comprehension

1. How is a new chemical product designed?
2. List the methods of purifying a product.
3. Describe what happens during the formulation process.

Critical Thinking

1. **Summarizing** Name the steps in developing a new chemical product.
2. **Contrasting** Explain the difference between batch, or unit, processing and continuous processing.

QUICK ACTIVITY

Using three clean, clear drinking glasses, fill each one with 1" of water. Add approximately 1/4" of vegetable oil to each glass and stir the mixture. What happens after you stop stirring? Now slowly add a teaspoon of liquid laundry detergent to the first glass, liquid hand soap to the second, and dishwashing detergent to the third. What happens? Now stir each of the glasses. What happens?
For more related Design Activities, see pages 536–539.

Chapter Summary

- Everything we see, smell, touch, and taste is made up of chemicals. Many of the things we use every day are made using chemical technology.

- A chemical is a substance that contains any of the atoms found in the periodic table of elements. Atoms bond together to form molecules in different arrangements for different substances.

- Natural products are made of chemicals found in nature. Those that are made by humans are called synthetic products. Chemists and chemical technologists figure out how to produce synthetic products with the same molecules as in the natural product.

- Chemical technology systems combine the seven technological resources to produce more than 70,000 commercial items.

- Chemical technology has improved many aspects of our everyday lives; however, it poses risks that need to be taken into account as well.

- Chemical technology can be broken down into several categories, including agriculture and food, medicine, personal care, electronics, energy and fuels, housing, paper and printing, textiles, and polymers.

- The development of chemical products follows a predictable cycle: market analysis, product design and synthesis, characterization and purification, formulation and testing, and scaling up and processing.

Building Vocabulary

Your teacher may give you a crossword puzzle. Complete the puzzle using the following words and phrases from this chapter. Exchange puzzles with a classmate to check each other's answers.

1. batch process
2. catalyst
3. chemical
4. compound
5. continuous process
6. cracking
7. endothermic reaction
8. exothermic reaction
9. formulation
10. gas chromatograph
11. natural product
12. periodic table
13. pharmaceutical
14. purify
15. rate of reaction
16. spectrometer
17. surfactant
18. synthesize
19. synthetic product

See your teacher for the Crosstech puzzle.

Reviewing Content

1. What is the relationship between an element and an atom?

2. Name three tools or machines used to process chemicals, and explain the function of each.

3. What is the first step in developing a new chemical product?

4. How did the biochemical engineering industry get started?

5. Give ten examples of products that are produced through chemical technology.

Applying Your Knowledge

1. Ozone depletion in the atmosphere is a serious concern. Investigate the role that chemist Paul J. Crutzen had in alerting us to this problem.

2. Investigate catalysts and explain their role in chemical reactions.

3. How might you find information to help you develop a formulation for a hair gel? What design criteria would you use for the gel?

4. Read the labels on the clothes you wear. Which textiles were used in making them? Which ones are natural and which ones are synthetic?

5. Examine the food labels on cookies and soda. Were any chemicals added to the food product? What is their purpose?

Critical Thinking

1. **Inferring** How would you use the periodic table of elements to determine the molecular weight of table salt (NaCl) and of carbon dioxide (CO_2)?

2. **Defending** Defend a pro or con position about the following statement: Natural products are safer than synthetic products.

3. **Predicting** Look at the table of chemical technologies in Section 3 of this chapter. Predict three new or improved products in different categories, and describe the properties and function of each.

4. **Analyzing** Why were so many synthetic and other chemical products developed during World War II? Use specific examples in your answer.

5. **Making a Judgment** Do you think that it is safe to eat foods that have been produced using pesticides? Explain your answer.

 Connecting to STEM

science • technology • engineering • math

Scanning Tunneling Microscopes

Atoms are far too small to be seen with the naked eye, or even with conventional microscopes. A special high-tech device called a scanning tunneling microscope can be used to view atoms. Do research on the Internet to learn more about these microscopes. See if you can find any images of atoms that were taken using a scanning tunneling microscope.

Design Activity 33

LIQUID CRYSTALS

Problem Situation

Chemical technology has provided many new compounds that have improved our lives and allowed us to create new products. Liquid crystals are one such compound. A liquid crystal behaves both as a liquid and as a solid. In some liquid crystals, the orientation of molecules varies with their temperature. The wavelengths—hence the colors—of the light reflected from the liquid crystal surface will vary because of the molecular orientation, so we can use this to determine temperature.

Materials

You will need:

- beakers
- black paper
- cholesteryl nonanoate/perargonate (CN)
- cholesteryl oleyl carbonate (COC)
- hot plate
- plastic gloves
- plastic transparencies
- small plastic or glass tubes with screw tops
- tape
- test-tube holder
- test tubes
- thermometer
- wooden stirrers

Your Challenge

You and your team members are to construct a device that will make use of temperature-dependent liquid crystals.

> Go to your **Student Activity Guide, Design Activity 33.** Complete this activity in your Guide, and state the design challenge in your own words.

① Clarify the Design Specifications and Constraints

To solve the problem, your design must meet the following specifications and constraints:

- The device will use a liquid-crystal display to indicate high temperatures or a temperature range.

> In your Guide, state the design specifications and constraints. Add any others that your team or your teacher included.

② Research and Investigate

To better complete the design challenge, you need to first gather information to help you build a knowledge base.

> In your Guide, complete Knowledge and Skill Builder 1: Investigating Liquid Crystals.

> In your Guide, complete Knowledge and Skill Builder 2: Creating Temperature-Dependent Liquid Crystals.

In your Guide, complete Knowledge and Skill Builder 3: Application Investigation.

③ Generate Alternative Designs

In your Guide, describe two possible solutions that your team has created for the problem.

④ Choose and Justify the Optimal Solution

Refer to your Guide. Explain why you selected the solution you did, and why it was the better choice.

⑤ Develop a Prototype

Construct your device and make the liquid crystal that functions with it. Put a sketch or photograph of the device with the liquid crystal attached in your Guide, as well as a notation about the liquid-crystal formulation you used.

In any technological activity, you will use seven resources: people, capital, time, information, energy, materials, and tools and machines. In your Guide, indicate which resources were most important in this activity, and how you made trade-offs among them.

⑥ Test and Evaluate

How will you test and evaluate your design? In your Guide, describe the testing procedure you will use. Explain how the results show that the design solves the problem and meets the specifications and constraints.

⑦ Redesign the Solution

Respond to the questions in your Guide about how you would redesign your solution. Your redesign should be based on the knowledge and information that you gained during the activity.

⑧ Communicate Your Achievements

In your Guide, describe the plan you will use to present your solution to your class. Show what handouts and/or PowerPoint slides you will use.

Design Activity 34
MAKING SLIME

Problem Situation

Slime is a gooey substance sold at many toy stores. In its pure state, it is a clear colorless gel that resembles hair gel. Slime can be shaped into a ball and bounced, or flattened to make imprints of comic strips. It can also be stretched if pulled slowly, or it can "fracture" if pulled apart quickly. It has properties such as tensile strength, flexibility, and elasticity. It can be stored in the refrigerator in a tightly sealed plastic bag for about 2 weeks before becoming moldy. Slime is made from a polymer called PVA, polyvinyl alcohol. A polymer is a very large molecule that is made up of many smaller parts.

Materials

You will need:

- 10 ml white glue
- 5 ml borax solution (4 gm borax dissolved in 100 ml of water)
- 50 ml polymer solution (4% polyvinyl alcohol, PVA)
- disposable plastic cup
- food coloring
- maximum of 3 ml of borax per formulation
- mixing rod or tongue depressor

Your Challenge

You and your team members are to design your own slime that will bounce and stretch.

> Go to your **Student Activity Guide, Design Activity 34.** Complete this activity in your Guide, and state the design challenge in your own words.

① Clarify the Design Specifications and Constraints

To solve the problem, your design must meet the following specifications and constraints:

- When formed into a ball, the slime should bounce at least 60 cm high.
- When pulled apart, the same ball of slime should stretch at least 20 cm.
- All slime formulations must total 10 ml.

> In your Guide, state the design specifications and constraints. Add any others that your team or your teacher included.

② Research and Investigate

To better complete the design challenge, you need to first gather information to help you build a knowledge base.

> In your Guide, complete Knowledge and Skill Builder 1: Cross-Linking Polymers.

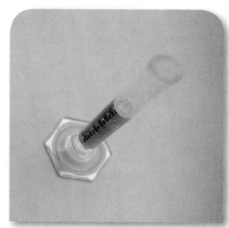

In your Guide, complete Knowledge and Skill Builder 2: Preliminary Slime and Glue Formulations.

In your Guide, complete Knowledge and Skill Builder 3: Factor Analysis.

③ Generate Alternative Designs

In your Guide, describe two possible solutions to the problem.

④ Choose and Justify the Optimal Solution

Refer to your Guide. Explain why you selected the solution you did, and why it was the better choice.

⑤ Develop a Prototype

Make your final formulation of slime. Include a copy of the formulation you used in your Guide.

In any technological activity, you will use seven resources: people, capital, time, information, energy, material, and tools and machines. In your Guide, indicate which resources were most important in this activity, and how you made trade-offs among them.

⑥ Test and Evaluate

How will you test and evaluate your design? In your Guide, describe the testing procedures you will use. Indicate how the results will show that the design solves the problem and meets the specifications and constraints.

⑦ Redesign the Solution

Respond to the questions in your Guide about how you would redesign your solution. Your redesign should be based on the knowledge and information that you gained during the activity.

⑧ Communicate Your Achievements

In your Guide, describe the plan you will use to present your solution to your class. Show what handouts and/or PowerPoint slides you will use.

UNIT 7

The Future of Technology in Society

Unit Outline

Chapter 18
The Future of Technology

"I want to encourage the readers of your book to be studious and to stay away from alcohol and drugs. It's a much bigger adventure to get into technology."

—Harvey Severson

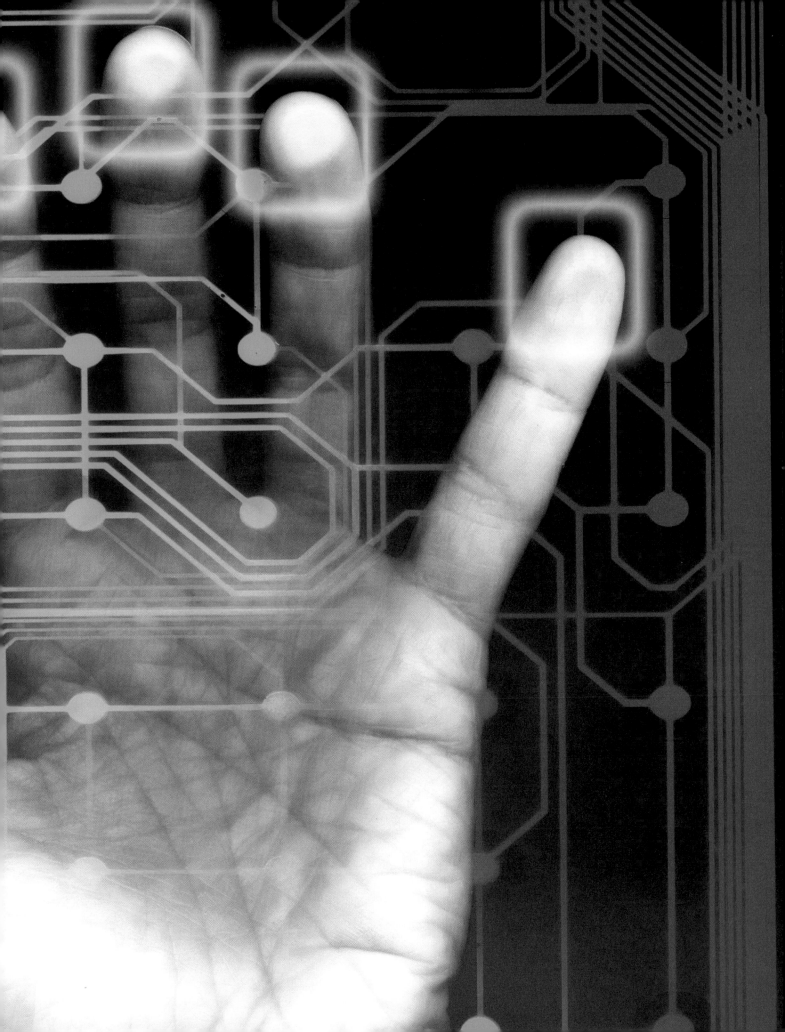

CHAPTER 18

The Future of Technology

SECTION 1   The Wireless Revolution
SECTION 2   Personal Manufacturing
SECTION 3   Forecasting and Living with the Future

In this chapter, you will explore the following aspects of the future of technology:

- the wireless revolution
- artificial intelligence
- virtual reality
- personal manufacturing
- technostress

What effects will technology have on your life in the future? What amazing new inventions will you see in the next few years, or by the end of your lifetime? Here are some possibilities for the near future.

- Computer screens will be lightweight, flexible, and rolled into a metal tube for storage.

- Chemists will combine elements to create new foods that are as healthful and tasty as the foods we eat today.

- Holographic telephones will project life-size images of the caller and the person being called.

- Fossil fuels will run out. Renewable energy sources will fill most of our energy needs.

- People will manufacture items they need in their homes.

As technology changes and becomes more complex, these possibilities may become a reality. Part of preparing for the future will be deciding how we want technology to change.

The Wireless Revolution

Benchmarks for Learning

- Wireless technology will create a new communications revolution.
- Nanotechnology will allow us to build computers and medical devices using atoms and molecules.
- Artificial intelligence will be used in robotics, computer games, and virtual reality.

Reading Strategy

Listing   As you read this section, think about the effects the Wireless Revolution will have on our lives. Make a list of as many different effects as you can.

Effects of the Wireless Revolution
1. Technology will be more convenient.
2. Computers will perform many more functions.

Vocabulary

wireless personal area
  network
nanotechnology

artificial
  intelligence
speech recognition

speech synthesis
robotics
virtual reality

Looking to the Future

You learned in Chapter 1 that existing ideas are used as building blocks for newer and more powerful technologies. Future technologies will become more and more complex. Space shuttles, satellites, robotic factories, and maglev trains all use ideas from a variety of technological systems—electrical, transportation, manufacturing, and scientific. This confluence, or flowing together, of systems means that new technologies will not be separate and distinct. Many systems will be combined to produce new technologies.

Because of the rapid improvements in computers and electronics, new communication technologies are emerging every day. What will technology be like in the future? Bill Gates, head of Microsoft Corporation, says computers that can talk, see, and listen will soon be developed. Faster, more powerful chips and new software will enable computers to recognize and converse with people. Gates notes, "When people look back on the computers of today, they'll say: 'What did they do? They couldn't see, they couldn't listen, they couldn't speak.'" Computers will be getting smaller and becoming even more powerful.

Nanotechnology and Computers

Linking all of these wireless networks will be faster—and much smaller—computers. Computers of the future will be based on nanotechnology. **Nanotechnology** is the creation

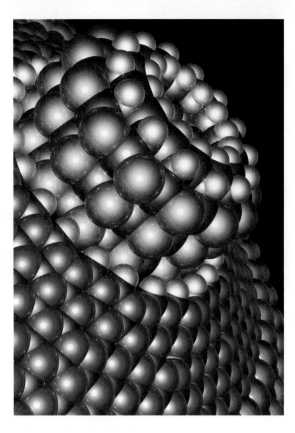

Figure 18.1 These interlocking gears were assembled using many individual carbon atoms.

Summarizing *Why is nanotechnology such an important technological development?*

of materials, devices, and systems using individual atoms and molecules. The word comes from the prefix *nano-*, which means one billionth. Nanotechnology uses particles that are 1/80,000 the diameter of a human hair. At such a small scale, new physical, chemical, and biological properties become evident.

Instead of using transistors and silicon chips, quantum computers will follow the laws of physics that govern subatomic particles. For example, electrons will be able to "tunnel" through a barrier that would normally block their path. Quantum computers may be able to solve extremely complex problems, such as breaking codes for government work.

Like today's computers, electronic nanocomputers will represent information as the storage and movement of electrons. Future transistors will be much smaller, allowing up to 10,000 transistors to fit in the space taken up by one current transistor. Mechanical nanocomputers will calculate by using moving molecule-size rods and rotating gears that spin on shafts and bearings (Figure 18.1). These tiny machines and computers will be assembled by mechanically positioning atoms one at a time.

Artificial Intelligence

Artificial intelligence (AI) is any use of machines to imitate human intelligence. AI can be used to recognize and

Technology in the Real World

From the Film Studio to the Living Room

Motion-sensing, or motion capture, technology is any technology that translates motion into a digital signal that can be interpreted by a computer. If you have seen almost any movie with special effects in the last five years, you have seen one of the things that can be created with motion capture technology.

In film, motion capture technology enables an actor's movements to be tracked by computers. The actor wears an outfit that has reflective dots stuck to it. These dots are tracked by cameras. The cameras communicate the motions to the computer, which can then create an animated character that moves just like the actor. This results in more lifelike animated characters and saves the film studio a great deal of time and money. They don't have to draw or have a computer create every single element of a motion. Gollum in *The Lord of the Rings* movies and King Kong in the 2005 version of the movie are both animated characters that were created using motion capture technology.

Another recent use of motion-sensing technology that many people are familiar with is Nintendo's Wii, a gaming console with a remote, motion-sensing controller. This controller lets players interact with their games by moving the controller through space, swinging it like a bat, pointing it like a gun, or turning it like a steering wheel, for example.

The Wii controller uses infrared (IR), just like a regular television remote control, as well as Bluetooth wireless technology to communicate with a sensor bar that must be placed near the television. The controller can track its own distance and angle from the sensor bar. There is also a device inside the controller, called the accelerometer, that can sense how fast the controller is moving through space, and another one, called a gyrometer, that can sense how the controller is being tilted. Finally, it all shows up on the television because the controller uses Bluetooth to communicate all that location and motion information to the game console, which interprets it as shooting the bad guy or hitting the baseball.

Manufacturers are using motion-sensing technology to make product design and testing more efficient. It allows designers to interact with computer simulations and models.

Marketers are using it to create interactive advertising, such as billboards that react to people walking close by or moving their hands. Corporations are also using motion-sensing computer controls to run dramatic sales or information presentations. Instead of clicking and dragging with a mouse, special glove and camera systems can allow people to move things around on a computer screen using hand gestures. This can add an exciting new element of showmanship to presentations.

Companies are finding new ways to use motion-sensing technology every day. Many of those discoveries will be finding their ways into homes very soon, just like the Wii already has.

re-create speech, to make humanlike robots, and to create complete virtual reality environments.

Voice Recognition Technologies

As it continues to improve, voice recognition will go hand in hand with a wireless environment. New voice recognition technologies will allow speech recognition (converting speech into text) and speech synthesis (converting text into speech).

In **speech recognition** applications, the sound waves of speech are digitized and matched to coded waves. These matches are then converted to text, as in typed language. Speech recognition technology allows you to speak into devices when it is not practical to enter data using a keyboard, such as when using handheld computers and car phones. Also, computer games can become more friendly and realistic if players can speak to the characters.

Speech synthesis applications convert text into speech. This involves breaking down the written words into the basic sounds of a language and then converting the sounds into digital audio. Because the sound is generated electronically, it does not sound as natural as a real human voice. One important use of speech synthesis is to help people who are visually impaired get information from text materials.

Robotics

Robotics is the application of AI in the design and construction of robots. Several different types of problems must be addressed when designing a robot. In many cases, the robot must be programmed to "think" so it can develop a plan for doing a job. Mechanical problems might include the need to design a robot's arm so that it can move strongly in all directions. Sensory problems include the need to design a robot so that it can pick up fragile objects without breaking them.

Another design challenge is robotic vision (Figure 18.2). Vision systems must be developed to allow robots to recognize objects and interpret visual images. When you look around, recognizing desks or chairs is easy, but it is very difficult for a robot. Another problem that is easy for you but hard for a robot is finding a path from one place to another. Knowing how to avoid bumping into objects must be programmed into the robot.

Figure 18.2 Artificial intelligence is being used to make robots communicate more like people.

Applying *What applications might very lifelike robots be useful for?*

Virtual Reality

Virtual reality uses AI to create a three-dimensional computer-generated environment into which a person is projected using special eyewear or a glove. In virtual reality,

Technology in the Real World

Nanotechnology in Medicine

You don't need to look into the future to see technology in use in the human body. Many electronic devices are already being used in the human body to help it function better. Heart pacemakers keep the heart beating regularly. Hearing aids are implanted in the ear to aid in hearing. What will be different in the future is that these devices will become much smaller.

Technology as Small as a Pill

With nanotechnology, new medical sensors will improve hearing, smell, and even the sense of taste. They will connect directly to the nervous system and deliver information directly to the human brain.

Other types of sensors could provide people with valuable information about their environments. Small sensors on the body might detect information about the chemical makeup of your surroundings. This type of information could provide a type of security system and also help people avoid unhealthy or toxic environments. Some years from now, doctors may provide you with swallowable capsules containing millions of sensors that have been programmed to find and kill cancer cells.

Nanobots

In the future, nanobots—robots of microscopic size built using nanotechnology—will travel through the bloodstream to provide a picture of the way the brain's circuitry works.

To make examinations of the stomach and intestinal area easier, patients will be able to swallow a pill-sized capsule that moves smoothly and painlessly through the body. As the capsule passes through the stomach and intestinal area, it will transmit signals to an array of antennas placed on the outside of the patient's body. Data will be stored in a data recorder worn on the person's belt. A computer will process the data and produce a short video clip of the results.

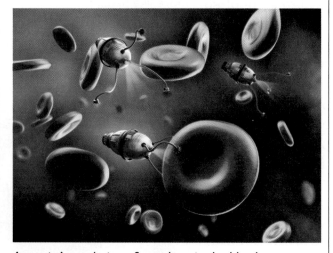

An artist's rendering of nanobots in the blood stream.

Critical Thinking

1. **Summarizing** How can a simple change in size lead to new technologies?

2. **Inferring** Why might it be helpful for people to have better sensory evaluation of their environment?

you can go where you want to go, experience what you want to experience, and be who you want to be (Figure 18.3). Soon, people will be able to "virtually" touch each other. People will be using virtual reality glasses to put images directly on their retinas. Virtual reality will allow you to drive your dream car and look under its hood. It will take you on tours through foreign countries. It will allow you to view the inner workings of molecules.

"Smart" Houses and Cars

In the future, "smart" houses will have built-in computers that can control systems. These houses will have bathrooms in which you can choose a shower, a spring rain, or a warm breeze. Bedrooms will have voice-activated radio and television systems.

"Smart" cars will be able to travel on autopilot. Communication systems will allow cars to exchange information about speed, direction, and location with an automated highway control network and with other vehicles. Superhighways will send signals to computers in cars, keeping them on the road until the correct exit is reached. Magnetic markers will be installed along the center or edges of highways. The markers will contain digital information to warn drivers of exits or curves.

Figure 18.3 Virtual reality games are one application of artificial intelligence.

Hypothesizing *Why are virtual reality games more popular than traditional games?*

SECTION 1 Assessment

Recall and Comprehension

1. What is a wireless personal area network? How is such a network used?
2. Describe a nanocomputer. How is it different from today's computers?
3. Explain how homes will be "smart" in the future.

Critical Thinking

1. **Contrasting** What is the difference between speech recognition and speech synthesis?
2. **Summarizing** How might artificial intelligence be used to make machines imitate human intelligence?
3. **Analyzing** What problems must be addressed when designing a robot?

QUICK ACTIVITY

Accurate speech recognition programs must be able to distinguish between words that sound alike but have different meanings. Some examples are: by, buy, bye; see, sea; hi, high, Hy; two, too, to. Make a list of as many words as you can that could cause problems in speech recognition software. Compare your list with those of other students in the class.
For more related Design Activities, see pages 564–567.

Personal Manufacturing

Benchmarks for Learning

- People will be able to manufacture products that are customized to their individual needs.
- People's own cells can be used to produce replacement tissue and bones for them.

Reading Strategy

Listing Make a list of different products that one day might be manufactured by people in their homes. Indicate why you think these products would benefit from customization, rather than mass production.

Personal Fabrication
1. Manufacturing
   - Make cell phone

Vocabulary

Fabber   Bioprinting   Self replication

Personal Manufacturing

Imagine that you need a new cell phone. Now you can purchase one online or go to a nearby store and select a model. What if you built your own? What if there were a whole variety of cell phone designs, from simple to complex, you downloaded the design, sent it to your personal fabricator and in a couple of hours you had your own custom-designed, and personally manufactured, cell phone? Not really so far-fetched. The word that is used to describe this is Fabbers—machines that quickly create objects from specifications in a CAD file.

How Fabrication Works

Figure 18.4 illustrates a generic Fabber. In this system there is a syringe with a gel substance that is accurately extruded

Figure 18.4 A personal fabrication device, Fabber, can be used to manufacture objects that are personalized by and for you.

Summarizing *What are some advantages to personal manufacturing?*

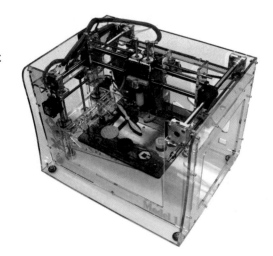

to form the object from the CAD files, which are converted to stereolithographic files. The object, such as the propeller, or gear, shown in the picture, is formed on the plate and the gear mechanisms raise the extruder as the object is built up. Thus, the extruder needs to move only vertically in the Z direction, but in a variety of patterns in the X-Y directions. Other designs used an extruder that is essentially like a hot glue gun, where a material is fed continuously through the nozzle and heated, so it melts, and forms the object. There are situations where the table base moves in the Z direction, and the nozzle only moves in the X-Y directions.

There are Fabbers that have two or more syringe/extruders, so different materials can be used in the fabrication process. For instance, teams have fabbed batteries that were embedded in devices, such as a flashlight, which you could pick up and turn on.

Biological Designs

One of the difficulties facing surgeons is replacing bones that have degenerated or been badly fractured. What if an image of the bone was sent to the Fabber and a custom-designed replacement was created with the precise size and

Figure 18.5 Schematics of building a tubular organ module by bioprinting. The blue sheets represent the biopaper or scaffolding biocompatible gel. The gel and the bio-ink particles are deposited/printed layer–by layer. The cells then naturally connect with each other in a multicellular self-assembly.

Identifying *What do you think will be some challenges in fabricating biological tissues?*

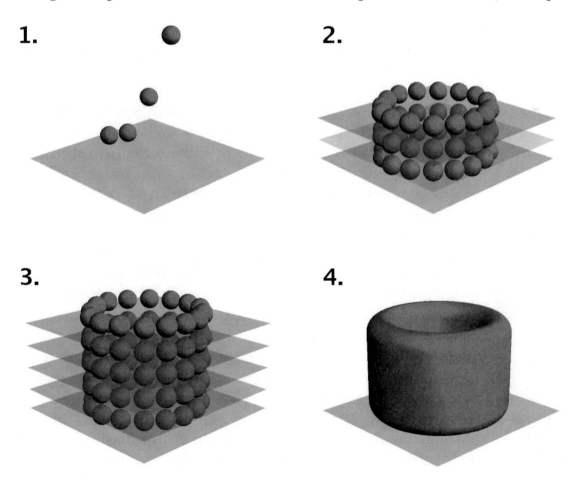

shape, tailored to the patient? Teams are working on bones and cartilage that were created from the patient's own cells, thus avoiding the problem of rejection by the body. Cells are taken from the patient and grown in laboratories, so they multiply millions of times, and then these are used to form the bone or cartilage.

Bone and cartilage do not require a blood supply, but tissue does. There is great interest in creating organs, such as a kidney or the liver. The concept involves using a bio-ink, a mixture created from a person's cells. An organ is created as the bio-ink forms a three-dimensional structure with the cells self-assembling to form a functioning organ. This will become more of a medical issue as the population ages and there are fewer and fewer organ donations available. A significant hurdle is that the tissue requires blood vessels so that it can receive vital nutrients for its survival. How to form these and provide the nutritive blood has not been solved yet.

Self-Replicating Robots

One of the challenges in space exploration is to have devices that can repair themselves and their environment automatically, independent of human intervention. Thus, thinking robots that can self-replicate, can repair parts damaged by radiation and the hostile environment, are being developed. Researchers are developing electro-mechanical systems that can self-reproduce, analogously as biological systems reproduce.

Current efforts, such as creating a flashlight, which needs structural and mechanical elements, indicate that advanced fabbers could be used for this type of manufacturing. Other work has demonstrated that the artificial intelligence used in robots allows for self-replication. Other studies have indicated that robots can design and produce devices to meet specifications independent of human intervention. Many of the elements needed for self-replication exist. However, there may be dangers involved—because the robots do not need human intervention, people cannot control the outcomes.

3-D Printing and Art

Technology and art have been intertwined from the beginnings of humankind. There are many types of 3-D printing devices that can be used to create large objects using a different technology than used by Fabbers. It is more expensive and can use one type of ingredient, for instance metal powder. In this type of printer, illustrated in Figure 18.6, the

Figure 18.6 A 3-D printer that uses powder layers as a means to build an object.

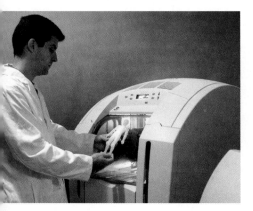

Connecting to STEM

Energy from Space: Using Helium-3 from the Moon

As fossil fuels become more and more depleted, we need to explore alternative sources of energy. One new source of energy may be nuclear fusion. In this process, atoms are forced to combine, releasing huge amounts of energy.

Although these reactions produce a lot of energy, they also require huge amounts of energy to start. Researchers are trying to find ways to get at least as much energy *from* the

reaction as is needed to start it. Once this has been achieved, the excess energy can be used to generate electricity.

Fuel for the Future?

Because nuclear fusion is such a rich source of energy, some researchers predict that fusion reactors will be built by the year 2020. Energy from atomic fusion might well become a major energy source in the decades ahead. The simplest fuels used for fusion are two types of hydrogen atoms, deuterium and tritium.

Although deuterium and tritium produce a lot of energy, the perfect source of fuel for fusion may be helium-3, a variation of the helium used to blow up balloons. A fusion reactor using helium-3 and deuterium would produce very little radioactive waste. Therefore, you could safely build a helium-3 plant in the middle of a populated area.

An Abundant Resource

The problem is that helium-3 is very rare on Earth. However, it is abundant on the moon. Scientists estimate that there are about 1 million tons of helium-3 on the moon, enough to power the world for thousands of years. One Space Shuttle load of 25 tons could satisfy the energy needs of the United States for an entire year.

Critical Thinking

1. **Predicting** Helium-3 is found in the powdery soil of the moon, which means people would need to mine the surface of the moon to recover it. Do you foresee any negative impact of collecting the soil in this way?

2. **Making a Judgment** Would you feel comfortable having a helium-3 plant built in your neighborhood? Why or why not?

Figure 18.7

Figure 18.8

Figure 18.7 Artwork designed and fabricated in a 3-D printer.

Figure 18.8 A sugary treat designed and fabricated using 3-D printing.

CAD image is again sliced into different layers. The printer itself puts down a thin layer of powder and the print head squirts a binder where the slice image is. Another layer of powder is put in place, the binder is directed again at the powder in the appropriate locations, and the object is built.

The renderings can be quite complex and interesting, as in Figure 18.7. People are using this technique in creating food objects, as they do with Fabbers. Sugar is a popular ingredient and elaborate renderings can be created (Figure 18.8).

SECTION 2 **Assessment**

Recall and Comprehension

1. Name some products that could be manufactured at home.
2. Explain bio-printing and how cells join together.
3. What is a Fabber?

Critical Thinking

1. **Inferring** How might Fabbers change manufacturing industries?
2. **Predicting** What might be some dangers associated with self-replicating robots?
3. Imagine that you wanted to set up a business that assisted people in custom manufacturing objects.
4. What aspects of your business would you feature?

QUICK ACTIVITY

Working as a team with your classmates, decide on a fairly simple project, e.g. a flashlight, that you would like to fabricate. Sketch your design and include all the materials that are required. How many different syringes or extruders will be needed to complete your design?

For more related Design Activities, see pages 564–567.

For: Alternative Engine Activity
Visit: www.mytechedkit.com

What Will You Be Driving?
Alternative Engines Online

Technology is everywhere in today's cars. Global positioning systems, DVD players, and cellphones are common features, but what about technology under the hood?

Cleaner burning fuels may be the next development to affect the way cars are manufactured. Go online to learn more about one advance: the hydrogen engine.

Forecasting and Living with the Future

Benchmarks for Learning

- Using different forecasting techniques, people can anticipate the consequences of new technologies.
- In modern society, we must deal with technostress.
- Future employers will seek workers with technical skills, problem-solving abilities, and the ability to work on a team.
- Technology interacts with society causing it to change, which often lead to new wants and needs.

Reading Strategy

Listing Make a list to help you organize different methods for forecasting the future. Each item of the list should note a different forecasting technique, along with a brief description.

Forecasting the Future
1. Futures Wheel
   - Ideas are represented as spokes on a wheel.

Vocabulary

telecommuting   technostress   information overload

The Changing Face of Work

Technology affects society, sometimes causing changes in industry, which ripple through the economy and job opportunities, politics, and how we as people interact with one another. In the workplace of the future, jobs will require workers with a great deal of knowledge. As machines gradually replace laborers, education and training will become more important. This is even true today. The highest paying jobs are those that require people to use their brains, not their muscles. Web designers, engineers, and computer programmers are paid more than clerks and fast-food workers.

Telecommuting

In the future, more and more business will be conducted over the World Wide Web. Because of the Internet and other communication technologies, the number of small businesses operating from a home or a community environment will increase. As technology continues to improve, more workers will reap the benefits of telecommuting (Figure 18.9). **Telecommuting** is working at home using an electronic linkup with a central office. Telecommuters interact with co-workers and customers via computers, fax machines, and telephones, and transmit work in digital form over the Internet to a company's computer.

People who telecommute make their own schedules. They do not have to travel to and from an office every day. Parents are especially interested in telecommuting because they can stay at home with their children. Businesses save money

Figure 18.9 Telecommuting presents new challenges for workers' productivity and well-being.

Summarizing *Why might telecommuting become more common in the future?*

because they do not typically have to provide desks, workspace, or health benefits.

As with most innovation, there are some drawbacks as well as benefits to consider. A telecommuting worker has little face-to-face communication with other workers, so he or she may miss the human contact. In addition, many employers believe that better ideas and products are generated by people working together and interacting creatively, which is difficult for telecommuters.

Skills for the Future

With the growth of technology, people with technical skills who are good problem solvers—capable of finding creative solutions and looking at things in new ways—will be valued increasingly as employees. Future jobs will require employees to use their abilities creatively and in harmony with people having different skills.

Differences are prized because individuals bring unique strengths to a team and complement other team members' abilities. People from different racial, ethnic, and cultural backgrounds add a great deal to a company. They represent

Figure 18.10 Technostress is any form of stress caused by working with technology.

Analyzing *What causes technostress for you?*

a variety of views and ideas that often lead to innovative solutions. In a well-run company, everybody counts.

Technostress

The more technology surrounds us, the more we depend on it. Our dependence on complex machines creates a type of stress called **technostress** (Figure 18.10). For example, how would you feel if the automated teller machine (ATM) swallowed your ATM card, did not return it, and the bank was closed? Have you ever walked into a public restroom to rinse your hands and found that there were no faucets, only electronically controlled spouts? In these spouts, the water flow is activated by a microprocessor-controlled sensor, which responds to nearby motion. This greatly reduces the amount of water being wasted, but it can be frustrating to figure out how the device works.

It takes time to adjust to new technologies, and that also creates stress. As technology changes more and more rapidly, people must adjust to even more new devices and methods of doing things. People are constantly bombarded with messages from cellular and landline phones, instant messaging, email, and fax machines. The Internet provides great opportunities to receive and search for vast amounts of information, but so much information may be overwhelming. Having to deal with too much information is known as **information overload.**

As events occur more quickly, it becomes difficult to slow down. Have you ever talked on the phone while working on

LIVING GREEN
Calculating Your Carbon Footprint

In the future, our ability to sustain resources and a healthy environment will certainly become more and more important as those resources dwindle. As humans, we must pay particular attention to our own individual effects on the environment, becoming aware of how our actions have a direct effect on biodiversity, pollution, and climate change, to name a few. One way that we can do this is to be aware of our individual carbon footprint, or the total amount of carbon dioxide (the most common greenhouse gas) we are directly responsible for emitting into the environment.

MAKE A CHANGE
There are many factors involved in calculating your carbon footprint, and many calculators available for free on the Internet. One calculator, sponsored by the Environmental Protection Agency, utilizes information about how much carbon a class of students emits. Most of these tools ask you about your usage of water and energy. You input your own data, and the calculator gives you an estimate of how many pounds of carbon dioxide you currently emit into the environment.

TRY THIS
Once you know about how much carbon dioxide your daily routines produce, it's critical that you take the next step to reduce those emissions. Turning off the tap while brushing your teeth, walking or riding your bike whenever possible, using less energy and responsibly recycling whenever possible are all important steps to lightening your individual impact on global climate change. However, the most crucial step to sustaining and improving the environment is to be aware and informed of the issues relevant to living green. What will you do differently? How will you learn more, and what changes are you willing to make?

the computer on an unrelated topic? Computers are made for multitasking. However, too much multitasking in too short a time also leads to technostress.

Forecasting the Future

You need to prepare for the future and the technologies that will change your life. People who study the future help

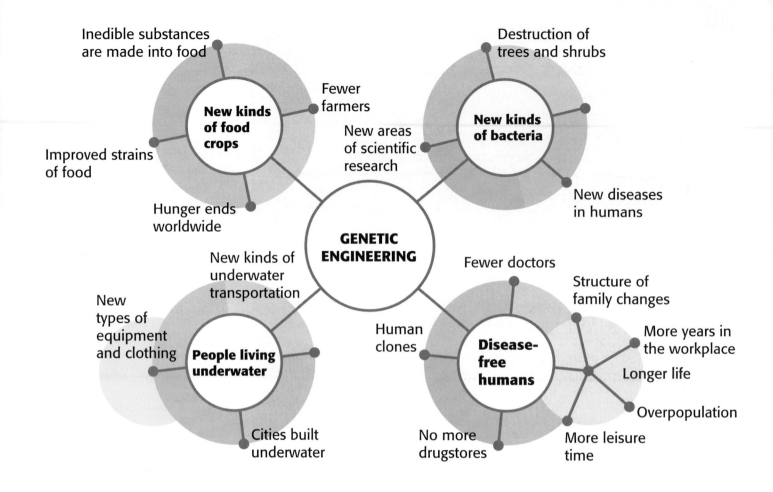

Inedible substances are made into food

New kinds of food crops

Fewer farmers

Improved strains of food

Hunger ends worldwide

New areas of scientific research

New kinds of bacteria

Destruction of trees and shrubs

New diseases in humans

GENETIC ENGINEERING

New kinds of underwater transportation

New types of equipment and clothing

People living underwater

Cities built underwater

Human clones

Disease-free humans

Fewer doctors

Structure of family changes

More years in the workplace

Longer life

Overpopulation

No more drugstores

More leisure time

Figure 18.11 A futures wheel helps us determine possible outcomes of a new idea.

Extending *What are some other consequences of genetic engineering?*

decide which technologies should be developed. They consider the impact of new technologies on the way people live and work. They consider ethical issues and how these issues can be resolved.

One way of forecasting the future is to use a futures wheel. A futures wheel is a diagram with one main idea at the center of a circle and other ideas on the spokes (Figure 18.11). It shows how the central idea leads to other ideas. Each of these ideas, in turn, has its own likely future outcomes.

Another way of forecasting the future is through a Delphi survey. In a Delphi survey, experts rank a list of ideas for future development. Each member of the group first lists ten kinds of changes he or she expects to see in the future. The lists are combined and then sent to each group member, who then arranges the ideas from most likely to happen to least likely to happen. The final ranked list is the group's forecast of the future.

A trend analysis forecasts the future by looking at the past. For example, the cost of computers has dropped at an increasing rate over the last ten years. Forecasters might suggest that prices will continue to drop in this way. Computer prices in the year 2020 might be so low that computers could be used everywhere and anywhere. On the other hand, simply because events occurred a certain way in the past does not mean that a trend will continue. Trend analysis does not always work.

Conclusion

One thing is certain: Technology will continue to grow at an ever-increasing rate. Because of this, people must be ready to plan and direct technology.

Life in the future will depend on how well technology is controlled. The quality of life on Earth should be our main concern. People must think more in terms of the human race and less in terms of individual countries or groups. People worldwide must work together to ensure their own survival as well as that of the living things with which they share our planet.

Technology is not magic. As you learn more about science, mathematics, and technology, you will see that technology must be controlled—for the present good and for the good of future generations.

SECTION 3 Assessment

Recall and Comprehension

1. Describe trend analysis.
2. What is information overload? Why is this becoming more and more common?
3. In the future, will training and education become more or less desirable?
4. What is a futures wheel? How is it used?

Critical Thinking

1. **Predicting** Draw a futures wheel. At the center, place a skin cream that brings back youth. Predict the possible outcomes.
2. **Taking a Position** Who should make decisions about developing and using a new technology? Should it be scientists, religious leaders, or the government? Explain your position.

QUICK ACTIVITY

Have you ever been techno-stressed out? How about the time you lost all of your homework because your computer crashed? Or the time you *really* wanted to see something on television and the cable went out? Make a list of some of the most frustrating technostress situations you have experienced. Share the list with your classmates.
For more related Design Activities, see pages 564–567.

18 Review and Assessment

Chapter Summary

- The next revolution in communication will be the wireless revolution. Wireless personal area networks make it possible for many devices to share information without having to be physically connected. The Intelligent Wireless Web combines speech recognition, WPANs, integrated wireless and wired networks, and wireless protocols.

- Nanotechnology will allow devices to be built using atoms or molecules as building blocks. Nanotechnology will be applied to computers, medicine, and other areas of technology.

- Future computers will be smaller and even more powerful than the computers of today. They will use nanotechnology to solve more and more complex problems.

- Artificial intelligence can be applied to robotics, computer games, virtual reality, and text and voice recognition.

- "Smart" houses and cars will have built-in computers to control many systems and provide wireless access to the Internet.

- "Manufacturing" will become more automated and will use new materials and manufacturing methods.

- Personal manufacturing will become commonplace.

- More and more business will be conducted over the World Wide Web. People will do more telecommuting instead of using vehicles to commute to work.

- People who study the future will help determine which technologies should be developed. They will predict how new technologies will change lives.

Building Vocabulary

Use the chapter vocabulary words and phrases listed below to create a crossword puzzle. Exchange puzzles with a classmate. Complete the puzzles and then check each other's answers.

1. artificial intelligence
2. Delphi survey
3. Fabber
4. futures wheel
5. information overload
6. nanotechnology
7. personal manufacturing
8. robotics
9. speech recognition
10. speech synthesis
11. technostress
12. telecommuting
13. trend analysis
14. virtual reality
15. wireless personal area network
16. wireless revolution

See your teacher for the Crosstech puzzle.

Reviewing Content

1. How will wireless technology create a new communications revolution?

2. Which two products would you choose to manufacture in space? Why?

3. Describe three ways that future travel will differ from present-day travel.

4. Name some medical applications of nanotechnology.

5. Describe "smart" technology. How might it help reduce traffic accidents?

Applying Your Knowledge

1. Make a presentation on three emerging and innovative technologies. As a class, select the three best presentations, and display these in your classroom.

2. Draw a design for a futuristic communication system that would allow you to communicate with a class of students in Europe.

3. Do research on nanotechnology. What kinds of devices might be built from atoms and molecules?

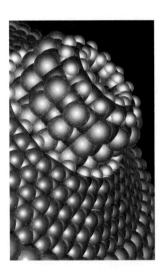

4. Create a futures wheel for living in space. What are the different possible outcomes of establishing communities and cities in a space station?

5. Do you ever feel technostress? Describe one situation that caused technostress and your reaction to it.

6. Describe several career opportunities available in technology based on what you have read throughout the text. Investigate the entry level and advancement requirements for each career.

7. Compare the impacts of genetic engineering of crops on an industrialized society and an agrarian society, indicating what aspects are favorable and sustainable in one, but not in the other.

Critical Thinking

1. **Predicting** How will technology affect the way we live in the future?

2. **Comparing** One possible new energy source is nuclear fusion, especially using helium-3 from the moon. What are the advantages and disadvantages of using this energy source?

3. **Analyzing** How will nanotechnology affect different technological systems or a combination of systems?

4. **Extending** How might virtual reality be used to provide experiences for people with disabilities?

5. **Synthesizing** Give an example of how a future technology will require inputs from several existing technological systems. (Hint: Think of how biotechnical, communication, construction, manufacturing, and transportation systems act together to produce new technologies.)

Connecting to STEM
science · technology · engineering · math

Living in an Undersea World

Oceans are a comparatively unexplored space on Earth. Imagine that you are an aquanaut living underwater in an underwater space station. What would life be like? Where would you obtain an air supply? What would the pressure be? Could you go outside the station? Would fabrication play an important role?

Design Activity 35
FUTURE FORECASTING

Problem Situation

New technologies are often complex, with many ramifications. Technologies also interact with one another. For example, satellite communication can be used by ships at sea to determine location, as well as for tracking trucks on the highway.

Futurists are people who study and forecast the future. They try to suggest things that might happen so people can plan ahead. Futurists often use graphic or physical models to describe the relationships among their ideas, making them easier to understand.

Materials

You will need:
- colored paper
- compasses
- foam board display
- index cards, 3" x 5"
- markers

Your Challenge

You and your team members are to design and create a futures wheel.

> Go to your **Student Activity Guide, Design Activity 35.** Complete this activity in your Guide, and state the design challenge in your own words.

① Clarify the Design Specifications and Constraints

To solve the problem, your design must meet the following specifications and constraints:
- The wheel should include a minimum of three technical areas, such as genetic engineering, telecommunications, and nanotechnology.
- The wheel should include areas in which the technology areas influence one another.
- You should be able to draw at least three conclusions about possible new technologies from analyzing your futures wheel.

> In your Guide, state the design specifications and constraints. Add any others that your team or your teacher included.

② Research and Investigate

To better complete the design challenge, you need to first gather information to help you build a knowledge base.

In your Guide, complete Knowledge and Skill Builder 1: Investigating Technologies.

In your Guide, complete Knowledge and Skill Builder 2: Selecting a Futures Wheel Theme.

In your Guide, complete Knowledge and Skill Builder 3: In-Depth Investigations of Technologies.

③ Generate Alternative Designs

In your Guide, describe two possible solutions that your team has created for the problem.

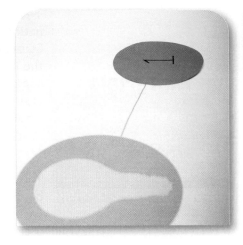

④ Choose and Justify the Optimal Solution

Refer to your Guide. Explain why you selected the solution you did, and why it was the better choice.

⑤ Develop a Prototype

Design your futures wheel. Put a sketch of your final design in your Guide.

In any technological activity, you will use seven resources: people, capital, time, information, energy, materials, and tools and machines. In your Guide, indicate which resources were most important in this activity, and how you made trade-offs among them.

⑥ Test and Evaluate

How will you test and evaluate your design? In your Guide, describe the testing procedure you will use. Explain how the results show that the design solves the problem and meets the specifications and constraints.

⑦ Redesign the Solution

Respond to the questions in your Guide about how you would redesign your solution. Your redesign should be based on the knowledge and information that you gained during the activity.

⑧ Communicate Your Achievements

In your Guide, describe the plan you will use to present your solution to your class. Show any handouts and/or PowerPoint slides you will use.

Design Activity 36

INTERNATIONAL SPACE STATION

Problem Situation

The International Space Station (ISS) is the largest and most complex international science project ever undertaken. Sixteen nations, led by the United States, each have different components to complete for the fabrication of the ISS. The United States will develop and operate the major components aboard the station. The ISS is an orbiting low-gravity laboratory, so many interesting scientific experiments can be performed in the station's six laboratories.

Your Challenge

You and your team members will design and construct a model of the ISS to be displayed in your technology class or at another location in your school. Each team will be assigned a different component, just as the nations are creating different components, that will fit together to form the entire ISS.

Go to your **Student Activity Guide, Design Activity 36.** Complete this activity in your Guide, and state the design challenge in your own words.

① Clarify the Design Specifications and Constraints

To solve the problem, your design must meet the following specifications and constraints:

- The overall size of the ISS will be determined by your teacher. For planning purposes, the model should not exceed 8 ft. in length and 4 ft. in width and height.
- The component(s) you are designing are subsystems of the total ISS system. You must make certain that the scale is the same among components and that you and other teams agree on how to connect the components to one another.

    In your Guide, state the design specifications and constraints. Add any others that your team or your teacher included.

Materials

You will need:

- aluminum foil (heavy duty)
- foam board
- mylar
- paint
- Plexiglas
- PVC and conduit couplings
- PVC piping (various diameters)
- sheet metal
- small-diameter metal tubing or conduit

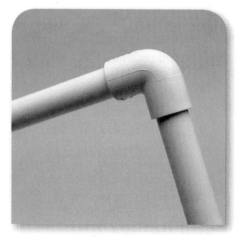

② Research and Investigate

In order to better complete the design challenge, you need to first gather information to help you build a knowledge base.

> In your Guide, complete Knowledge and Skill Builder 1: Visiting NASA.

> In your Guide, complete Knowledge and Skill Builder 2: Team Meeting.

> In your Guide, complete Knowledge and Skill Builder 3: Investigating Your Component.

③ Generate Alternative Designs

> In your Guide, describe two possible solutions that your team has created for the problem.

④ Choose and Justify the Optimal Solution

> Refer to your Guide. Explain why you selected the solution you did, and why it was the better choice.

⑤ Develop a Prototype

> Design and construct your ISS component. Put a sketch of your final design or a photograph of your component in your Guide.

> In any technological activity, you will use seven resources: people, capital, time, information, energy, materials, and tools and machines. In your Guide, indicate which resources were most important in this activity, and how you made trade-offs among them.

⑥ Test and Evaluate

> How will you test and evaluate your design? In your Guide, describe the testing procedure you will use. Explain how the results show that the design solves the problem and meets the specifications and constraints.

⑦ Redesign the Solution

> Respond to the questions in your Guide about how you would redesign your solution. Your redesign should be based on the knowledge and information that you gained during the activity.

⑧ Communicate Your Achievements

> In your Guide, describe the plan you will use to present your solution to your class. Show any handouts and/or PowerPoint slides you will use.

Technology Student Association

TSA is the only student organization designed specifically to meet the needs of middle and high school students who are enrolled in technology and engineering classes. TSA was founded in1978 and has grown significantly since then. Today there are more than 2,500 TSA chapters in 48 states and several other countries. A total of 150,000 students participate in TSA activities each year. Approximately 45% of the members are females, 35% are minorities and about 75% are college-bound.

The motto of TSA is *Learning to live in a technical world.* The motto supports the two-part mission of the organization: technological literacy and leadership development. As you have learned in this book, technological literacy is important for all citizens. TSA activities are designed to integrate science, technology, engineering and mathematics (STEM) concepts that lead to technological literacy through classroom activities, conferences and competitive events. Leadership is developed through public speaking, cooperation with others and through leadership conferences held at the national, state and local levels. The competitive events enhance 10 core leadership skills that help prepare members to become leaders while they are students and in the future.

If your school does not have a chapter, you will need to locate a teacher to serve as chapter advisor and enroll 10 or more students. Students who have completed or are currently taking a technology education or engineering class are eligible to join. Most chapter advisors are technology education or pre-engineering teachers. In some schools teachers of other subjects serve as advisors. Large chapters may have several advisors. Advisors enjoy being part of a national organization dedicated to helping students develop the skills needed for success in today's technical world.

The Organization

The National TSA office is located in Reston, Virginia. All memberships are processed through the national office. Two types of memberships are available, individual and CAP. Small schools and new chapters often select individual membership because a minimum of 10 members are needed to affiliate. CAP (Chapter Affiliation Program) is a good choice for large schools because it allows an unlimited number of students to join TSA for a flat fee. Most chapters serve middle or high school students. Schools that include middle and high school students can have a separate chapter for each level. TSA has a unified membership structure that requires all members to affiliate and pay dues for the local, state and national level. Dues are reasonable and many local chapters cover the costs through fund-raising activities.

National TSA has an executive director, a board of directors, office staff and a team of national officers. The national officer team includes a president, vice-president, secretary, treasurer, reporter and sergeant-at-arms. The officers are elected at the annual conference and serve for one school year. Most national officers have had leadership experience at the local and state levels. During the year they serve, the officers continue to build their leadership skills and represent the national organization at state TSA and other professional conferences. The national officers conduct a daily general session at the national conference and the award program which most students consider to be the highlight of the conference.

Most states have a leadership structure that includes elected student officers, a state advisor who coordinates chapter affiliations and conferences and a corporate member who serves as the official state representative to the national organization. In some states the same person serves as state advisor and corporate member. Corporate members participate in an annual meeting held at the national TSA conference. Their responsibilities include representing the interests of their state organizations and voting to elect members to the national board of directors. State advisors gather at the national conference to receive updates from the executive director and exchange ideas with other state advisors about activities within their home states.

TSA National Conference

The national TSA conference is an exciting event. It is held over a 5-day period in late June each year and moves around the country to cities such as Nashville, Orlando, Dallas and Denver. 4,000 to 5,000 people including students, advisors and parents attend the conference each year. Major activities at the conference include a daily general session, competitive events, a leadership academy, special interest sessions, an education fair, advisor updates and time to enjoy local recreational activities. The first evening of the conference features a mixer with a DJ. During the mixer students exchange pins that have been designed in the various states. Some states hold an annual design competition and have the winning design made into pins for their members to wear and exchange.

The typical state TSA holds several conferences each year. State officers play an important role in running the state organizations and conducting the conferences. Many states hold a leadership conference in the fall. Although the conferences vary, they generally invite a select audience such as the officers of each affiliated chapter to attend. Leadership conferences feature a motivational speaker, an opportunity to interact with students from other schools and hands-on activities that encourage teamwork. Students who attend these conferences continue to build their leadership skills by sharing the activities with members of their local chapters.

In the spring, states hold their annual conferences. Some are held on a single day while others last 2–3 days. The conferences are similar to the national TSA conference and include general sessions, competitive events and an award program. Some states also hold one or more regional conferences. Regional conferences are common in states that have a large number of chapters and in states that are large geographically. In many cases, winners of certain events at the regional level advance to compete in those same events at the state level. Winners at the state level are usually offered the opportunity to represent their state at the national conference.

Local chapters meet regularly to plan events for the school year, elect officers and prepare for the state and national competitions. In some schools TSA activities are co-curricular and conducted by chapter advisors during regular technology and engineering classes. This enables all students to learn about and participate in TSA activities. In other schools preparation for state and national competitions takes place after-school. Many chapters conduct service projects in their communities. Some raise funds to support the work of non-profit organizations. For example, during recent years many chapters have conducted activities to raise funds for the American Cancer Society. Thousands of dollars have been donated and participating schools are recognized at state and national conferences.

Types of Activities

Every two years TSA publishes a new middle school competitive event guide on a CD. Every chapter advisor receives a copy. The guide includes approximately 30 individual and team competitions which focus on various areas of technology as well as leadership. The guide includes a complete description of each event including an overview, the number of students from each state that may participate in that event at the national conference, the procedures and regulations and a description of how entries will be judged. Some of the most popular middle school events are described below.

For the **Agriculture and Biotechnology Issues** event members conduct research on a current topic and create a display to report their findings. **Challenging Technology Issues** requires a team of two members to explain opposing viewpoints on a contemporary topic in a debate-style format. The **Digital Photography** competition has two parts. All students submit an album of digital photographs which address a theme that changes each year. Semifinalists are given an additional challenge which requires taking and editing three additional photos at the conference site.

A large audience always gathers for the **Dragster** competition. For this event members prepare working drawings and then build a CO-2 dragster. Vehicles which qualify are then raced in an exciting head-to-head competition. For the **Flight** event members construct a glider from the materials provided and submit a design notebook. The gliders are launched from a catapult and the goal is to have the glider remain airborne as long as possible. The **Prepared Speech** event requires members to prepare and deliver a 3–5 minute talk on an assigned topic which changes every year. In **Problem Solving** teams of two members are provided with limited materials and develop the best possible solution to as problem that is specified on site. The **Tech Bowl** is a 3-member team event that has two parts. Members first take a STEM-based written test. Semifinalist teams then compete against other teams in an exciting quiz bowl competition. The **Write Now! Technical Writing** event requires advance preparation and an onsite performance. Members conduct research on a topic and three related subtopics. At the conference members prepare a written report on one of the subtopics which is specified on site. There are 20 additional middle school events that are briefly described on the TSA web site (www.tsaweb.org).

Each event in the guide has a common set of features which will help you prepare for the competitions. The **Procedure** section provides detailed information about how to prepare for the event. The **Regulations** section describes competition rules and specifies what needs to be brought to the event. The **Evaluation** section covers specific details about how entries will be judged and a rating form that will be used by the judges. Take a close look at the rating form because it will help you understand what is expected and the relative importance of each component of the event. For example, many events require documentation in addition to the finished product. A common element of winning entries is carefully prepared documentation.

Honors and Awards

TSA has several recognition programs. The TSA Achievement Program which awards bronze, silver and gold pins is a non-competitive program designed to recognize student members for outstanding achievement in a school's engineering and/or technology education program. Students who complete the required activities for the bronze level award may then begin to work toward the silver award and eventually toward the gold award. TSA usually awards two scholarships each year. One goes to member who demonstrates outstanding academic success and the other goes to a member planning to become a technology education teacher. TSA also has an honor society open to middle and high school students. The honor society recognizes TSA members who excel in academics, leadership and service to the school and community. One additional recognition program encourages students to nominate their teachers for an Advisor of the Year award.

Advantages of TSA Membership

TSA conferences are exciting. Members get to travel and see new places and meet others with common interests. As members prepare for their competitive events they have an opportunity to improve their knowledge of a variety of technology and engineering topics. Many events are team-based and require that members work closely with others toward a common goal. **Chapter Team** is an example of an event that requires extensive preparation as well as an on-site performance. For this event each member of the six-person team must take a written parliamentary procedures test. Semifinalists then conduct an opening ceremony, use parliamentary procedures to handle routine business and perform the closing ceremony all within a specified period of time. For many years Chapter Team has been one of TSA's most popular and challenging team events.

TSA recognizes the importance our nation's need for effective leaders. That's why leadership skill development is an integral part of the TSA mission and every competitive event. By participating in TSA you will have an opportunity to develop skills in the areas of problem solving, critical thinking, decision making, communication, teamwork, evaluation and ethics. Ask your teacher about the leadership activities that are included on the CD that they receive each year. The activities are fun and help develop skills that can be used in other classes and outside of school.

Leadership skills are developed through participation in team and individual events. For example, the Chapter Team event helps members develop problem solving skills as they conduct business. Oral communication skills are needed to perform the opening and closing ceremonies and teamwork is required throughout the event. For the Dragster competition which is an individual event, creative thinking is required to design a vehicle that meets specifications, problem solving is often needed to repair the vehicle and evaluation skills are needed to identify changes needed to improve performance of the vehicle after initial testing. For additional information be sure to visit the TSA web site (www.tsaweb.org).

Technology Timeline

Stone Age **1,000,000 B.C.**	Fire and the development of early stone tools
12,000 B.C.	Early oil lamps that burned fish and animal oil
Agricultural Age **3500 B.C.**	Writing (cuneiform and hieroglyphics)
Bronze Age **3000 B.C.**	Tools made from metal; development of the wheel
Iron Age **1200 B.C.**	Smelting of iron; development of the Greek alphabet
700 B.C.	Irrigation and sewers in Rome
221 B.C.	Construction of the Great Wall of China
144 B.C.	First high-level aqueduct in Rome; use of cement
100 B.C.	Water wheels
A.D. 500	Sailing ships that could travel with and against the wind
A.D. 1024	Chinese use first paper currency
A.D. 1200	Windmills
A.D. 1281	Chinese use gunpowder in war against Mongols
A.D. 1400–1600	Age of Exploration
A.D. 1450	Gutenberg invents movable type

1,000,000 B.C.

3500 B.C.

221 B.C.

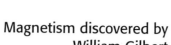

Magnetism discovered by William Gilbert — **1600**

Galileo's telescope — **1610**

William Harvey discovers circulation of the blood — **1628**

Newton's Laws — **1700**

Newcomen's steam engine — **1712**

Fahrenheit thermometer invented — **1724**

Benjamin Franklin proves lightning is electricity — **1752**

Industrial Revolution in Britain — **Industrial Age 1760–1840**

James Watt's steam engine — **1769**

Hargreaves's spinning jenny patented — **1770**

Wrought-iron process by Henry Cort — **1784**

Cartwright's loom — **1785**

LaVoisier's theory of chemical combustion — **1789**

Eli Whitney's cotton — **1793**

Jacquard's loom — **1804**

1807   Robert Fulton's steamboat

1822   First textile mill in United
States, in Rhode Island

1822

1829   The steam locomotive

1835   Bessemer process for
making steel

1839   Photography invented

1843   Morse begins telegraph line between
Baltimore, MD, and Washington, DC

1876

1876   Alexander Graham Bell
invents the telephone

1885   First successful gasoline-driven
car, by Carl Benz in Germany

1887   Daimler's internal combustion
engine automobile

1888   First Kodak handheld
camera

1903

1893   Diesel engine invented by
Dr. Rudolf Diesel

1901   Marconi sends first
transatlantic radio signals

1903   Wright brothers' airplane

1905   Albert Einstein proposes
his theory of relativity

1908   Model T Ford built using
mass production

1908

1928

Beginning of age of plastics	**1909**
SONAR invented	**1915**
Hybrid corn greatly improves crop yields	**1921**
Automatic traffic light patented by Garrett A. Morgan	**1923**
First round-the-world flight; Kleenex tissues	**1924**
First successful experiments with hydroponics	**1925**
First television demonstration; pop-up toaster	**1926**
Penicillin invented by Alexander Fleming	**1928**
Electroencephalograph (EEG); foam rubber	**1929**
Empire State Building in New York City	**1931**
The electron microscope	**1934**
Radar developed; nylon patented; B-17 bomber produced	**1935**
The ballpoint pen is patented	**1938**
First jet aircraft flies in Germany	**1939**
First color snapshots	**1942**

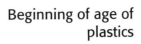

1944 First automatic digital computer, the Mark I

1945 ENIAC computer built using 18,000 vacuum tubes

1945

1947 Microwave oven; first Honda motorcycle; development of the transistor at Bell Laboratories

1950 Power steering; start of regular color TV broadcasts; first credit card (Diner's Club)

1952 Salk polio vaccine; 3-D movies

1954 Nuclear-powered submarine "Nautilus"; photocell developed

1955 Nuclear-powered electricity generation; fiber optics

1956 First desktop computer (Burroughs E-101)

1950

1957 Sputnik

Information Age
1958 First U.S. satellite

1959 Russia sends unmanned spacecraft to the moon (Luna2); Xerox copier

1960 Lasers; light-emitting diodes (LEDs); weather satellites

1957

1964 China tests nuclear bomb; IBM word processors

1965 First space walk

1966 Electronic fuel injection; Russians make successful soft landing on the moon (Luna 9)

1969

American manned moon landing; first Concorde SST flight — **1969**

Microprocessor developed by Intel; Mariner 9 orbits Mars — **1971**

CAT scan; photography from satellites (Landsat); videodiscs — **1972**

Genetic engineering; Skylab orbiting space station launched — **1973**

Toronto Communications Tower (tallest building in the world) — **1974**

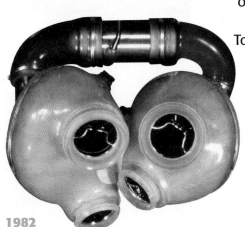

1982

Birth of the Apple computer; electronic cameras — **1976**

Trans-Alaska pipeline system completed — **1977**

Skylab, the orbiting space station, falls back to Earth — **1979**

Solar-powered aircraft — **1980**

First launch of a reusable space vehicle (space shuttle *Columbia*) — **1981**

Artificial heart — **1982**

Compact disc player — **1984**

DNA fingerprinting — **1986**

NASA's Hubble Space Telescope deployed by space shuttle *Discovery* — **1990**

Ebay online auction and shopping website — **1997**

1990

1999	First human chromosome completely sequenced
2001	Xbox video game console
2001	Worldwide introduction of the Toyota Prius hybrid car
2001	iPod and iTunes

2002

2002	First production of Segway electric vehicle
2003	MySpace website launched
2004	World of Warcraft massively multiplayer online game released
2004	Nintendo DS

2004

Facebook social networking
website launched — 2004

Sony PlayStation Portable
released in North America — 2005

YouTube.com launches — 2005

iPhone — 2007

Google search engine becomes
the most visited Web site — 2007

Mapping and sequencing
of eight human genomes — 2008

Tesla Roadster and Chevrolet
Volt electric cars developed — 2008

3D Television — 2010

2008

Glossary

A

advanced photo system (APS) a photo system in which a magnetic layer is applied to the film and information is recorded on the magnetic layer as the user takes pictures

aesthetics the way something looks and how that affects people's feelings

agile manufacturing an organizational strategy in which decisions are produced quickly and a greater variety of products is made faster and cheaper than in other types of manufacturing

agricultural era the period from 8,000 B.C. to the 1700s when people lived off the land

air-cushion vehicle (ACV) a vehicle that uses large fans to push air underneath it, lifting it on a cushion of air above the water

alloy a mixture of two or more metals

ampere the unit of measurement for electrical current

analog circuit an electronic circuit that works with analog signals

animation a series of still images in sequence

annealing heating steel to red hot and then allowing it to cool very slowly, making the steel softer

anthropometry the science of measuring people

applications program a program that gives a computer information for carrying out a specific task

aquaculture the farming of aquatic organisms, such as salmon, oysters, and plants

aqueduct a structure that carries water through channels carved into rock, allowing the water to pass over a valley or river

arch bridge a bridge that uses a curved structure above an opening to support the bridge's load

architects people who design buildings and other structures

Artificial Intelligence any use of machines to imitate human intelligence

assembly line a system in which an item is moved quickly from one workstation to the next

asymmetrical lacking symmetry; one side of an asymmetrical drawing is different from the other side

automation the process of controlling machines automatically

B

balance the way various parts of a design relate to one another

bar chart a chart that allows people to compare different categories; a chart in which different quantities are indicated by the lengths of the bars

basic system model describes any technological system by showing how the parts of a system work together

batch process a way of making a chemical formulation in which all of the ingredients are added and cooked in a chemical reactor until the formulation is done

beam bridge the simplest kind of bridge, using one beam to span a distance

binary a number system that uses only two numbers, 0 and 1

biomass vegetation and animal wastes

bionics the technology of making body parts

bioprocessing using living organisms to process materials, break them down, and change them into other materials

biotechnology the use of technology to make living things into products or new forms of life

bit short for **b**inary dig**it**, bits are represented by a 1 or a 0

bitmap (BMP) a common format in which computers store digital images, providing good clarity and detail but taking up considerable amounts of memory

blimp a helium-filled airship that is propelled forward by an engine

blow molding a process in which a bit of heated plastic is placed in the center of a mold, injected with air, and then expanded in a uniform thickness to form the desired shape

brainstorming a method of coming up with alternative designs

bridge a structure that crosses over water, a valley, or a road

brittle will not deform without breaking

Bronze Age starting around 3,000 B.C., the period during which people began to craft tools and weapons from bronze

browser a software program used to access Web pages

building any structure with walls and a roof that is made for permanent use

building permit a document allowing construction to take place on a specific piece of land

buoyancy the tendency of an object to float when submerged in a fluid

byte a group of eight bits

C

CAD/CAM computer-aided design/computer-aided manufacturing technology that allows the user to create a design on the computer screen and then send it directly to a machine tool

camera obscura the earliest camera-like device, first used in Italy in the 1500s

cantilever bridge a bridge whose two sections rise from the center, with each section firmly attached at its end

capital one of the seven technological resources; any source of wealth

carrier frequency in broadcast TV, each transmitter's own frequency on which it sends the video

casting a process by which a liquid is poured into a mold, allowed to harden, and then removed from the mold

CAT scan a cross-sectional image taken by a rotating X-ray device that takes 288 pictures as it moves around a patient

catalyst a material that speeds up chemical reactions

cell phone phone that uses radio waves to transmit signals over a large area

central processing unit (CPU) the "brain" of a computer, which reads programs and changes each program's instructions into actions

ceramics objects made from clay or similar inorganic (nonliving) materials

Certificate of Occupancy a certificate issued by the local building department stating that a building can be occupied

channel the medium through which a message moves

chemical a substance containing any of the atoms found in the periodic table of elements

chromosome part of a cell made up of DNA, which determines and transmits hereditary characteristics

CIM computer-integrated manufacturing technology that uses computers not only for design and manufacturing but also for business needs; used to store information about raw materials and parts

circle chart a chart allowing people to compare things relative to each other and to the whole by illustrating the fraction or percentage of the total for different products or categories; a pie chart

circuit a group of components connected together to do a specific job

cloning a nonsexual way of creating another organism

closed-loop system a system that can respond and change its inputs or outputs

coke a carbon-like substance that comes from coal, used to make iron pure and much stronger

combustion the process of burning a fossil fuel

communication the process of sending any type of message

communication medium the method used to communicate messages

compact disc a plastic disc containing digital information encoded as a pattern of pits

comparator an instrument that compares the actual output of a system with the desired output or result

composite a combination of several materials that improves on their individual properties

compound a substance formed by the combination of two or more elements

compression a force that pushes on or squeezes a material

computer any electronic machine that stores, processes, and retrieves information

computer-aided design (CAD) a tool that allows people to produce drawings using a computer instead of technical drawing tools

computer network many interconnected computers and computer devices, such as printers, scanners, and storage hardware

computer-to-plate printing a type of printing system in which data is used to generate text and images on a screen, producing an image that is reflected directly onto the paper

computer virus a software program that attaches itself to other programs

concrete a mixture of stone, sand, water, and cement (a mixture of limestone and clay)

conductors materials whose atoms give up some electrons easily

constraints the limits imposed on a design solution

construction the process of building a structure where it will be used

construction site the location where a structure is built

consumers people who buy products or services

continuous process a way of manufacturing a chemical formulation in which starting materials are constantly added to the reactor, mixed, and heated accordingly, thereby producing the desired product

contract a written agreement that states what the contractor will do for a given project and how much the owner will pay

control a device used to adjust a system

controller a device that turns a process on or off, or that changes it in some way

conveyer belt part of an assembly line, along which parts are moved from workstation to workstation

cracking a chemical process in which a complex chemical compound is broken (cracked) into simpler compounds

craft approach a method of manufacturing based on making products one at a time, from start to finish

crating the technique of working from geometric shapes to create new shapes

crossbreeding a form of biotechnology that combines the traits of one plant or animal with those of another

current the flow of electrons

D

darkroom a room used for developing photographs

data raw facts and figures

decoder the element that converts a message back to a form that can be understood by the receiver

Delphi survey a way of forecasting the future in which experts rank a list of ideas for future development

demodulation the process by which a modem converts the sounds sent from one end of an Internet connection back into data signals

design brief list of specifications for a design

designing a process of creating and planning

desktop publishing a system that uses a computer, special software, a mouse, and a computer printer to turn out a book or newsletter, page by page

developer a chemical used in photography that turns the exposed silver bromide grains black

diesel engine an engine in which diesel fuel is injected into the cylinder at the end of the compression step

digital camera camera in which images are recorded on an electronic image sensor rather than on film

digital circuit circuit in which information is coded into a series of 0s and 1s

digital image an image that stores visual information as numbers; can be created using a digital camera or a scanner, as well as on the computer using special software

Digital Video Discs (DVDs) plastic discs that carry visual and audio information

diode a component made using semiconductors, allowing current to flow in one direction but not the other

dividends shares of a company's profits that are paid to investors

DNA long, threadlike molecules made up of many genes that combine in different ways to give organisms their traits

downlink transmission of a changed signal from a satellite to Earth

drafting tablet a special surface and pen allowing the drafter to make lines and drawings that appear directly on the computer screen

drag wind resistance that tends to hold a plane back when it moves forward

drilling a separating process used to cut round holes in materials

drywall a sheet of plaster covered with heavy paper that comes in 4 x 8-foot panels and is fastened to the studs in a building using special drywall nails or screws

ductile able to be twisted, bent, or pressed without breaking

dye-sublimation printer printer that uses a page-sized printer ribbon, a heated print head, and specially coated printer paper; typically used by professional photographers, graphic artists, and scientists who process satellite images

E

elastic potential energy the energy stored in an object that is being stretched or compressed

elasticity the stiffness of a material

electrical generator a device that converts mechanical energy into electric energy

electrical panel the fixture through which electricity enters a house; it distributes electricity to different branch circuits in a house

electromotive force the force that produces a current

electron a particle that carries a negative charge

electronic communication a form of communication in which electrical signals carry the message

electronic components devices that control the flow of electricity

electroplating a process that uses electricity to form a thin metal coating on an object

element a pure substance that consists of only one type of atom

e-mail a text message that is sent using electronic signals

encoder an element that converts information into a form that can be sent, or transmitted, easily

endothermic reaction a reaction that absorbs energy in the form of heat

energy the ability to do work

engineers people who, by preparing exact drawings and plans for a structure's framework and foundation, work with architects to make sure that a bridge, road, or building is structurally sound

entrepreneur a person who comes up with a good idea and uses that idea to make money

ethical dilemma a difficult decision about whether something is right or wrong

ergonomics also called *human factors engineering;* deals with designing products so they can be used easily and comfortably

exothermic reaction a reaction that releases energy in the form of heat

exponential rate of change a faster and faster rate of change

exterior finishing painting or adding the final layer of materials to the outside walls of a house

external combustion fuel is burned in a chamber outside the engine, providing heat to another liquid or gas

extruding the process of squeezing softened materials through an opening so they take the shape of that opening

F

facsimile (FAX) electronic transmission of text and pictures

factory system a method of making products by machine rather than by hand

feedback information about the output of a system that can be used to adjust it

feedback control using feedback to adjust the way a machine works

fermentation a biological process that takes place when microorganisms act on other materials

ferrous metals types of alloys made of more than 50 percent iron

fertilizers chemicals added to soil to improve plant growth

fiber optic cables cables made of very thin strands of coated glass fibers

film an acetate (plastic) base that is covered with silver and other substances

film-based photography photography that uses light to record an image on film and then onto paper

filtering a method of separating solids from liquids in a mixture

fixer a chemical preservative used to fix, or set, a photographic image

flexible manufacturing the efficient production of small amounts of products

floppy disks thin, flexible magnetic storage disks that can be easily inserted into or removed from the computer

footing the base of a foundation

forging heating (but not melting) a metal part and then hammering it into shape

forming changing the shape of a material without cutting it

formulation a product developed by adding chemicals to create a product with the desired properties

fossil fuels energy-rich substances from the remains of plants and animals that lived millions of years ago

foundation an underground structure that spreads the weight of a building over a larger area

foundation wall a wall built on top of the footing to support the weight of a house; often made of concrete blocks that are attached to the footing and to each other with mortar

freehand how sketching is usually done, without drawing tools or instruments

frequency the number of times that an electric signal changes each second

functionality the ability of a product, system, or process to fulfill its intended purpose over its desired life span

futures wheel a diagram with one big idea at the center of a circle and other ideas on the spokes

G

Gantt chart a scheduling tool that shows when each part of a project begins and ends

gas chromatograph a piece of equipment that analyzes by heating a product past its boiling point to make a gas

gasoline engine an engine in which a spark ignites the fuel

gasoline turbine engine an engine that uses pressurized gas to spin a turbine for power

general contractor a person or company with overall responsibility for a construction project

genetic engineering taking a gene from one cell and splicing (moving) it into another cell, giving it the traits of the spliced gene

genome an organism's complete set of DNA

geosynchronous describing a satellite that moves around Earth at the same speed at which Earth rotates around its axis

geothermal energy thermal energy that is stored below Earth's surface

glaze a glasslike material that protects the surface of ceramics and gives them color

graph a way of presenting information visually that usually shows how something changes over a period of time

graphic communication a form of communication in which printed pictures or words carry the message

graphic techniques ways of communicating ideas visually that include sketching, drawing, and using recognizable symbols

gravitation the force of attraction between any two objects with mass

gravitational potential energy energy that is stored in an object because of its position relative to the ground

gravure printing a method of printing in which a line is scratched into the surface of a piece of metal, which is then inked and wiped dry, leaving the ink in the scratch; when a piece of paper is pressed against the metal surface, the paper pulls the ink out

grinding using small particles to sharpen or sand the surface of materials

H

hard disk a disk drive that is usually built into the computer

hardening heating steel red hot and quickly cooling it in water, resulting in harder steel

hardness a material's ability to resist being scratched or dented

hardware the physical components inside a computer system

hardwood wood from trees such as maples, oaks, and poplars

heat thermal energy transferred from something warmer to something cooler

herbicide a chemical used to kill weeds so that plants and crops will grow better

heredity the passing of traits from parents to offspring

human-made world all the systems and manufactured goods that are the products of human creativity

hybrid a crossbred plant or animal

hybrid electric vehicle (HEV) a car that combines the engine of a conventional vehicle with the battery and electric motor of an electric vehicle, allowing it to achieve twice the fuel economy of a conventional car

hydrofoil a fin with a flat surface that is attached to the bottom of a boat to lift it above the water when a certain speed is reached

hydroponics growing plants in water, sand, or gravel rather than in soil

I

industrial materials materials that have been processed

Industrial Revolution the period from the late 1700s through the mid-1900s, in which human and animal muscle power was replaced by machines

information age the period from the 1950s to today, in which inventions are based on electronics and computers

information overload more information than one can handle

informed design a process providing a way to make decisions without complete knowledge and then reassess those decisions later on

inkjet printer a type of printer in which small ink droplets are electrostatically sprayed from a nozzle onto the paper

innovation an improvement made to an invention

input the command given to a system

Instant Messaging a private e-mail connection between two people included on a buddy (or contact) list

insulator a material binding electrons to an atom so the electrons cannot move freely between atoms

integrated circuit often less than one tenth of an inch long by one tenth of an inch wide, it provides a complete circuit on a tiny piece of semiconductor

interchangeable parts parts manufactured to be exactly alike so they may be substituted for each other

interest the amount of money charged by a lender that must be paid in addition to the amount borrowed

interior finishing completing the inside walls and ceilings of a house

intermodal transportation system a system in which freight is hauled by trucks, trains, and/or ships and does not have to be unloaded along the way

internal combustion engine an engine in which the combustion process takes place inside the engine; when the fuel is ignited, chemical energy is converted to thermal energy and then to mechanical energy

International Space Station construction project in space being built by the United States and 15 other nations, for which 45 rockets and space shuttle missions will carry 460 tons of parts to the construction site

Internet a worldwide network of computer servers

Internet phone a telephone with which the user contacts another user through an e-mail address

Internet Service Provider a company providing access to the Internet

invention a new device, method, or process that has been developed as a result of technology

investigating examining; how scientists do their work

Iron Age the period from around 1200 B.C. in the Middle East and about 450 B.C. in Britain when iron came into common use

irrigation the artificial application of water to help plant growth

isometric drawing a pictorial drawing that is done within a framework of three lines, or an isometric axis

J

jet engine gas turbine attached to the wing of an airplane

joule the amount of energy required to move an object one meter using one Newton of force

just-in-time manufacturing a type of manufacturing that uses careful scheduling to keep materials and products in storage for as little time as possible

K

kinetic energy energy that results from motion

L

labor cost the amount of money required to hire people to work

landline phone a telephone that sends signals through wires or cables on land

laser a very strong burst of light energy

laser printer a computer printer that uses a laser beam to reproduce images or text on a photosensitive drum to which a black powder called toner, made of iron grains and plastic resin, is applied

law of conservation of energy the principle that energy cannot be created or destroyed

lift the upward force equaling the weight of the air displaced by an object floating in air

limits of controllability minimum and maximum outputs a system can produce

linear rate of change increasing by the same rate; changing proportionally; on a graph, this rate is represented by a straight line

lithography a method of printing in which a greasy print of the image to be printed is placed on a metal sheet, which is wrapped around a wet metal cylinder; the plate is inked and turns onto another cylinder that is covered with a rubber blanket; the rubber blanket turns and transfers the image onto the paper as it goes through the press

local area network (LAN) a small data network that allows computers throughout an office to share data

M

macadam material used as a surface for many roads, in which a layer of stone is placed on a base of hard soil and then topped by a layer of tar

Mach 1 the speed of sound (approximately 700 miles per hour)

machine-to-machine communication a form of communication in which machines communicate with one another

maglev magnetic levitation, used to lift, propel, and guide trains on special tracks at speeds up to 300 miles per hour

mainframe computers the large computers used by companies, government agencies, and universities

manufacturing the process of making finished products

market research a type of research in which companies survey a group of people to find out what customers want in a product

market share the percentage of sales for a particular product that can be claimed by one company

mass media forms of communication that reach large numbers of people

mass production the production of goods in large quantities by groups of workers in factories

mathematical model a model that uses equations to describe how a product will function or perform

memory the means by which a computer stores programs

message the information sent through a communication system

MiniDisc a two-inch-wide disc that holds audio recordings

modem a device that converts data signals into sounds that a telephone line can carry

modular construction a construction technique in which buildings or other structures are made from many similar units

modulation a process by which a modem converts data signals into sounds that the telephone line can carry

module a section of uniform size that can include one or more rooms and usually measures 12 feet by 20 feet

mortar a mixture of cement, lime, sand, and water

mortgage a bank loan for the purchase of a house

motherboard a printed circuit board containing all of the system hardware

movable type individual letters of type that were molded out of metal and then coated with ink

MP3 a digital recording technology that reduces the size of computer files containing music so they can be more easily moved or stored

N

nail a pointed, tapered piece of metal used to fasten together two pieces of wood

nanosecond billionth of a second

nanotechnology the creation of materials, devices, and systems using individual atoms and molecules

natural product a product made up of chemicals derived from plants

natural world the Earth and all living things

network interface card (NIC card) a computer circuit board or card installed in a computer

networking operating system software that manages all the other programs in a server

neutron a particle that has the same mass as a proton but does not carry any charge

nonferrous metal metal other than iron, as well as any alloy without a large amount of iron

nonrenewable energy source a source of energy that cannot be replaced once it has been used up

nonrenewable raw material a raw material that cannot be grown or replaced

nuclear fission the splitting of an atom's nucleus

nuclear fusion a process in which the nuclei of two atoms are forced together to form a new nucleus, releasing a large amount of energy

nucleus the center, or core, of an atom

O

oblique drawing a pictorial drawing that shows a straight-on view of one surface of an object

ohm the unit by which resistance is measured; a measure of how well a material conducts electricity

Ohm's law an equation that describes the relationship between current, voltage, and resistance;

$$\text{Current (amps)} = \frac{\text{Voltage (volts)}}{\text{Resistance (ohms)}}$$

open-loop system a system that has no way of monitoring and adjusting itself

operating system a program that allows a user to control the computer and its components

optical properties a material's ability to transmit or reflect light

optimization the process of making alternatives work as well as they can to get the best possible solution

orthographic drawing two-dimensional (2-D) drawing

outline a list of topics or concepts

output what is produced by a system; the actual result

P

peer-to-peer networks networks that involve two stand-alone computers connected together to share information, or peripheral devices like drives or printers

periodic table a table listing all of the known elements in such a way that elements with similar properties are in the same column

person-to-person communication the sending and receiving of messages between two people

personal computers general-purpose, single-user computers designed to be operated by one person at a time

perspective drawing the most realistic type of pictorial drawing, in which the parts of the object that are farther away appear smaller

pesticides chemicals used by farmers to kill pests and help plants and animals grow better

pharmaceutical describing products that include medicines, vaccines, and drugs used for healthcare

photocopier a device that creates copies using photography and static electricity

photoresistor an electronic component whose resistance changes with the amount of light hitting it

photovoltaic cell a solar cell that can turn sunlight into electricity

pictorial drawing a three-dimensional (3-D) drawing, which is more realistic than a 2-D drawing because it gives a clearer picture of the object

pie chart a chart that allows the comparison of things relative to each other and to the whole by illustrating the fraction or percentage of the total for different products or categories; a circle chart

pipeline a type of transportation system used to move crude oil or natural gas; extends from the field where these resources are pumped to the place where they are refined or to trucks or ships for further transport

pixel the basic picture element of an image displayed on a video screen

plastic a material made of long, chainlike molecules called polymers

plasticity the ability to be reshaped under pressure without breaking

plotter a large-scale printer used to print mechanical and architectural drawings; output device that uses one or more pens to make a drawing

point-to-point transmission the sending of a message from a single transmitter to a single receiver

potential energy energy that is stored

power the amount of work done during a given period of time

power system a machine that uses energy to do work

pressing a process similar to casting, but which uses force to change the shape of a material

primary material any material that is taken directly from Earth

printed circuit board a thin board made of an insulating material such as fiberglass

probability the likelihood that something will take place

process the action part of a system

processing changing resources into desired results

productivity the rate at which goods or services are produced, especially output per unit of labor

program control a method of automation in which a list of instructions is used to direct a machine's activity

project manager a person who oversees the contracts, scheduling, material deliveries, and overall progress of a large construction job

proportion different sizes within a design

proton a positively charged particle

prototype a full-scale, fully operational version of a solution

purify to make a product pure by filtering, distilling, or subliming, so that it is made of only one kind of matter

Q

quality degree of excellence

quality control system for ensuring the maintenance of proper standards during manufacturing

R

radio broadcasting communication system that uses one transmitter to send out a signal to many listeners

radioactive waste waste material containing unstable atoms that give off radiation

RAM **r**andom-**a**ccess **m**emory; the main memory in a computer

rate of reaction the length of time for a chemical reaction to take place, dependent on the temperature and the concentration of the starting materials

raw material any natural resource that is used to make finished products

relief printing a method of printing in which only the raised surfaces of the letters are inked; when the letter is pressed against a piece of paper, only the inked surface prints

rendering a way of making a drawing look more realistic, often by using shading and textures

renewable energy source any source of energy that can be replaced

renewable raw material a raw material that can be grown and replaced

renovation the process of rebuilding an existing building

requirements the features that define the performance of a system

resistance opposition to the flow of current through a material

resource one component in the production of goods, such as capital, time, or energy

rhythm the way the eye of the viewer moves around an entire design

risk/benefit trade-off a common trade-off made in solving large problems; to obtain desired benefits, designers must allow for risk while trying to keep the risk as low as possible

rivet a metal bolt or pin used to hold pieces of sheet metal or other materials together

robotics the area of artificial intelligence that designs and builds robots

rocket engine an engine that is similar to a jet engine but that carries its own oxygen supply to burn fuel

ROM **r**ead-**o**nly **m**emory; fixed memory that has been built into chips on a computer's motherboard

roof trusses large wooden triangles that make up most of a roof

S

satellite anything that orbits some other, larger item

satellite dish a special antenna that can pick up the transmission of uplinks and downlinks

sawing separating materials by using a metal blade with teeth

scale model a model that is either larger or smaller than the final product

scanner an input device that can transfer input from the printed page directly to the computer

scientific inquiry explaining the natural world based on the evidence gotten as a result of investigation

screen printing a method of printing in which a stencil is used to print designs using ink

screw a tapered and pointed metal pin with ridges used to pull one piece of material tightly against another

secondary storage computer memory that stores information for later use

semiconductors materials that are neither good insulators nor good conductors

sensor a monitor of a system's output

separating removing part of a material, usually through cutting or grinding

series circuit a circuit that connects components in one continuous path

shading using lighter and darker tones to show how light falls on an object

shaping a process used to change the shape of a piece of material

shareware software that you use on the honor system, whereby you download it for free from the Internet and then pay a small registration fee if you decide to use it

shearing using a knifelike blade for separating

shearing stress a force causing two contacting parts to slide upon each other in opposite directions

sheathing a waterproof layer that encloses a building and protects it from the weather

short circuit an unintended flow of current

siding a layer of aluminum, wood, or vinyl boards that waterproofs and protects a house

silicon the most common semiconductor material

single lens reflex camera a reflex camera in which a single lens supplies both the image on the film and the one in the viewfinder

sintering the process of heating powdered metal to make the particles fuse together

sketch a simplified view of an object or place

smelting the process of making iron from iron ore

software program a list of instructions according to which a computer does its work

softwood wood that comes from trees with needle-like leaves, such as pines and firs

solar energy energy from the sun

solder a metal that melts easily and makes a good connection

soldering joining two metals with heat and soft solder, which is an alloy made from lead and tin

source an element that produces the information or message to be sent

space factory a factory in space that is powered by solar energy and run by robots

space tourism traveling into space for a large fee

specifications the performance requirements, or output requirements, that a solution must fulfill

spectrometer a piece of equipment, used in processing chemicals, that measures the light energy absorbed or transmitted through a sample

speech recognition digitizing the sound waves of speech and then converting them to the basic sounds that are used in language

speech synthesis converting text to speech by breaking down the written words into basic sounds of the language and then converting the sounds into digital audio

stock shares of a company that allow investors to become part owners of it

Stone Age the period from about 1,000,000 B.C. to about 3,000 B.C. when people used stone, bones, and wood for tools

stop bath a chemical used in photography to stop the developing process

strength the ability of a material to keep its own shape when a force is applied to it

studs vertical pieces of 8-foot lengths of 2" x 4" lumber that frame most walls, spaced 16 inches apart to allow room for insulation, electrical wiring, and plumbing

subcontractors workers, such as carpenters, plumbers, and masons, who do the actual construction on a project

submarine a large ship that can operate above or below the surface of the water, used primarily by governments as part of their navy

submersible a small vehicle that can stay underwater for short periods of time

subsystems smaller systems that often make up other, larger systems

supercomputers the fastest and largest computers, most often used for research, for analyzing huge amounts of data, or for other very large jobs

superstructure the part of a structure that is aboveground

surfactant a surface-active agent, such as soap

suspension bridge a bridge that uses steel cables to hang the deck, or roadbed, from towers

symmetry exact correspondence of both sides of a design; mirror imaging

synthesize make; often used to describe the creation of molecules by combining two or more substances in a chemical reaction

synthetic material any material made in a factory

synthetic products entirely human-made products; often called artificial products

system a group of interrelated components designed collectively to achieve a desired goal

T

T square a long straightedge with a cross-piece at one end, allowing the user to draw horizontal lines that are exactly parallel to each other

technical drawing production of highly accurate drawings using special instruments and tools

technologically literate possessing an understanding of what technology can and cannot do for us

technology the process of using human knowledge to turn resources into goods and services that people need

technostress a type of stress that results from our dependence on the technology that surrounds us

telecommuting working at home using an electronic linkup with a central office

telephone switching the connection of one telephone to another

telnet an Internet communications protocol that allows a computer user remote access to all the files on another computer

tempering heating steel, after hardening, to a temperature that is not quite red hot and cooling it quickly, making it less brittle

tension the force that pulls on a material

texture the look and feel of the surface of an object

thermal energy the energy of the atoms that make up a substance

thermal glass window made from two or three panes of glass that are separated by a gas or vacuum, whose purpose is to reduce the flow of heat and to conserve energy

thermal properties a material's ability to conduct heat

thermistor an electronic component having a resistance that changes with the temperature

thermoplastics plastics that soften when heated, allowing them to be melted and shaped

thermoset plastics plastics that do not soften when heated, but char and burn instead

thrust the forward force produced by engines that move an airplane

toner negatively charged black powder used in photocopiers

torsion the twisting of a material

total market an estimate of how many products will be bought in a year

toughness the ability of a material to absorb energy without breaking

trade-off an exchange of the benefits and disadvantages of one solution for those of another solution

tradespeople workers who do the actual construction of a project

transducer a device that changes information in one form of energy to information in another form

transistor a type of resistor that lets a small amount of current control the flow of a much larger amount of current

transmission the gears or belts that connect a motor to the wheels

trend analysis forecasting the future by looking at the past

turbine a circular device with blades

tunnel a covered passageway through or under an obstruction

turning a process in which the material to be cut (the workpiece) is spun by a machine called a lathe

twin lens reflex camera a camera that uses two lenses, one for viewing and the other for focusing light on the film

U

ultrasound a medical technology that uses sound waves too high for humans to hear to view images of the human body

union a labor organization that works with companies to set pay and working conditions for its members

unity the concept that all parts of a design should work together to produce a single general effect

uplink transmission of a signal on one frequency from Earth to a satellite

V

vacuum forming a process of stretching a sheet of warm, soft plastic down and letting it cling to whatever it is drawn against

values the principles we consider worthwhile that influence all of our decisions

variables the different factors that affect the performance of a design

vehicle a container that carries people and objects from one place to another

venture capitalist a person who supplies money to finance the start-up of a new company

view camera the simplest type of film-based camera, with a lens at the front and film at the back

virtual reality using artificial intelligence to create a computer-generated environment

volt the unit by which electromotive force is measured

voltage a measure of the force with which electric current is pushed through wires

W

warranty the terms under which a company will repair or replace a defective product

Web server a computer that serves Web pages to the Internet

weight the force of gravity acting upon a body or object that pulls it toward Earth

welding joining metals by heating them enough to fuse them together

wide area network (WAN) a network that connects a large number of computers that are far apart

wind drift the distance a building moves from its vertical center because of the effect of wind

wire cables copper wires used to connect the computers in a network

Wireless Personal Area Network a short-range, wireless network that links personal devices together without the use of cables

Wireless Revolution the next revolution in communication, in which computer and radio technologies will merge to fundamentally change how we live and work

word processor a device that combines the functions of a typewriter with those of a computer

work something produced or accomplished through effort or activity

World Wide Web (WWW or "the Web") a network linking millions of computers worldwide

Z

zoning board a group in municipal government that controls what types of construction take place and how land is used within the municipality

Index

shutter, 393
sidewalks, 454
siding, 263
sign language, 283
silk screening, 388
sill plate, 257
SimCity 3000, 370
simple machines, 120
single-lens reflex (SLR) camera, 393, 394
sketching, 86, 87
skills, 557, 558
Skype, 350
slip ring, 442
Smalley, Richard E., 177, 179
smart car, 549
smart house, 549
smelting, 32
soap, 528
social concerns, 75
social networking, 352–354
software, 331–333
software driver, 330
softwood, 167
soil, 252
solar cell, 431
solar energy, 430, 431
solar panels, 115
soldering, 190, 191
Sony's PlayStation Portable (PSP), 370
sound, 347
sound barrier, 466
sound waves, 345, 347, 466
source, 287
space travel, 467, 468
Spacewar, 370
speaker, 345
speaking, 280, 281
specifications, 62, 63
spectrometer, 521
speech recognition, 547
speech synthesis, 547
speed of sound, 465
spoken conversation, 285, 287
Sputnik, 467, 576
steam, 441
steam engine, 36, 573
steam generator, 441
steel, 176, 244
stereolithography, 196, 197
stock, 111
stone, 252
stone age, 30, 31, 572
Stonehenge, 112
stop bath, 392
storage media, 372
strength, 173, 174
Strowger, Almon, 347
structures, 247–253
student association, 568–571
studs, 256, 257
subcontractor, 242
subfloor, 256, 257
submarine, 462
submersible, 462
subsystem, 141
Sunshine Skyway Bridge, 248
supercoils, 487
supercomputer, 321
superstructure, 256

surface-active agent, 527
surfactant, 527–528
surgery, 492–494
suspension bridge, 247, 248
sustainability, 371
switchboard, 347
switching office, 348
symbols, 86, 87
symmetry, 92
synthesis, 531, 532
synthesize, 520
synthetic chemicals, 514, 515
synthetic clothing, 171
synthetic fibers, 171
synthetic materials, 118
synthetic product, 514
synthetics, 523, 524
system
    bicycle as a, 137
    closed-loop, 143
    controllability of a, 151, 152
    controlling a, 147–154
    controls, types of, 150
    defined, 16
    intermodal transportation, 471
    living green, 152
    nontechnological, 153 ,154
    nonvehicle transportation, 472, 473
    open-loop, 142
    parts of a, 138–141
    production, 225–231
    and subsystems, online, 151
    transportation, 452–457
    universal system model, 138

T

T square, 87, 88
tables, 97
tactile feedback, 145
Tagged Image File Format (TIFF), 408
tapes, 329
teamwork, 214
Tech Bowl event, 570
technical drawing, 87
technological world, 56–83
technologically literate, 14
technologist, 8, 10
technology. *See also* online technology
    advances in, 7, 8
    car, build your own, 19
    challenges, 12, 13
    change, 36–40
    chemical, 510–539
    combining, 38
    culture and, 42–44
    in daily life, 5–7
    defined, 7, 8
    digital photography, 397, 399
    environmental, 44–47, 525
    ethics and, 47, 48
    flight, 465
    history and, 30–35
    human body, 494–496
    impact of, 42–48
    improvements from, 11, 12
    information, 340–381
    jobs and, 42, 43
    key ideas, 16–21

landscape and, 43, 44
literacy, 14, 568
medical, 492–496
nanotechnology, 544, 545, 548
nature of, 2–27
people in, 15, 114, 212
printing, 386
in the real world, 49
resources, 108–135
science and engineering, 6–10
in society, 540–567
society and, 28–55
study of, 11–16
systems, 136–161
then and now, 29
timeline, 572–579
Technology Student Association (TSA), 568–571
technostress, 558, 559
telecommuting, 556, 557
telegraph, 29
telephone
    cell, 348, 349
    cordless, 348
    design and performance, 346, 347
    information technology, 346–352
    internet free calling, 350
    landline, 346
    satellite, 356
    switching, 347
    technology, 29
telescope, 573
television, 354–356, 576
telnet, 365
tempering, 193, 194
tension, 174
Tesia Roadster, 579
testing, 532, 533
Texas Instrument, 315
text symbols, 86, 87
textile mill, 574
textiles, 525
texture, 91
thermal energy, 422
thermal glass, 259
thermal property, 175
thermal windows, 245
thermistor, 310
thermoplastic, 170
thermoset plastic, 170
3-D printing, 552, 554
3-dimensional drawing, 88, 89
thrust, 469
thumb drive, 326, 327
tiles, 395
time, 109, 112, 215, 292, 293, 454, 517, 518
tissue culture, 500, 501
titanium, 495
tools, 109, 118, 217, 224, 294, 456, 521, 572
Tools, Robert, 495
torsion, 174
total market, 229
touch screen, 329
touch sensor, 329
toughness, 174
trade-off, 18, 125
tradesperson, 242
Trans-Alaska pipeline, 472
transcontinental railroad, 457

Credits

2 traffic_analyzer/iStockphoto.com; **4** *b.* 2K Sports/Newscom, Nordling/Shutterstock.com; **5** *b.l.* Michel Setboun/Sygma/Corbis News/CORBIS, *t.r.* Michelle D. Milliman/Shutterstock.com; **6** traffic_analyzer/iStockphoto.com; **7** t.r. Steve Lovegrove/Shutterstock.com, *b.l.* 2happy/Fotolia.com; **8** *t.l.* Bajinda/Dreamstime.com, *b.r.* Gabriel Blaj/Fotolia.com; **9** Igabriela/Dreamstime.com; **11** Monkey Business Images/Dreamstime.com; **12** Carroteater/Dreamstime.com; **15** *t.r.* Jank1000/Dreamstime.com, *b.l.* Jim West/Alamy; lisegagne/iStockphoto.com; mecaleha/iStockphoto.com **16** Gjermund Alsos/Shutterstock.com; **17** © dbox for the Lower Manhattan Development Corporation/CORBIS; **19** lushik/iStockphoto.com; **21** browndogstudios/iStockphoto.com; mecaleha/iStockphoto.com; **23** Carl & Ann Purcell/Documentary Value/CORBIS; **24–25** traffic_analyzer/iStockphoto.com; **26** Steveheap/Dreamstime.com; **27** TOMO/Shutterstock.com; **28** Defun/Dreamstime.com; **29** *t.r.* Danylchenko Iaroslav/Shutterstock.com, ostill/Shutterstock.com; **30** Publiphoto/Photo Researchers, Inc.; **31** Bronze Age kettle, Iran (bronze),/Ashmolean Museum, University of Oxford, UK/The Bridgeman Art Library; **32** Bettmann/CORBIS; **33** *t.* InavanHateren/Shutterstock.com, *c.* I. Quintanilla/Shutterstock.com; **34** *t.r.* Michael Busselle/Encyclopedia/CORBIS; **36** Baptist/Shutterstock.com; **37** Power loom weaving, 1834 (engraving), Allom, Thomas (1804–72) (after)/Private Collection/The Bridgeman Art Library; **39** *t.l.* valdis torms/Shutterstock.com, Lanceb/Dreamstime.com; *t.r.* valdis torms/Shutterstock.com; **44** John Kropewnicki/Shutterstock.com; **45** Richard Packwood/PhotoLibrary; **46** Darren Brode/Shutterstock.com; **49** *t.r.* Robert Holmgren/Photolibrary, *b.l.* Nigel Cattlin/Alamy; mecaleha/iStockphoto.com; khz/iStockphoto.com; **51** Darren Brode/Shutterstock.com; **56–57** Michael Felix Photography/Shutterstock.com; **58** Stuartkey/Dreamstime.com; **59** *t.r.* 2K Sports/Newscom, *b.l.* Buena Vista Games/Newscom; **61** *b.r.* Jgroup/Dreamstime.com, *t.l.* Mauro Fermariello/Photo Researchers, Inc.; **64** Jeff Greenberg/Age Fotostock; **67** *t.l.* Cla78/Dreamstime LLC, *b.r.* Dmitry Vereshchagin/Fotolia.com; **70** lyf1/Shutterstock.com; **72** Grisho/Dreamstime.com; **74** *t.r.* MAYA Design, Inc., *b.l.* MAYA Design, Inc.; **75** Steve Mann/Shutterstock.com; **76** David H. Wells/CORBIS; **77** *t.l.* Gerald Holubowicz/Alamy, *t.r.* Xalanx/Dreamstime.com; **84** Creations/Shutterstock.com; **85** Jim Arbogast/Digital Vision/Getty Images; **91** Bharati Chaudhuri/SuperStock; **92** *t.r.* Javiergil/Dreamstime.com; *t.l.* Lisa Svara/Fotolia.com; **93** Michael Rosenfeld/Stone/Getty Images; **94** *t.l.* Pseudolongino/Dreamstime.com, *t.r.* Maximilian Stock Ltd./Photo Researchers, Inc.; **96** *t.r.* R.orecchia/Dreamstime.com, *b.l.* Elishabetka/Dreamstime.com;

97 Andresr/Shutterstock.com; **98** *b.c.* Monika Wisniewska/Fotolia.com, *t.l.* ssuaphotos/Shutterstock.com; **101** carlo dapino/Shutterstock.com; **108** dendron/Fotolia.com; **109** *b.l.* Andres Rodriguez/Fotolia.com, *t.r.* Brocreative/Fotolia.com; **110** NASA Headquarters; **111** Dennis Brack/Newscom; **112** *t.l.* Ray/Fotolia.com, *t.r.* Andrea Danti/Fotolia.com; **115** Yü Lan/Shutterstock.com; **118** Ambient Ideas/Shutterstock.com; **119** *t.l.* Kurhan/Shutterstock.com, *t.r.* prism68/Shutterstock.com; **121** AP Photo/Thomas Kienzle; **124** Annedave/Dreamstime.com; **126** Richard T. Nowitz/CORBIS; **131** prism68/Shutterstock.com; **136** s_oleg /Shutterstock.com; **137** *b.* djemphoto/Fotolia.com, *t.r.* Norbert von der Groeben/The Image Works; **139** JAUBERT IMAGES/Alamy; **140** Ted Spiegel/CORBIS; **145** Artlux/Fotolia.com; **147** Sergiy Zavgorodny/Shutterstock.com; **154** *t.l.* mdfiles/Fotolia.com, *t.r.* fred goldstein/Fotolia.com; **155** *b.l.* NASA Headquarters, *b.l.* NASA; **157** Darren Green/Shutterstock.com; **162–163** vichie81/Shutterstock.com; **164** omers/Shutterstock.com; **165** *b.l.* Johnnydevil/Dreamstime.com, *t.r.* WireImage/Getty Imagesl **167** CORBIS; **168** Kathy Burns-Millyard/Fotolia.com; **169** trailexplorers/Shutterstock.com; **170** Kurhan/Shutterstock.com; **171** Oleg Zabielin/Shutterstock.com; **172** Nicholas Moore/Shutterstock.com; **175** *t.r.* David R. Frazier/Photo Researchers, Inc; **175** *b.r.* sydeen/Shutterstock.com; **176** *t.* Stephen McBrady/PhotoEdit, *m.* UK Centre for Materials Education; **177** Colin Cuthbert/Photo Researchers, Inc.; **179** sgame/Shutterstock.com; **182** wandee007/Shutterstock.com; **183** Danny Lehman/CORBIS; **184** Annette Coolidge/PhotoEdit; **187** David R. Frazier Photolibrary, Inc./Alamy; **191** Vasyl Helevachuk/Shutterstock.com; **196** Charles O'Rear/CORBIS; **197** *t.l.* Juice Images/Alamy; **199** Vasyl Helevachuk/Shutterstock.com; **204** Toynutz/Dreamstime.com; **205** *t.r.* ARTEKI/Shutterstock.com, *b.* Small Town Studio/Shutterstock.com; **206** Tom Carter/PhotoEdit; **207** Marmi/Shutterstock.com; **208** Claudio Bravo/Shutterstock.com; **212** *t.r.* Andrew Woodley/Alamy, *b.l.* Danny Lehman/CORBIS; **214** AP Photo/Harry Cabluck; **215** Monkey Business Images/Shutterstock.com; **216** Sam Ogden/Photo Researchers, Inc.; **218** Hank Morgan-Rainbow/Science Faction/CORBIS; **219** Monty Rakusen/cultura/CORBIS; **221** Michael Rosenfeld/Stone/Getty Images; **223** Smalltownstudio/Dreamstime.com; **228** Roger Allyn Lee/SuperStock; **229** Losevsky Pavel/Shutterstock.com; **233** Hank Morgan - Rainbow/Science Faction/CORBIS; **238** Jordanrusev/Dreamstime.com; **239** *b.* Steve Lipofsky/CORBIS, *t.r.* Bjorn Heller/Shutterstock.com; **240** Francis Nief/Fotolia.com; **241** Ginasanders/Dreamstime.com; **242** *t.c.* Kadmy/Dreamstime.co, *b.l.* Greg Pickens/Fotolia.com; **244** *t.l.* Wave Royalty Free/Alamy, *b.l.* PSHAW-PHOTO/Shutterstock.com;

246 Keith Brofsky/Photodisc/Thinkstock.com; **247** Ray22/Dreamstime.com; **249** Chrishowey/Dreamstime.com; **251** jetsetmodels/Shutterstock.com; **253** Jacques Langevin/Sygma/CORBIS; **255** George Goodwin/SuperStock; **262** Andrei Nekrassov/Shutterstock.com; **263** Imagebroker.net/SuperStock; **264** benicce/Shutterstock.com; **265** *b.l.* Digital Vision/Getty Images, *t.r.* U.S. Department of Energy; **266** Christineg/Dreamstime.com; **268** Lisa F. Young/Shutterstock.com; **276–277** Mihai Simonia /Fotolia.com; **278** Paulfleet/Dreamstime.com; **279** *t.r.* Ajupp/Dreamstime.com; **281** krechet/www.Shutterstock.com; **282** Thomas Coex/AFP/Getty Images; **283** *b.l.* Pictorial Press Ltd/Alamy, *t.* Susan Kuklin/Photo Researchers, Inc.; **285** Yuri Arcurs/Shutterstock.com; **288** Ulrich Niehoff/PhotoLibrary; **290** maska82/Fotolia.com; **291** *b.l.* Andrew Brusso/CORBIS; **292** withGod/Fotolia.com; **293** Ron Brown/SuperStock; **297** Jvdwolf/Dreamstime.com; **300** *b.l.* EDHAR/Shutterstock.com, *b.r.* Leah-Anne Thompson/Shutterstock.com; **301** *b.r.* Blend Images/Hill Street Studios/Getty Images, *t.r.* Marc Dietrich/Shutterstock.com; **302** risteski goce/Shutterstock.com; **303** Monkey Business Images/Shutterstock.com, Poco_bw/Fotolia.com; **306** Jhaz Photography/Shutterstock.com; **307** Annette Shaff/Shutterstock.com; **309** Martin Green/Fotolia.com; **310** John McGrail/Taxi/Getty Images; **313** Amgun/Shutterstock.com; **315** *t.r.* Fotosearch/Getty Images, *b.l.* Hermann Danzmayr/Shutterstock.com; **316** Dave King/DK Limited/Documentary Value/CORBIS; **317** National Archives; **319** Chad Slattery/Stone/Getty Images; **320** *t.l.* Vishwakiran/Dreamstime.com, *b.c.* Lesa/Dreamstime.com; **321** Paul Shambroom/Photo Researchers, Inc.; **322** Kayros Studio "Be Happy!"/Shutterstock.com; **324** *b.r.* Lesa/Dreamstime.com, *b.l.* Feng Yu/Fotolia.com; **325** *b.l.* Draghicich/Dreamstime.com, *l.* Norman Chan/Fotolia.com, *r.* Gudellaphoto /Fotolia.com, *b.r.* Péter Gudella/Shutterstock.com, Wam1975/ Dreamstime.com; **327** Bestshortstop/Dreamstime.com; **329** Taiga/Shutterstock.com; **330** Alessandrozocc/Dreamstime.com; **331** AP Photo/Paul Sakuma; **332** Stockbyte/Getty Images; **335** John McGrail/Taxi/Getty Images; **338** Darren Hubley/Shutterstock.com, Elnur/Shutterstock.com; **339** *r.* Bojan Pavlukovic/Shutterstock.com; **340** rossco/Fotolia.com; **341** *b.* YAKOBCHUK VASYL/Shutterstock.com, *t.r.* koh sze kiat/Shutterstock.com; **343** *t.r.* Bettmann/CORBIS, *b.* Rob Cousins/Alamy; **346** Monkey Business/Fotolia.com; **347** SuperStock/SuperStock; **349** cobalt88/Shutterstock.com; **355** Neo Edmund/Fotolia.com; **360** Paul Taylor/Riser/Getty Images; **361** Sergei Devyatkin/Shutterstock.com; **363** Toria/Shutterstock.com; **366** Mitch010/itradu/Dreamstime.com; **367** Exactostock/SuperStock; **368** Reuters/CORBIS, *t.r.* Eugene Berman/Shutterstock.com; **369** Toy Alan King/Alamy;